OpenIntro Statistics
Fourth Edition

David Diez
Data Scientist
OpenIntro

Mine Çetinkaya-Rundel
Associate Professor of the Practice, Duke University
Professional Educator, RStudio

Christopher D Barr
Investment Analyst
Varadero Capital

Table of Contents

Preface

OpenIntro Statistics covers a first course in statistics, providing a rigorous introduction to applied statistics that is clear, concise, and accessible. This book was written with the undergraduate level in mind, but it's also popular in high schools and graduate courses.

We hope readers will take away three ideas from this book in addition to forming a foundation of statistical thinking and methods.

- Statistics is an applied field with a wide range of practical applications.
- You don't have to be a math guru to learn from real, interesting data.
- Data are messy, and statistical tools are imperfect. But, when you understand the strengths and weaknesses of these tools, you can use them to learn about the world.

Textbook overview

The chapters of this book are as follows:

1. **Introduction to data.** Data structures, variables, and basic data collection techniques.
2. **Summarizing data.** Data summaries, graphics, and a teaser of inference using randomization.
3. **Probability.** Basic principles of probability.
4. **Distributions of random variables.** The normal model and other key distributions.
5. **Foundations for inference.** General ideas for statistical inference in the context of estimating the population proportion.
6. **Inference for categorical data.** Inference for proportions and tables using the normal and chi-square distributions.
7. **Inference for numerical data.** Inference for one or two sample means using the t-distribution, statistical power for comparing two groups, and also comparisons of many means using ANOVA.
8. **Introduction to linear regression.** Regression for a numerical outcome with one predictor variable. Most of this chapter could be covered after Chapter 1.
9. **Multiple and logistic regression.** Regression for numerical and categorical data using many predictors.

OpenIntro Statistics supports flexibility in choosing and ordering topics. If the main goal is to reach multiple regression (Chapter 9) as quickly as possible, then the following are the ideal prerequisites:

- Chapter 1, Sections 2.1, and Section 2.2 for a solid introduction to data structures and statistical summaries that are used throughout the book.
- Section 4.1 for a solid understanding of the normal distribution.
- Chapter 5 to establish the core set of inference tools.
- Section 7.1 to give a foundation for the t-distribution
- Chapter 8 for establishing ideas and principles for single predictor regression.

Examples and exercises

Examples are provided to establish an understanding of how to apply methods

EXAMPLE 0.1

This is an example. When a question is asked here, where can the answer be found?

The answer can be found here, in the solution section of the example!

When we think the reader should be ready to try determining the solution to an example, we frame it as Guided Practice.

GUIDED PRACTICE 0.2

The reader may check or learn the answer to any Guided Practice problem by reviewing the full solution in a footnote.[1]

Exercises are also provided at the end of each section as well as review exercises at the end of each chapter. Solutions are given for odd-numbered exercises in Appendix A.

Additional resources

Video overviews, slides, statistical software labs, data sets used in the textbook, and much more are readily available at

openintro.org/os

We also have improved the ability to access data in this book through the addition of Appendix B, which provides additional information for each of the data sets used in the main text and is new in the Fourth Edition. Online guides to each of these data sets are also provided at **openintro.org/data** and through a companion R package.

We appreciate all feedback as well as reports of any typos through the website. A short-link to report a new typo or review known typos is **openintro.org/os/typos**.

For those focused on statistics at the high school level, consider *Advanced High School Statistics*, which is a version of *OpenIntro Statistics* that has been heavily customized by Leah Dorazio for high school courses and AP® Statistics.

Acknowledgements

This project would not be possible without the passion and dedication of many more people beyond those on the author list. The authors would like to thank the OpenIntro Staff for their involvement and ongoing contributions. We are also very grateful to the hundreds of students and instructors who have provided us with valuable feedback since we first started posting book content in 2009.

We also want to thank the many teachers who helped review this edition, including Laura Acion, Matthew E. Aiello-Lammens, Jonathan Akin, Stacey C. Behrensmeyer, Juan Gomez, Jo Hardin, Nicholas Horton, Danish Khan, Peter H.M. Klaren, Jesse Mostipak, Jon C. New, Mario Orsi, Steve Phelps, and David Rockoff. We appreciate all of their feedback, which helped us tune the text in significant ways and greatly improved this book.

---

[1]Guided Practice problems are intended to stretch your thinking, and you can check yourself by reviewing the footnote solution for any Guided Practice.

Chapter 1

Introduction to data

Scientists seek to answer questions using rigorous methods and careful observations. These observations – collected from the likes of field notes, surveys, and experiments – form the backbone of a statistical investigation and are called **data**. Statistics is the study of how best to collect, analyze, and draw conclusions from data, and in this first chapter, we focus on both the properties of data and on the collection of data.

For videos, slides, and other resources, please visit
www.openintro.org/os

1.1 Case study: using stents to prevent strokes

Section 1.1 introduces a classic challenge in statistics: evaluating the efficacy of a medical treatment. Terms in this section, and indeed much of this chapter, will all be revisited later in the text. The plan for now is simply to get a sense of the role statistics can play in practice.

In this section we will consider an experiment that studies effectiveness of stents in treating patients at risk of stroke. Stents are devices put inside blood vessels that assist in patient recovery after cardiac events and reduce the risk of an additional heart attack or death. Many doctors have hoped that there would be similar benefits for patients at risk of stroke. We start by writing the principal question the researchers hope to answer:

Does the use of stents reduce the risk of stroke?

The researchers who asked this question conducted an experiment with 451 at-risk patients. Each volunteer patient was randomly assigned to one of two groups:

Treatment group. Patients in the treatment group received a stent and medical management. The medical management included medications, management of risk factors, and help in lifestyle modification.

Control group. Patients in the control group received the same medical management as the treatment group, but they did not receive stents.

Researchers randomly assigned 224 patients to the treatment group and 227 to the control group. In this study, the control group provides a reference point against which we can measure the medical impact of stents in the treatment group.

Researchers studied the effect of stents at two time points: 30 days after enrollment and 365 days after enrollment. The results of 5 patients are summarized in Figure 1.1. Patient outcomes are recorded as "stroke" or "no event", representing whether or not the patient had a stroke at the end of a time period.

Patient	group	0-30 days	0-365 days
1	treatment	no event	no event
2	treatment	stroke	stroke
3	treatment	no event	no event
⋮	⋮	⋮	
450	control	no event	no event
451	control	no event	no event

Figure 1.1: Results for five patients from the stent study.

Considering data from each patient individually would be a long, cumbersome path towards answering the original research question. Instead, performing a statistical data analysis allows us to consider all of the data at once. Figure 1.2 summarizes the raw data in a more helpful way. In this table, we can quickly see what happened over the entire study. For instance, to identify the number of patients in the treatment group who had a stroke within 30 days, we look on the left-side of the table at the intersection of the treatment and stroke: 33.

	0-30 days		0-365 days	
	stroke	no event	stroke	no event
treatment	33	191	45	179
control	13	214	28	199
Total	46	405	73	378

Figure 1.2: Descriptive statistics for the stent study.

GUIDED PRACTICE 1.1

Of the 224 patients in the treatment group, 45 had a stroke by the end of the first year. Using these two numbers, compute the proportion of patients in the treatment group who had a stroke by the end of their first year. (Please note: answers to all Guided Practice exercises are provided using footnotes.)[1]

We can compute summary statistics from the table. A **summary statistic** is a single number summarizing a large amount of data. For instance, the primary results of the study after 1 year could be described by two summary statistics: the proportion of people who had a stroke in the treatment and control groups.

Proportion who had a stroke in the treatment (stent) group: $45/224 = 0.20 = 20\%$.

Proportion who had a stroke in the control group: $28/227 = 0.12 = 12\%$.

These two summary statistics are useful in looking for differences in the groups, and we are in for a surprise: an additional 8% of patients in the treatment group had a stroke! This is important for two reasons. First, it is contrary to what doctors expected, which was that stents would *reduce* the rate of strokes. Second, it leads to a statistical question: do the data show a "real" difference between the groups?

This second question is subtle. Suppose you flip a coin 100 times. While the chance a coin lands heads in any given coin flip is 50%, we probably won't observe exactly 50 heads. This type of fluctuation is part of almost any type of data generating process. It is possible that the 8% difference in the stent study is due to this natural variation. However, the larger the difference we observe (for a particular sample size), the less believable it is that the difference is due to chance. So what we are really asking is the following: is the difference so large that we should reject the notion that it was due to chance?

While we don't yet have our statistical tools to fully address this question on our own, we can comprehend the conclusions of the published analysis: there was compelling evidence of harm by stents in this study of stroke patients.

Be careful: Do not generalize the results of this study to all patients and all stents. This study looked at patients with very specific characteristics who volunteered to be a part of this study and who may not be representative of all stroke patients. In addition, there are many types of stents and this study only considered the self-expanding Wingspan stent (Boston Scientific). However, this study does leave us with an important lesson: we should keep our eyes open for surprises.

---

[1]The proportion of the 224 patients who had a stroke within 365 days: $45/224 = 0.20$.

Exercises

1.1 Migraine and acupuncture, Part I. A migraine is a particularly painful type of headache, which patients sometimes wish to treat with acupuncture. To determine whether acupuncture relieves migraine pain, researchers conducted a randomized controlled study where 89 females diagnosed with migraine headaches were randomly assigned to one of two groups: treatment or control. 43 patients in the treatment group received acupuncture that is specifically designed to treat migraines. 46 patients in the control group received placebo acupuncture (needle insertion at non-acupoint locations). 24 hours after patients received acupuncture, they were asked if they were pain free. Results are summarized in the contingency table below.[2]

		Pain free		
		Yes	No	Total
Group	Treatment	10	33	43
	Control	2	44	46
	Total	12	77	89

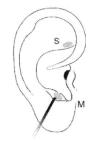

Figure from the original paper displaying the appropriate area (M) versus the inappropriate area (S) used in the treatment of migraine attacks.

(a) What percent of patients in the treatment group were pain free 24 hours after receiving acupuncture?

(b) What percent were pain free in the control group?

(c) In which group did a higher percent of patients become pain free 24 hours after receiving acupuncture?

(d) Your findings so far might suggest that acupuncture is an effective treatment for migraines for all people who suffer from migraines. However this is not the only possible conclusion that can be drawn based on your findings so far. What is one other possible explanation for the observed difference between the percentages of patients that are pain free 24 hours after receiving acupuncture in the two groups?

1.2 Sinusitis and antibiotics, Part I. Researchers studying the effect of antibiotic treatment for acute sinusitis compared to symptomatic treatments randomly assigned 166 adults diagnosed with acute sinusitis to one of two groups: treatment or control. Study participants received either a 10-day course of amoxicillin (an antibiotic) or a placebo similar in appearance and taste. The placebo consisted of symptomatic treatments such as acetaminophen, nasal decongestants, etc. At the end of the 10-day period, patients were asked if they experienced improvement in symptoms. The distribution of responses is summarized below.[3]

		Self-reported improvement in symptoms		
		Yes	No	Total
Group	Treatment	66	19	85
	Control	65	16	81
	Total	131	35	166

(a) What percent of patients in the treatment group experienced improvement in symptoms?

(b) What percent experienced improvement in symptoms in the control group?

(c) In which group did a higher percentage of patients experience improvement in symptoms?

(d) Your findings so far might suggest a real difference in effectiveness of antibiotic and placebo treatments for improving symptoms of sinusitis. However, this is not the only possible conclusion that can be drawn based on your findings so far. What is one other possible explanation for the observed difference between the percentages of patients in the antibiotic and placebo treatment groups that experience improvement in symptoms of sinusitis?

---

[2]G. Allais et al. "Ear acupuncture in the treatment of migraine attacks: a randomized trial on the efficacy of appropriate versus inappropriate acupoints". In: *Neurological Sci.* 32.1 (2011), pp. 173–175.

[3]J.M. Garbutt et al. "Amoxicillin for Acute Rhinosinusitis: A Randomized Controlled Trial". In: *JAMA: The Journal of the American Medical Association* 307.7 (2012), pp. 685–692.

1.2 Data basics

Effective organization and description of data is a first step in most analyses. This section introduces the *data matrix* for organizing data as well as some terminology about different forms of data that will be used throughout this book.

1.2.1 Observations, variables, and data matrices

Figure 1.3 displays rows 1, 2, 3, and 50 of a data set for 50 randomly sampled loans offered through Lending Club, which is a peer-to-peer lending company. These observations will be referred to as the `loan50` data set.

Each row in the table represents a single loan. The formal name for a row is a **case** or **observational unit**. The columns represent characteristics, called **variables**, for each of the loans. For example, the first row represents a loan of $7,500 with an interest rate of 7.34%, where the borrower is based in Maryland (MD) and has an income of $70,000.

GUIDED PRACTICE 1.2

What is the grade of the first loan in Figure 1.3? And what is the home ownership status of the borrower for that first loan? For these Guided Practice questions, you can check your answer in the footnote.[4]

In practice, it is especially important to ask clarifying questions to ensure important aspects of the data are understood. For instance, it is always important to be sure we know what each variable means and the units of measurement. Descriptions of the `loan50` variables are given in Figure 1.4.

	loan_amount	interest_rate	term	grade	state	total_income	homeownership
1	7500	7.34	36	A	MD	70000	rent
2	25000	9.43	60	B	OH	254000	mortgage
3	14500	6.08	36	A	MO	80000	mortgage
⋮	⋮	⋮	⋮	⋮	⋮	⋮	⋮
50	3000	7.96	36	A	CA	34000	rent

Figure 1.3: Four rows from the `loan50` data matrix.

variable	description
loan_amount	Amount of the loan received, in US dollars.
interest_rate	Interest rate on the loan, in an annual percentage.
term	The length of the loan, which is always set as a whole number of months.
grade	Loan grade, which takes a values A through G and represents the quality of the loan and its likelihood of being repaid.
state	US state where the borrower resides.
total_income	Borrower's total income, including any second income, in US dollars.
homeownership	Indicates whether the person owns, owns but has a mortgage, or rents.

Figure 1.4: Variables and their descriptions for the `loan50` data set.

The data in Figure 1.3 represent a **data matrix**, which is a convenient and common way to organize data, especially if collecting data in a spreadsheet. Each row of a data matrix corresponds to a unique case (observational unit), and each column corresponds to a variable.

---

[4]The loan's grade is A, and the borrower rents their residence.

When recording data, use a data matrix unless you have a very good reason to use a different structure. This structure allows new cases to be added as rows or new variables as new columns.

GUIDED PRACTICE 1.3

The grades for assignments, quizzes, and exams in a course are often recorded in a gradebook that takes the form of a data matrix. How might you organize grade data using a data matrix?[5]

GUIDED PRACTICE 1.4

We consider data for 3,142 counties in the United States, which includes each county's name, the state where it resides, its population in 2017, how its population changed from 2010 to 2017, poverty rate, and six additional characteristics. How might these data be organized in a data matrix?[6]

The data described in Guided Practice 1.4 represents the county data set, which is shown as a data matrix in Figure 1.5. The variables are summarized in Figure 1.6.

---

[5]There are multiple strategies that can be followed. One common strategy is to have each student represented by a row, and then add a column for each assignment, quiz, or exam. Under this setup, it is easy to review a single line to understand a student's grade history. There should also be columns to include student information, such as one column to list student names.

[6]Each county may be viewed as a case, and there are eleven pieces of information recorded for each case. A table with 3,142 rows and 11 columns could hold these data, where each row represents a county and each column represents a particular piece of information.

	name	state	pop	pop_change	poverty	homeownership	multi_unit	unemp_rate	metro	median_edu	median_hh_income
1	Autauga	Alabama	55504	1.48	13.7	77.5	7.2	3.86	yes	some_college	55317
2	Baldwin	Alabama	212628	9.19	11.8	76.7	22.6	3.99	yes	some_college	52562
3	Barbour	Alabama	25270	-6.22	27.2	68.0	11.1	5.90	no	hs_diploma	33368
4	Bibb	Alabama	22668	0.73	15.2	82.9	6.6	4.39	yes	hs_diploma	43404
5	Blount	Alabama	58013	0.68	15.6	82.0	3.7	4.02	yes	hs_diploma	47412
6	Bullock	Alabama	10309	-2.28	28.5	76.9	9.9	4.93	no	hs_diploma	29655
7	Butler	Alabama	19825	-2.69	24.4	69.0	13.7	5.49	no	hs_diploma	36326
8	Calhoun	Alabama	114728	-1.51	18.6	70.7	14.3	4.93	yes	some_college	43686
9	Chambers	Alabama	33713	-1.20	18.8	71.4	8.7	4.08	no	hs_diploma	37342
10	Cherokee	Alabama	25857	-0.60	16.1	77.5	4.3	4.05	no	hs_diploma	40041
⋮	⋮			⋮	⋮	⋮	⋮	⋮	⋮	⋮	⋮
3142	Weston	Wyoming	6927	-2.93	14.4	77.9	6.5	3.98	no	some_college	59605

Figure 1.5: Eleven rows from the county data set.

variable	description
name	County name.
state	State where the county resides, or the District of Columbia.
pop	Population in 2017.
pop_change	Percent change in the population from 2010 to 2017. For example, the value 1.48 in the first row means the population for this county increased by 1.48% from 2010 to 2017.
poverty	Percent of the population in poverty.
homeownership	Percent of the population that lives in their own home or lives with the owner, e.g. children living with parents who own the home.
multi_unit	Percent of living units that are in multi-unit structures, e.g. apartments.
unemp_rate	Unemployment rate as a percent.
metro	Whether the county contains a metropolitan area.
median_edu	Median education level, which can take a value among below_hs, hs_diploma, some_college, and bachelors.
median_hh_income	Median household income for the county, where a household's income equals the total income of its occupants who are 15 years or older.

Figure 1.6: Variables and their descriptions for the county data set.

1.2.2 Types of variables

Examine the `unemp_rate`, `pop`, `state`, and `median_edu` variables in the `county` data set. Each of these variables is inherently different from the other three, yet some share certain characteristics.

First consider `unemp_rate`, which is said to be a **numerical** variable since it can take a wide range of numerical values, and it is sensible to add, subtract, or take averages with those values. On the other hand, we would not classify a variable reporting telephone area codes as numerical since the average, sum, and difference of area codes doesn't have any clear meaning.

The `pop` variable is also numerical, although it seems to be a little different than `unemp_rate`. This variable of the population count can only take whole non-negative numbers (0, 1, 2, ...). For this reason, the population variable is said to be **discrete** since it can only take numerical values with jumps. On the other hand, the unemployment rate variable is said to be **continuous**.

The variable `state` can take up to 51 values after accounting for Washington, DC: `AL`, `AK`, ..., and `WY`. Because the responses themselves are categories, `state` is called a **categorical** variable, and the possible values are called the variable's **levels**.

Finally, consider the `median_edu` variable, which describes the median education level of county residents and takes values `below_hs`, `hs_diploma`, `some_college`, or `bachelors` in each county. This variable seems to be a hybrid: it is a categorical variable but the levels have a natural ordering. A variable with these properties is called an **ordinal** variable, while a regular categorical variable without this type of special ordering is called a **nominal** variable. To simplify analyses, any ordinal variable in this book will be treated as a nominal (unordered) categorical variable.

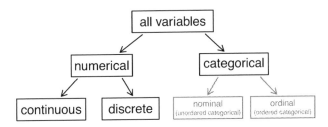

Figure 1.7: Breakdown of variables into their respective types.

EXAMPLE 1.5

Data were collected about students in a statistics course. Three variables were recorded for each student: number of siblings, student height, and whether the student had previously taken a statistics course. Classify each of the variables as continuous numerical, discrete numerical, or categorical.

The number of siblings and student height represent numerical variables. Because the number of siblings is a count, it is discrete. Height varies continuously, so it is a continuous numerical variable. The last variable classifies students into two categories – those who have and those who have not taken a statistics course – which makes this variable categorical.

GUIDED PRACTICE 1.6

An experiment is evaluating the effectiveness of a new drug in treating migraines. A `group` variable is used to indicate the experiment group for each patient: treatment or control. The `num_migraines` variable represents the number of migraines the patient experienced during a 3-month period. Classify each variable as either numerical or categorical?[7]

---

[7]There `group` variable can take just one of two group names, making it categorical. The `num_migraines` variable describes a count of the number of migraines, which is an outcome where basic arithmetic is sensible, which means this is numerical outcome; more specifically, since it represents a count, `num_migraines` is a discrete numerical variable.

1.2.3 Relationships between variables

Many analyses are motivated by a researcher looking for a relationship between two or more variables. A social scientist may like to answer some of the following questions:

(1) If homeownership is lower than the national average in one county, will the percent of multi-unit structures in that county tend to be above or below the national average?

(2) Does a higher than average increase in county population tend to correspond to counties with higher or lower median household incomes?

(3) How useful a predictor is median education level for the median household income for US counties?

To answer these questions, data must be collected, such as the `county` data set shown in Figure 1.5. Examining summary statistics could provide insights for each of the three questions about counties. Additionally, graphs can be used to visually explore data.

Scatterplots are one type of graph used to study the relationship between two numerical variables. Figure 1.8 compares the variables `homeownership` and `multi_unit`, which is the percent of units in multi-unit structures (e.g. apartments, condos). Each point on the plot represents a single county. For instance, the highlighted dot corresponds to County 413 in the `county` data set: Chattahoochee County, Georgia, which has 39.4% of units in multi-unit structures and a homeownership rate of 31.3%. The scatterplot suggests a relationship between the two variables: counties with a higher rate of multi-units tend to have lower homeownership rates. We might brainstorm as to why this relationship exists and investigate each idea to determine which are the most reasonable explanations.

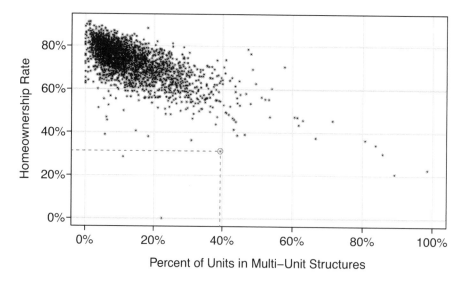

Figure 1.8: A scatterplot of homeownership versus the percent of units that are in multi-unit structures for US counties. The highlighted dot represents Chattahoochee County, Georgia, which has a multi-unit rate of 39.4% and a homeownership rate of 31.3%.

The multi-unit and homeownership rates are said to be associated because the plot shows a discernible pattern. When two variables show some connection with one another, they are called **associated** variables. Associated variables can also be called **dependent** variables and vice-versa.

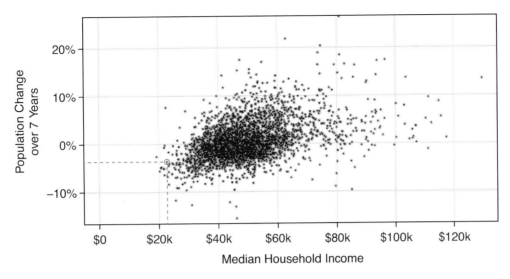

Figure 1.9: A scatterplot showing `pop_change` against `median_hh_income`. Owsley County of Kentucky, is highlighted, which lost 3.63% of its population from 2010 to 2017 and had median household income of $22,736.

GUIDED PRACTICE 1.7

Examine the variables in the `loan50` data set, which are described in Figure 1.4 on page 12. Create two questions about possible relationships between variables in `loan50` that are of interest to you.[8]

EXAMPLE 1.8

This example examines the relationship between a county's population change from 2010 to 2017 and median household income, which is visualized as a scatterplot in Figure 1.9. Are these variables associated?

The larger the median household income for a county, the higher the population growth observed for the county. While this trend isn't true for every county, the trend in the plot is evident. Since there is some relationship between the variables, they are associated.

Because there is a downward trend in Figure 1.8 – counties with more units in multi-unit structures are associated with lower homeownership – these variables are said to be **negatively associated**. A **positive association** is shown in the relationship between the `median_hh_income` and `pop_change` in Figure 1.9, where counties with higher median household income tend to have higher rates of population growth.

If two variables are not associated, then they are said to be **independent**. That is, two variables are independent if there is no evident relationship between the two.

ASSOCIATED OR INDEPENDENT, NOT BOTH

A pair of variables are either related in some way (associated) or not (independent). No pair of variables is both associated and independent.

---

[8]Two example questions: (1) What is the relationship between loan amount and total income? (2) If someone's income is above the average, will their interest rate tend to be above or below the average?

1.2.4 Explanatory and response variables

When we ask questions about the relationship between two variables, we sometimes also want to determine if the change in one variable causes a change in the other. Consider the following rephrasing of an earlier question about the county data set:

> *If there is an increase in the median household income in a county, does this drive an increase in its population?*

In this question, we are asking whether one variable affects another. If this is our underlying belief, then *median household income* is the **explanatory** variable and the *population change* is the **response** variable in the hypothesized relationship.[9]

EXPLANATORY AND RESPONSE VARIABLES

When we suspect one variable might causally affect another, we label the first variable the explanatory variable and the second the response variable.

$$\text{explanatory variable} \xrightarrow{\text{might affect}} \text{response variable}$$

For many pairs of variables, there is no hypothesized relationship, and these labels would not be applied to either variable in such cases.

Bear in mind that the act of labeling the variables in this way does nothing to guarantee that a causal relationship exists. A formal evaluation to check whether one variable causes a change in another requires an experiment.

1.2.5 Introducing observational studies and experiments

There are two primary types of data collection: observational studies and experiments.

Researchers perform an **observational study** when they collect data in a way that does not directly interfere with how the data arise. For instance, researchers may collect information via surveys, review medical or company records, or follow a **cohort** of many similar individuals to form hypotheses about why certain diseases might develop. In each of these situations, researchers merely observe the data that arise. In general, observational studies can provide evidence of a naturally occurring association between variables, but they cannot by themselves show a causal connection.

When researchers want to investigate the possibility of a causal connection, they conduct an **experiment**. Usually there will be both an explanatory and a response variable. For instance, we may suspect administering a drug will reduce mortality in heart attack patients over the following year. To check if there really is a causal connection between the explanatory variable and the response, researchers will collect a sample of individuals and split them into groups. The individuals in each group are *assigned* a treatment. When individuals are randomly assigned to a group, the experiment is called a **randomized experiment**. For example, each heart attack patient in the drug trial could be randomly assigned, perhaps by flipping a coin, into one of two groups: the first group receives a **placebo** (fake treatment) and the second group receives the drug. See the case study in Section 1.1 for another example of an experiment, though that study did not employ a placebo.

ASSOCIATION $\neq$ CAUSATION

In general, association does not imply causation, and causation can only be inferred from a randomized experiment.

---

[9]Sometimes the explanatory variable is called the **independent** variable and the response variable is called the **dependent** variable. However, this becomes confusing since a *pair* of variables might be independent or dependent, so we avoid this language.

Exercises

1.3 Air pollution and birth outcomes, study components. Researchers collected data to examine the relationship between air pollutants and preterm births in Southern California. During the study air pollution levels were measured by air quality monitoring stations. Specifically, levels of carbon monoxide were recorded in parts per million, nitrogen dioxide and ozone in parts per hundred million, and coarse particulate matter (PM_{10}) in $\mu g/m^3$. Length of gestation data were collected on 143,196 births between the years 1989 and 1993, and air pollution exposure during gestation was calculated for each birth. The analysis suggested that increased ambient PM_{10} and, to a lesser degree, CO concentrations may be associated with the occurrence of preterm births.[10]

(a) Identify the main research question of the study.

(b) Who are the subjects in this study, and how many are included?

(c) What are the variables in the study? Identify each variable as numerical or categorical. If numerical, state whether the variable is discrete or continuous. If categorical, state whether the variable is ordinal.

1.4 Buteyko method, study components. The Buteyko method is a shallow breathing technique developed by Konstantin Buteyko, a Russian doctor, in 1952. Anecdotal evidence suggests that the Buteyko method can reduce asthma symptoms and improve quality of life. In a scientific study to determine the effectiveness of this method, researchers recruited 600 asthma patients aged 18-69 who relied on medication for asthma treatment. These patients were randomly split into two research groups: one practiced the Buteyko method and the other did not. Patients were scored on quality of life, activity, asthma symptoms, and medication reduction on a scale from 0 to 10. On average, the participants in the Buteyko group experienced a significant reduction in asthma symptoms and an improvement in quality of life.[11]

(a) Identify the main research question of the study.

(b) Who are the subjects in this study, and how many are included?

(c) What are the variables in the study? Identify each variable as numerical or categorical. If numerical, state whether the variable is discrete or continuous. If categorical, state whether the variable is ordinal.

1.5 Cheaters, study components. Researchers studying the relationship between honesty, age and self-control conducted an experiment on 160 children between the ages of 5 and 15. Participants reported their age, sex, and whether they were an only child or not. The researchers asked each child to toss a fair coin in private and to record the outcome (white or black) on a paper sheet, and said they would only reward children who report white. The study's findings can be summarized as follows: "Half the students were explicitly told not to cheat and the others were not given any explicit instructions. In the no instruction group probability of cheating was found to be uniform across groups based on child's characteristics. In the group that was explicitly told to not cheat, girls were less likely to cheat, and while rate of cheating didn't vary by age for boys, it decreased with age for girls."[12]

(a) Identify the main research question of the study.

(b) Who are the subjects in this study, and how many are included?

(c) How many variables were recorded for each subject in the study in order to conclude these findings? State the variables and their types.

[10]B. Ritz et al. "Effect of air pollution on preterm birth among children born in Southern California between 1989 and 1993". In: *Epidemiology* 11.5 (2000), pp. 502–511.

[11]J. McGowan. "Health Education: Does the Buteyko Institute Method make a difference?" In: *Thorax* 58 (2003).

[12]Alessandro Bucciol and Marco Piovesan. "Luck or cheating? A field experiment on honesty with children". In: *Journal of Economic Psychology* 32.1 (2011), pp. 73–78.

1.6 Stealers, study components. In a study of the relationship between socio-economic class and unethical behavior, 129 University of California undergraduates at Berkeley were asked to identify themselves as having low or high social-class by comparing themselves to others with the most (least) money, most (least) education, and most (least) respected jobs. They were also presented with a jar of individually wrapped candies and informed that the candies were for children in a nearby laboratory, but that they could take some if they wanted. After completing some unrelated tasks, participants reported the number of candies they had taken.[13]

(a) Identify the main research question of the study.

(b) Who are the subjects in this study, and how many are included?

(c) The study found that students who were identified as upper-class took more candy than others. How many variables were recorded for each subject in the study in order to conclude these findings? State the variables and their types.

1.7 Migraine and acupuncture, Part II. Exercise 1.1 introduced a study exploring whether acupuncture had any effect on migraines. Researchers conducted a randomized controlled study where patients were randomly assigned to one of two groups: treatment or control. The patients in the treatment group received acupuncture that was specifically designed to treat migraines. The patients in the control group received placebo acupuncture (needle insertion at non-acupoint locations). 24 hours after patients received acupuncture, they were asked if they were pain free. What are the explanatory and response variables in this study?

1.8 Sinusitis and antibiotics, Part II. Exercise 1.2 introduced a study exploring the effect of antibiotic treatment for acute sinusitis. Study participants either received either a 10-day course of an antibiotic (treatment) or a placebo similar in appearance and taste (control). At the end of the 10-day period, patients were asked if they experienced improvement in symptoms. What are the explanatory and response variables in this study?

1.9 Fisher's irises. Sir Ronald Aylmer Fisher was an English statistician, evolutionary biologist, and geneticist who worked on a data set that contained sepal length and width, and petal length and width from three species of iris flowers (*setosa*, *versicolor* and *virginica*). There were 50 flowers from each species in the data set.[14]

(a) How many cases were included in the data?

(b) How many numerical variables are included in the data? Indicate what they are, and if they are continuous or discrete.

(c) How many categorical variables are included in the data, and what are they? List the corresponding levels (categories).

Photo by Ryan Claussen (http://flic.kr/p/6QTcuX) CC BY-SA 2.0 license

1.10 Smoking habits of UK residents. A survey was conducted to study the smoking habits of UK residents. Below is a data matrix displaying a portion of the data collected in this survey. Note that "£" stands for British Pounds Sterling, "cig" stands for cigarettes, and "N/A" refers to a missing component of the data.[15]

	sex	age	marital	grossIncome	smoke	amtWeekends	amtWeekdays
1	Female	42	Single	Under £2,600	Yes	12 cig/day	12 cig/day
2	Male	44	Single	£10,400 to £15,600	No	N/A	N/A
3	Male	53	Married	Above £36,400	Yes	6 cig/day	6 cig/day
⋮	⋮	⋮	⋮	⋮	⋮	⋮	⋮
1691	Male	40	Single	£2,600 to £5,200	Yes	8 cig/day	8 cig/day

(a) What does each row of the data matrix represent?

(b) How many participants were included in the survey?

(c) Indicate whether each variable in the study is numerical or categorical. If numerical, identify as continuous or discrete. If categorical, indicate if the variable is ordinal.

[13]P.K. Piff et al. "Higher social class predicts increased unethical behavior". In: *Proceedings of the National Academy of Sciences* (2012).

[14]R.A Fisher. "The Use of Multiple Measurements in Taxonomic Problems". In: *Annals of Eugenics* 7 (1936), pp. 179–188.

[15]National STEM Centre, Large Datasets from stats4schools.

1.11 US Airports. The visualization below shows the geographical distribution of airports in the contiguous United States and Washington, DC. This visualization was constructed based on a dataset where each observation is an airport.[16]

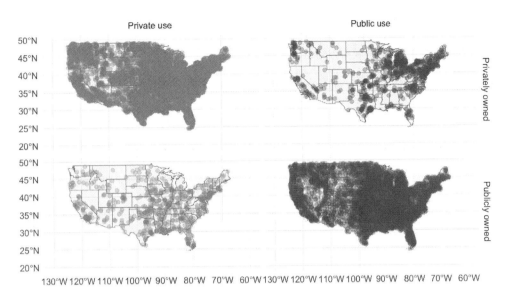

(a) List the variables used in creating this visualization.

(b) Indicate whether each variable in the study is numerical or categorical. If numerical, identify as continuous or discrete. If categorical, indicate if the variable is ordinal.

1.12 UN Votes. The visualization below shows voting patterns in the United States, Canada, and Mexico in the United Nations General Assembly on a variety of issues. Specifically, for a given year between 1946 and 2015, it displays the percentage of roll calls in which the country voted yes for each issue. This visualization was constructed based on a dataset where each observation is a country/year pair.[17]

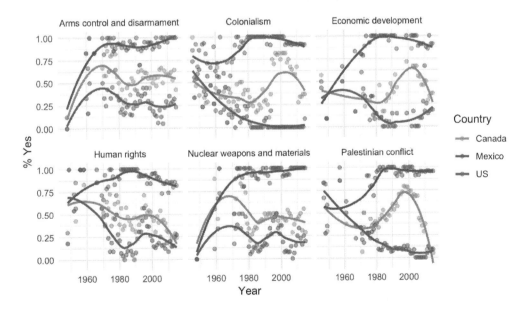

(a) List the variables used in creating this visualization.

(b) Indicate whether each variable in the study is numerical or categorical. If numerical, identify as continuous or discrete. If categorical, indicate if the variable is ordinal.

---

[16]Federal Aviation Administration, www.faa.gov/airports/airport_safety/airportdata_5010.

[17]David Robinson. *unvotes: United Nations General Assembly Voting Data*. R package version 0.2.0. 2017. URL: https://CRAN.R-project.org/package=unvotes.

1.3   Sampling principles and strategies

The first step in conducting research is to identify topics or questions that are to be investigated. A clearly laid out research question is helpful in identifying what subjects or cases should be studied and what variables are important. It is also important to consider *how* data are collected so that they are reliable and help achieve the research goals.

1.3.1   Populations and samples

Consider the following three research questions:

1. What is the average mercury content in swordfish in the Atlantic Ocean?
2. Over the last 5 years, what is the average time to complete a degree for Duke undergrads?
3. Does a new drug reduce the number of deaths in patients with severe heart disease?

Each research question refers to a target **population**. In the first question, the target population is all swordfish in the Atlantic ocean, and each fish represents a case. Often times, it is too expensive to collect data for every case in a population. Instead, a sample is taken. A **sample** represents a subset of the cases and is often a small fraction of the population. For instance, 60 swordfish (or some other number) in the population might be selected, and this sample data may be used to provide an estimate of the population average and answer the research question.

GUIDED PRACTICE 1.9

For the second and third questions above, identify the target population and what represents an individual case.[18]

1.3.2   Anecdotal evidence

Consider the following possible responses to the three research questions:

1. A man on the news got mercury poisoning from eating swordfish, so the average mercury concentration in swordfish must be dangerously high.
2. I met two students who took more than 7 years to graduate from Duke, so it must take longer to graduate at Duke than at many other colleges.
3. My friend's dad had a heart attack and died after they gave him a new heart disease drug, so the drug must not work.

Each conclusion is based on data. However, there are two problems. First, the data only represent one or two cases. Second, and more importantly, it is unclear whether these cases are actually representative of the population. Data collected in this haphazard fashion are called **anecdotal evidence**.

> **ANECDOTAL EVIDENCE**
>
> Be careful of data collected in a haphazard fashion. Such evidence may be true and verifiable, but it may only represent extraordinary cases.

---

[18](2) The first question is only relevant to students who complete their degree; the average cannot be computed using a student who never finished her degree. Thus, only Duke undergrads who graduated in the last five years represent cases in the population under consideration. Each such student is an individual case. (3) A person with severe heart disease represents a case. The population includes all people with severe heart disease.

Figure 1.10: In February 2010, some media pundits cited one large snow storm as valid evidence against global warming. As comedian Jon Stewart pointed out, "It's one storm, in one region, of one country."

Anecdotal evidence typically is composed of unusual cases that we recall based on their striking characteristics. For instance, we are more likely to remember the two people we met who took 7 years to graduate than the six others who graduated in four years. Instead of looking at the most unusual cases, we should examine a sample of many cases that represent the population.

1.3.3 Sampling from a population

We might try to estimate the time to graduation for Duke undergraduates in the last 5 years by collecting a sample of students. All graduates in the last 5 years represent the *population*, and graduates who are selected for review are collectively called the *sample*. In general, we always seek to *randomly* select a sample from a population. The most basic type of random selection is equivalent to how raffles are conducted. For example, in selecting graduates, we could write each graduate's name on a raffle ticket and draw 100 tickets. The selected names would represent a random sample of 100 graduates. We pick samples randomly to reduce the chance we introduce biases.

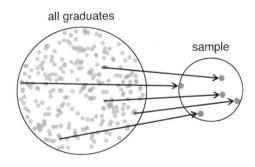

Figure 1.11: In this graphic, five graduates are randomly selected from the population to be included in the sample.

EXAMPLE 1.10

Suppose we ask a student who happens to be majoring in nutrition to select several graduates for the study. What kind of students do you think she might collect? Do you think her sample would be representative of all graduates?

Perhaps she would pick a disproportionate number of graduates from health-related fields. Or perhaps her selection would be a good representation of the population. When selecting samples by hand, we run the risk of picking a **biased** sample, even if their bias isn't intended.

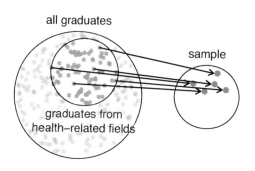

Figure 1.12: Asked to pick a sample of graduates, a nutrition major might inadvertently pick a disproportionate number of graduates from health-related majors.

If someone was permitted to pick and choose exactly which graduates were included in the sample, it is entirely possible that the sample could be skewed to that person's interests, which may be entirely unintentional. This introduces **bias** into a sample. Sampling randomly helps resolve this problem. The most basic random sample is called a **simple random sample**, and which is equivalent to using a raffle to select cases. This means that each case in the population has an equal chance of being included and there is no implied connection between the cases in the sample.

The act of taking a simple random sample helps minimize bias. However, bias can crop up in other ways. Even when people are picked at random, e.g. for surveys, caution must be exercised if the **non-response rate** is high. For instance, if only 30% of the people randomly sampled for a survey actually respond, then it is unclear whether the results are **representative** of the entire population. This **non-response bias** can skew results.

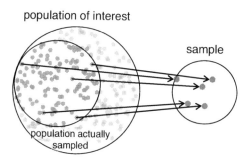

Figure 1.13: Due to the possibility of non-response, surveys studies may only reach a certain group within the population. It is difficult, and often times impossible, to completely fix this problem.

Another common downfall is a **convenience sample**, where individuals who are easily accessible are more likely to be included in the sample. For instance, if a political survey is done by stopping people walking in the Bronx, this will not represent all of New York City. It is often difficult to discern what sub-population a convenience sample represents.

GUIDED PRACTICE 1.11

Ⓖ

We can easily access ratings for products, sellers, and companies through websites. These ratings are based only on those people who go out of their way to provide a rating. If 50% of online reviews for a product are negative, do you think this means that 50% of buyers are dissatisfied with the product?[19]

---

[19]Answers will vary. From our own anecdotal experiences, we believe people tend to rant more about products that fell below expectations than rave about those that perform as expected. For this reason, we suspect there is a negative bias in product ratings on sites like Amazon. However, since our experiences may not be representative, we also keep an open mind.

1.3.4 Observational studies

Data where no treatment has been explicitly applied (or explicitly withheld) is called **observational data**. For instance, the loan data and county data described in Section 1.2 are both examples of observational data. Making causal conclusions based on experiments is often reasonable. However, making the same causal conclusions based on observational data can be treacherous and is not recommended. Thus, observational studies are generally only sufficient to show associations or form hypotheses that we later check using experiments.

GUIDED PRACTICE 1.12

Suppose an observational study tracked sunscreen use and skin cancer, and it was found that the more sunscreen someone used, the more likely the person was to have skin cancer. Does this mean sunscreen *causes* skin cancer?[20]

Some previous research tells us that using sunscreen actually reduces skin cancer risk, so maybe there is another variable that can explain this hypothetical association between sunscreen usage and skin cancer. One important piece of information that is absent is sun exposure. If someone is out in the sun all day, she is more likely to use sunscreen *and* more likely to get skin cancer. Exposure to the sun is unaccounted for in the simple investigation.

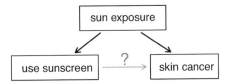

Sun exposure is what is called a **confounding variable**,[21] which is a variable that is correlated with both the explanatory and response variables. While one method to justify making causal conclusions from observational studies is to exhaust the search for confounding variables, there is no guarantee that all confounding variables can be examined or measured.

GUIDED PRACTICE 1.13

Figure 1.8 shows a negative association between the homeownership rate and the percentage of multi-unit structures in a county. However, it is unreasonable to conclude that there is a causal relationship between the two variables. Suggest a variable that might explain the negative relationship.[22]

Observational studies come in two forms: prospective and retrospective studies. A **prospective study** identifies individuals and collects information as events unfold. For instance, medical researchers may identify and follow a group of patients over many years to assess the possible influences of behavior on cancer risk. One example of such a study is The Nurses' Health Study, started in 1976 and expanded in 1989. This prospective study recruits registered nurses and then collects data from them using questionnaires. **Retrospective studies** collect data after events have taken place, e.g. researchers may review past events in medical records. Some data sets may contain both prospectively- and retrospectively-collected variables.

1.3.5 Four sampling methods

Almost all statistical methods are based on the notion of implied randomness. If observational data are not collected in a random framework from a population, these statistical methods – the estimates and errors associated with the estimates – are not reliable. Here we consider four random sampling techniques: simple, stratified, cluster, and multistage sampling. Figures 1.14 and 1.15 provide graphical representations of these techniques.

---

[20]No. See the paragraph following the exercise for an explanation.
[21]Also called a **lurking variable**, **confounding factor**, or a **confounder**.
[22]Answers will vary. Population density may be important. If a county is very dense, then this may require a larger fraction of residents to live in multi-unit structures. Additionally, the high density may contribute to increases in property value, making homeownership infeasible for many residents.

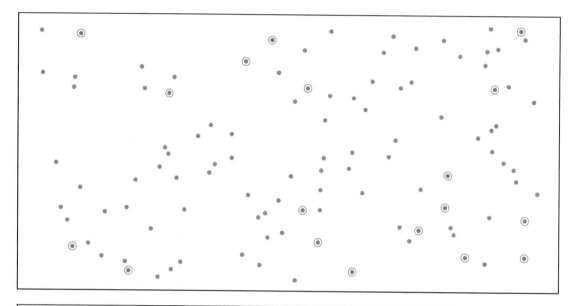

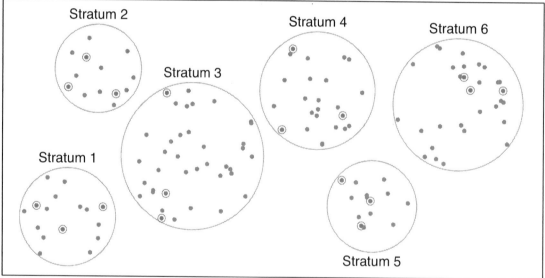

Figure 1.14: Examples of simple random and stratified sampling. In the top panel, simple random sampling was used to randomly select the 18 cases. In the bottom panel, stratified sampling was used: cases were grouped into strata, then simple random sampling was employed within each stratum.

Simple random sampling is probably the most intuitive form of random sampling. Consider the salaries of Major League Baseball (MLB) players, where each player is a member of one of the league's 30 teams. To take a simple random sample of 120 baseball players and their salaries, we could write the names of that season's several hundreds of players onto slips of paper, drop the slips into a bucket, shake the bucket around until we are sure the names are all mixed up, then draw out slips until we have the sample of 120 players. In general, a sample is referred to as "simple random" if each case in the population has an equal chance of being included in the final sample *and* knowing that a case is included in a sample does not provide useful information about which other cases are included.

Stratified sampling is a divide-and-conquer sampling strategy. The population is divided into groups called **strata**. The strata are chosen so that similar cases are grouped together, then a second sampling method, usually simple random sampling, is employed within each stratum. In the baseball salary example, the teams could represent the strata, since some teams have a lot more money (up to 4 times as much!). Then we might randomly sample 4 players from each team for a total of 120 players.

Stratified sampling is especially useful when the cases in each stratum are very similar with respect to the outcome of interest. The downside is that analyzing data from a stratified sample is a more complex task than analyzing data from a simple random sample. The analysis methods introduced in this book would need to be extended to analyze data collected using stratified sampling.

EXAMPLE 1.14

Why would it be good for cases within each stratum to be very similar?

We might get a more stable estimate for the subpopulation in a stratum if the cases are very similar, leading to more precise estimates within each group. When we combine these estimates into a single estimate for the full population, that population estimate will tend to be more precise since each individual group estimate is itself more precise.

In a **cluster sample**, we break up the population into many groups, called **clusters**. Then we sample a fixed number of clusters and include all observations from each of those clusters in the sample. A **multistage sample** is like a cluster sample, but rather than keeping all observations in each cluster, we collect a random sample within each selected cluster.

Sometimes cluster or multistage sampling can be more economical than the alternative sampling techniques. Also, unlike stratified sampling, these approaches are most helpful when there is a lot of case-to-case variability within a cluster but the clusters themselves don't look very different from one another. For example, if neighborhoods represented clusters, then cluster or multistage sampling work best when the neighborhoods are very diverse. A downside of these methods is that more advanced techniques are typically required to analyze the data, though the methods in this book can be extended to handle such data.

EXAMPLE 1.15

Suppose we are interested in estimating the malaria rate in a densely tropical portion of rural Indonesia. We learn that there are 30 villages in that part of the Indonesian jungle, each more or less similar to the next. Our goal is to test 150 individuals for malaria. What sampling method should be employed?

A simple random sample would likely draw individuals from all 30 villages, which could make data collection extremely expensive. Stratified sampling would be a challenge since it is unclear how we would build strata of similar individuals. However, cluster sampling or multistage sampling seem like very good ideas. If we decided to use multistage sampling, we might randomly select half of the villages, then randomly select 10 people from each. This would probably reduce our data collection costs substantially in comparison to a simple random sample, and the cluster sample would still give us reliable information, even if we would need to analyze the data with slightly more advanced methods than we discuss in this book.

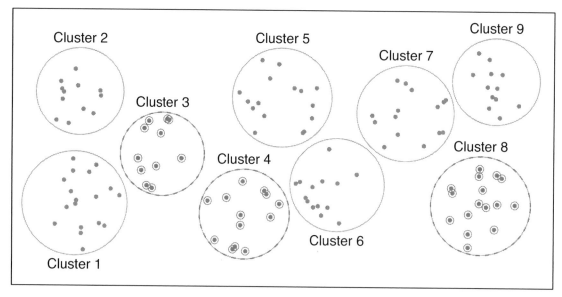

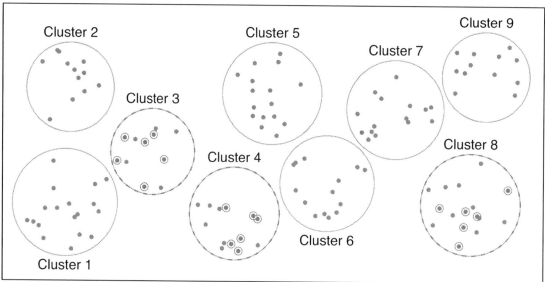

Figure 1.15: Examples of cluster and multistage sampling. In the top panel, cluster sampling was used: data were binned into nine clusters, three of these clusters were sampled, and all observations within these three cluster were included in the sample. In the bottom panel, multistage sampling was used, which differs from cluster sampling only in that we randomly select a subset of each cluster to be included in the sample rather than measuring every case in each sampled cluster.

Exercises

1.13 Air pollution and birth outcomes, scope of inference. Exercise 1.3 introduces a study where researchers collected data to examine the relationship between air pollutants and preterm births in Southern California. During the study air pollution levels were measured by air quality monitoring stations. Length of gestation data were collected on 143,196 births between the years 1989 and 1993, and air pollution exposure during gestation was calculated for each birth.

(a) Identify the population of interest and the sample in this study.

(b) Comment on whether or not the results of the study can be generalized to the population, and if the findings of the study can be used to establish causal relationships.

1.14 Cheaters, scope of inference. Exercise 1.5 introduces a study where researchers studying the relationship between honesty, age, and self-control conducted an experiment on 160 children between the ages of 5 and 15. The researchers asked each child to toss a fair coin in private and to record the outcome (white or black) on a paper sheet, and said they would only reward children who report white. Half the students were explicitly told not to cheat and the others were not given any explicit instructions. Differences were observed in the cheating rates in the instruction and no instruction groups, as well as some differences across children's characteristics within each group.

(a) Identify the population of interest and the sample in this study.

(b) Comment on whether or not the results of the study can be generalized to the population, and if the findings of the study can be used to establish causal relationships.

1.15 Buteyko method, scope of inference. Exercise 1.4 introduces a study on using the Buteyko shallow breathing technique to reduce asthma symptoms and improve quality of life. As part of this study 600 asthma patients aged 18-69 who relied on medication for asthma treatment were recruited and randomly assigned to two groups: one practiced the Buteyko method and the other did not. Those in the Buteyko group experienced, on average, a significant reduction in asthma symptoms and an improvement in quality of life.

(a) Identify the population of interest and the sample in this study.

(b) Comment on whether or not the results of the study can be generalized to the population, and if the findings of the study can be used to establish causal relationships.

1.16 Stealers, scope of inference. Exercise 1.6 introduces a study on the relationship between socio-economic class and unethical behavior. As part of this study 129 University of California Berkeley undergraduates were asked to identify themselves as having low or high social-class by comparing themselves to others with the most (least) money, most (least) education, and most (least) respected jobs. They were also presented with a jar of individually wrapped candies and informed that the candies were for children in a nearby laboratory, but that they could take some if they wanted. After completing some unrelated tasks, participants reported the number of candies they had taken. It was found that those who were identified as upper-class took more candy than others.

(a) Identify the population of interest and the sample in this study.

(b) Comment on whether or not the results of the study can be generalized to the population, and if the findings of the study can be used to establish causal relationships.

1.17 Relaxing after work. The General Social Survey asked the question, "After an average work day, about how many hours do you have to relax or pursue activities that you enjoy?" to a random sample of 1,155 Americans. The average relaxing time was found to be 1.65 hours. Determine which of the following is an observation, a variable, a sample statistic (value calculated based on the observed sample), or a population parameter.

(a) An American in the sample.

(b) Number of hours spent relaxing after an average work day.

(c) 1.65.

(d) Average number of hours all Americans spend relaxing after an average work day.

1.18 Cats on YouTube. Suppose you want to estimate the percentage of videos on YouTube that are cat videos. It is impossible for you to watch all videos on YouTube so you use a random video picker to select 1000 videos for you. You find that 2% of these videos are cat videos. Determine which of the following is an observation, a variable, a sample statistic (value calculated based on the observed sample), or a population parameter.

(a) Percentage of all videos on YouTube that are cat videos.

(b) 2%.

(c) A video in your sample.

(d) Whether or not a video is a cat video.

1.19 Course satisfaction across sections. A large college class has 160 students. All 160 students attend the lectures together, but the students are divided into 4 groups, each of 40 students, for lab sections administered by different teaching assistants. The professor wants to conduct a survey about how satisfied the students are with the course, and he believes that the lab section a student is in might affect the student's overall satisfaction with the course.

(a) What type of study is this?

(b) Suggest a sampling strategy for carrying out this study.

1.20 Housing proposal across dorms. On a large college campus first-year students and sophomores live in dorms located on the eastern part of the campus and juniors and seniors live in dorms located on the western part of the campus. Suppose you want to collect student opinions on a new housing structure the college administration is proposing and you want to make sure your survey equally represents opinions from students from all years.

(a) What type of study is this?

(b) Suggest a sampling strategy for carrying out this study.

1.21 Internet use and life expectancy. The following scatterplot was created as part of a study evaluating the relationship between estimated life expectancy at birth (as of 2014) and percentage of internet users (as of 2009) in 208 countries for which such data were available.[23]

(a) Describe the relationship between life expectancy and percentage of internet users.

(b) What type of study is this?

(c) State a possible confounding variable that might explain this relationship and describe its potential effect.

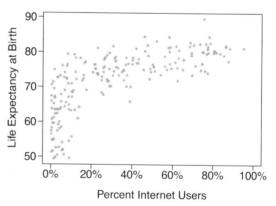

1.22 Stressed out, Part I. A study that surveyed a random sample of otherwise healthy high school students found that they are more likely to get muscle cramps when they are stressed. The study also noted that students drink more coffee and sleep less when they are stressed.

(a) What type of study is this?

(b) Can this study be used to conclude a causal relationship between increased stress and muscle cramps?

(c) State possible confounding variables that might explain the observed relationship between increased stress and muscle cramps.

1.23 Evaluate sampling methods. A university wants to determine what fraction of its undergraduate student body support a new $25 annual fee to improve the student union. For each proposed method below, indicate whether the method is reasonable or not.

(a) Survey a simple random sample of 500 students.

(b) Stratify students by their field of study, then sample 10% of students from each stratum.

(c) Cluster students by their ages (e.g. 18 years old in one cluster, 19 years old in one cluster, etc.), then randomly sample three clusters and survey all students in those clusters.

---

[23]CIA Factbook, Country Comparisons, 2014.

1.24   Random digit dialing. The Gallup Poll uses a procedure called random digit dialing, which creates phone numbers based on a list of all area codes in America in conjunction with the associated number of residential households in each area code. Give a possible reason the Gallup Poll chooses to use random digit dialing instead of picking phone numbers from the phone book.

1.25   Haters are gonna hate, study confirms. A study published in the *Journal of Personality and Social Psychology* asked a group of 200 randomly sampled men and women to evaluate how they felt about various subjects, such as camping, health care, architecture, taxidermy, crossword puzzles, and Japan in order to measure their attitude towards mostly independent stimuli. Then, they presented the participants with information about a new product: a microwave oven. This microwave oven does not exist, but the participants didn't know this, and were given three positive and three negative fake reviews. People who reacted positively to the subjects on the dispositional attitude measurement also tended to react positively to the microwave oven, and those who reacted negatively tended to react negatively to it. Researchers concluded that "some people tend to like things, whereas others tend to dislike things, and a more thorough understanding of this tendency will lead to a more thorough understanding of the psychology of attitudes."[24]

(a) What are the cases?

(b) What is (are) the response variable(s) in this study?

(c) What is (are) the explanatory variable(s) in this study?

(d) Does the study employ random sampling?

(e) Is this an observational study or an experiment? Explain your reasoning.

(f) Can we establish a causal link between the explanatory and response variables?

(g) Can the results of the study be generalized to the population at large?

1.26   Family size. Suppose we want to estimate household size, where a "household" is defined as people living together in the same dwelling, and sharing living accommodations. If we select students at random at an elementary school and ask them what their family size is, will this be a good measure of household size? Or will our average be biased? If so, will it overestimate or underestimate the true value?

1.27   Sampling strategies. A statistics student who is curious about the relationship between the amount of time students spend on social networking sites and their performance at school decides to conduct a survey. Various research strategies for collecting data are described below. In each, name the sampling method proposed and any bias you might expect.

(a) He randomly samples 40 students from the study's population, gives them the survey, asks them to fill it out and bring it back the next day.

(b) He gives out the survey only to his friends, making sure each one of them fills out the survey.

(c) He posts a link to an online survey on Facebook and asks his friends to fill out the survey.

(d) He randomly samples 5 classes and asks a random sample of students from those classes to fill out the survey.

1.28   Reading the paper. Below are excerpts from two articles published in the *NY Times*:

(a) An article titled *Risks: Smokers Found More Prone to Dementia* states the following:[25]

> "Researchers analyzed data from 23,123 health plan members who participated in a voluntary exam and health behavior survey from 1978 to 1985, when they were 50-60 years old. 23 years later, about 25% of the group had dementia, including 1,136 with Alzheimer's disease and 416 with vascular dementia. After adjusting for other factors, the researchers concluded that pack-a-day smokers were 37% more likely than nonsmokers to develop dementia, and the risks went up with increased smoking; 44% for one to two packs a day; and twice the risk for more than two packs."

Based on this study, can we conclude that smoking causes dementia later in life? Explain your reasoning.

(b) Another article titled *The School Bully Is Sleepy* states the following:[26]

> "The University of Michigan study, collected survey data from parents on each child's sleep habits and asked both parents and teachers to assess behavioral concerns. About a third of the students studied were identified by parents or teachers as having problems with disruptive behavior or bullying. The researchers found that children who had behavioral issues and those who were identified as bullies were twice as likely to have shown symptoms of sleep disorders."

A friend of yours who read the article says, "The study shows that sleep disorders lead to bullying in school children." Is this statement justified? If not, how best can you describe the conclusion that can be drawn from this study?

---

[24]Justin Hepler and Dolores Albarracín. "Attitudes without objects - Evidence for a dispositional attitude, its measurement, and its consequences". In: *Journal of personality and social psychology* 104.6 (2013), p. 1060.

[25]R.C. Rabin. "Risks: Smokers Found More Prone to Dementia". In: *New York Times* (2010).

[26]T. Parker-Pope. "The School Bully Is Sleepy". In: *New York Times* (2011).

1.4   Experiments

Studies where the researchers assign treatments to cases are called **experiments**. When this assignment includes randomization, e.g. using a coin flip to decide which treatment a patient receives, it is called a **randomized experiment**. Randomized experiments are fundamentally important when trying to show a causal connection between two variables.

1.4.1   Principles of experimental design

Randomized experiments are generally built on four principles.

Controlling. Researchers assign treatments to cases, and they do their best to **control** any other differences in the groups.[27] For example, when patients take a drug in pill form, some patients take the pill with only a sip of water while others may have it with an entire glass of water. To control for the effect of water consumption, a doctor may ask all patients to drink a 12 ounce glass of water with the pill.

Randomization. Researchers randomize patients into treatment groups to account for variables that cannot be controlled. For example, some patients may be more susceptible to a disease than others due to their dietary habits. Randomizing patients into the treatment or control group helps even out such differences, and it also prevents accidental bias from entering the study.

Replication. The more cases researchers observe, the more accurately they can estimate the effect of the explanatory variable on the response. In a single study, we **replicate** by collecting a sufficiently large sample. Additionally, a group of scientists may replicate an entire study to verify an earlier finding.

Blocking. Researchers sometimes know or suspect that variables, other than the treatment, influence the response. Under these circumstances, they may first group individuals based on this variable into **blocks** and then randomize cases within each block to the treatment groups. This strategy is often referred to as **blocking**. For instance, if we are looking at the effect of a drug on heart attacks, we might first split patients in the study into low-risk and high-risk blocks, then randomly assign half the patients from each block to the control group and the other half to the treatment group, as shown in Figure 1.16. This strategy ensures each treatment group has an equal number of low-risk and high-risk patients.

It is important to incorporate the first three experimental design principles into any study, and this book describes applicable methods for analyzing data from such experiments. Blocking is a slightly more advanced technique, and statistical methods in this book may be extended to analyze data collected using blocking.

1.4.2   Reducing bias in human experiments

Randomized experiments are the gold standard for data collection, but they do not ensure an unbiased perspective into the cause and effect relationship in all cases. Human studies are perfect examples where bias can unintentionally arise. Here we reconsider a study where a new drug was used to treat heart attack patients. In particular, researchers wanted to know if the drug reduced deaths in patients.

These researchers designed a randomized experiment because they wanted to draw causal conclusions about the drug's effect. Study volunteers[28] were randomly placed into two study groups. One group, the **treatment group**, received the drug. The other group, called the **control group**, did not receive any drug treatment.

---

[27]This is a different concept than a *control group*, which we discuss in the second principle and in Section 1.4.2.
[28]Human subjects are often called **patients**, **volunteers**, or **study participants**.

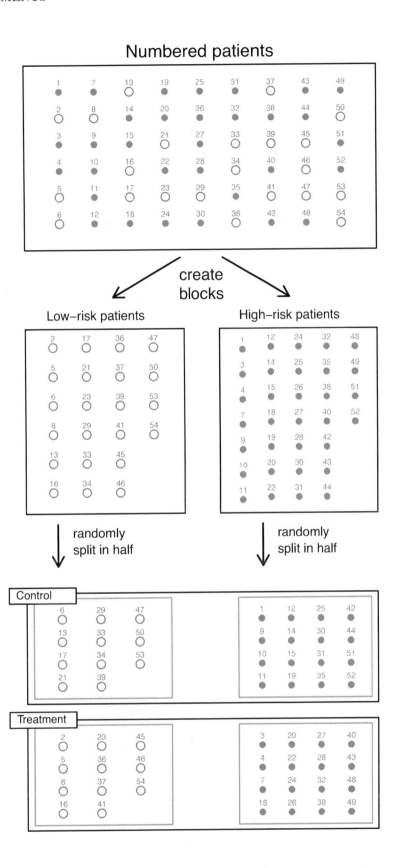

Figure 1.16: Blocking using a variable depicting patient risk. Patients are first divided into low-risk and high-risk blocks, then each block is evenly separated into the treatment groups using randomization. This strategy ensures an equal representation of patients in each treatment group from both the low-risk and high-risk categories.

Put yourself in the place of a person in the study. If you are in the treatment group, you are given a fancy new drug that you anticipate will help you. On the other hand, a person in the other group doesn't receive the drug and sits idly, hoping her participation doesn't increase her risk of death. These perspectives suggest there are actually two effects: the one of interest is the effectiveness of the drug, and the second is an emotional effect that is difficult to quantify.

Researchers aren't usually interested in the emotional effect, which might bias the study. To circumvent this problem, researchers do not want patients to know which group they are in. When researchers keep the patients uninformed about their treatment, the study is said to be **blind**. But there is one problem: if a patient doesn't receive a treatment, she will know she is in the control group. The solution to this problem is to give fake treatments to patients in the control group. A fake treatment is called a **placebo**, and an effective placebo is the key to making a study truly blind. A classic example of a placebo is a sugar pill that is made to look like the actual treatment pill. Often times, a placebo results in a slight but real improvement in patients. This effect has been dubbed the **placebo effect**.

The patients are not the only ones who should be blinded: doctors and researchers can accidentally bias a study. When a doctor knows a patient has been given the real treatment, she might inadvertently give that patient more attention or care than a patient that she knows is on the placebo. To guard against this bias, which again has been found to have a measurable effect in some instances, most modern studies employ a **double-blind** setup where doctors or researchers who interact with patients are, just like the patients, unaware of who is or is not receiving the treatment.[29]

GUIDED PRACTICE 1.16

Look back to the study in Section 1.1 where researchers were testing whether stents were effective at reducing strokes in at-risk patients. Is this an experiment? Was the study blinded? Was it double-blinded?[30]

GUIDED PRACTICE 1.17

For the study in Section 1.1, could the researchers have employed a placebo? If so, what would that placebo have looked like?[31]

You may have many questions about the ethics of sham surgeries to create a placebo after reading Guided Practice 1.17. These questions may have even arisen in your mind when in the general experiment context, where a possibly helpful treatment was withheld from individuals in the control group; the main difference is that a sham surgery tends to create additional risk, while withholding a treatment only maintains a person's risk.

There are always multiple viewpoints of experiments and placebos, and rarely is it obvious which is ethically "correct". For instance, is it ethical to use a sham surgery when it creates a risk to the patient? However, if we don't use sham surgeries, we may promote the use of a costly treatment that has no real effect; if this happens, money and other resources will be diverted away from other treatments that are known to be helpful. Ultimately, this is a difficult situation where we cannot perfectly protect both the patients who have volunteered for the study and the patients who may benefit (or not) from the treatment in the future.

---

[29]There are always some researchers involved in the study who do know which patients are receiving which treatment. However, they do not interact with the study's patients and do not tell the blinded health care professionals who is receiving which treatment.

[30]The researchers assigned the patients into their treatment groups, so this study was an experiment. However, the patients could distinguish what treatment they received, so this study was not blind. The study could not be double-blind since it was not blind.

[31]Ultimately, can we make patients think they got treated from a surgery? In fact, we can, and some experiments use what's called a **sham surgery**. In a sham surgery, the patient does undergo surgery, but the patient does not receive the full treatment, though they will still get a placebo effect.

Exercises

1.29 Light and exam performance. A study is designed to test the effect of light level on exam performance of students. The researcher believes that light levels might have different effects on males and females, so wants to make sure both are equally represented in each treatment. The treatments are fluorescent overhead lighting, yellow overhead lighting, no overhead lighting (only desk lamps).

(a) What is the response variable?

(b) What is the explanatory variable? What are its levels?

(c) What is the blocking variable? What are its levels?

1.30 Vitamin supplements. To assess the effectiveness of taking large doses of vitamin C in reducing the duration of the common cold, researchers recruited 400 healthy volunteers from staff and students at a university. A quarter of the patients were assigned a placebo, and the rest were evenly divided between 1g Vitamin C, 3g Vitamin C, or 3g Vitamin C plus additives to be taken at onset of a cold for the following two days. All tablets had identical appearance and packaging. The nurses who handed the prescribed pills to the patients knew which patient received which treatment, but the researchers assessing the patients when they were sick did not. No significant differences were observed in any measure of cold duration or severity between the four groups, and the placebo group had the shortest duration of symptoms.[32]

(a) Was this an experiment or an observational study? Why?

(b) What are the explanatory and response variables in this study?

(c) Were the patients blinded to their treatment?

(d) Was this study double-blind?

(e) Participants are ultimately able to choose whether or not to use the pills prescribed to them. We might expect that not all of them will adhere and take their pills. Does this introduce a confounding variable to the study? Explain your reasoning.

1.31 Light, noise, and exam performance. A study is designed to test the effect of light level and noise level on exam performance of students. The researcher believes that light and noise levels might have different effects on males and females, so wants to make sure both are equally represented in each treatment. The light treatments considered are fluorescent overhead lighting, yellow overhead lighting, no overhead lighting (only desk lamps). The noise treatments considered are no noise, construction noise, and human chatter noise.

(a) What type of study is this?

(b) How many factors are considered in this study? Identify them, and describe their levels.

(c) What is the role of the sex variable in this study?

1.32 Music and learning. You would like to conduct an experiment in class to see if students learn better if they study without any music, with music that has no lyrics (instrumental), or with music that has lyrics. Briefly outline a design for this study.

1.33 Soda preference. You would like to conduct an experiment in class to see if your classmates prefer the taste of regular Coke or Diet Coke. Briefly outline a design for this study.

1.34 Exercise and mental health. A researcher is interested in the effects of exercise on mental health and he proposes the following study: Use stratified random sampling to ensure representative proportions of 18-30, 31-40 and 41- 55 year olds from the population. Next, randomly assign half the subjects from each age group to exercise twice a week, and instruct the rest not to exercise. Conduct a mental health exam at the beginning and at the end of the study, and compare the results.

(a) What type of study is this?

(b) What are the treatment and control groups in this study?

(c) Does this study make use of blocking? If so, what is the blocking variable?

(d) Does this study make use of blinding?

(e) Comment on whether or not the results of the study can be used to establish a causal relationship between exercise and mental health, and indicate whether or not the conclusions can be generalized to the population at large.

(f) Suppose you are given the task of determining if this proposed study should get funding. Would you have any reservations about the study proposal?

---

[32]C. Audera et al. "Mega-dose vitamin C in treatment of the common cold: a randomised controlled trial". In: *Medical Journal of Australia* 175.7 (2001), pp. 359–362.

Chapter exercises

1.35   Pet names. The city of Seattle, WA has an open data portal that includes pets registered in the city. For each registered pet, we have information on the pet's name and species. The following visualization plots the proportion of dogs with a given name versus the proportion of cats with the same name. The 20 most common cat and dog names are displayed. The diagonal line on the plot is the $x = y$ line; if a name appeared on this line, the name's popularity would be exactly the same for dogs and cats.

(a) Are these data collected as part of an experiment or an observational study?

(b) What is the most common dog name? What is the most common cat name?

(c) What names are more common for cats than dogs?

(d) Is the relationship between the two variables positive or negative? What does this mean in context of the data?

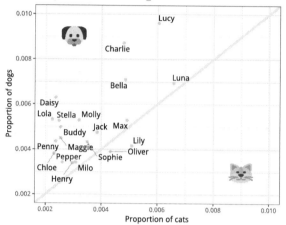

1.36   Stressed out, Part II. In a study evaluating the relationship between stress and muscle cramps, half the subjects are randomly assigned to be exposed to increased stress by being placed into an elevator that falls rapidly and stops abruptly and the other half are left at no or baseline stress.

(a) What type of study is this?

(b) Can this study be used to conclude a causal relationship between increased stress and muscle cramps?

1.37   Chia seeds and weight loss. Chia Pets – those terra-cotta figurines that sprout fuzzy green hair – made the chia plant a household name. But chia has gained an entirely new reputation as a diet supplement. In one 2009 study, a team of researchers recruited 38 men and divided them randomly into two groups: treatment or control. They also recruited 38 women, and they randomly placed half of these participants into the treatment group and the other half into the control group. One group was given 25 grams of chia seeds twice a day, and the other was given a placebo. The subjects volunteered to be a part of the study. After 12 weeks, the scientists found no significant difference between the groups in appetite or weight loss.[33]

(a) What type of study is this?

(b) What are the experimental and control treatments in this study?

(c) Has blocking been used in this study? If so, what is the blocking variable?

(d) Has blinding been used in this study?

(e) Comment on whether or not we can make a causal statement, and indicate whether or not we can generalize the conclusion to the population at large.

1.38   City council survey. A city council has requested a household survey be conducted in a suburban area of their city. The area is broken into many distinct and unique neighborhoods, some including large homes, some with only apartments, and others a diverse mixture of housing structures. For each part below, identify the sampling methods described, and describe the statistical pros and cons of the method in the city's context.

(a) Randomly sample 200 households from the city.

(b) Divide the city into 20 neighborhoods, and sample 10 households from each neighborhood.

(c) Divide the city into 20 neighborhoods, randomly sample 3 neighborhoods, and then sample all households from those 3 neighborhoods.

(d) Divide the city into 20 neighborhoods, randomly sample 8 neighborhoods, and then randomly sample 50 households from those neighborhoods.

(e) Sample the 200 households closest to the city council offices.

---

[33]D.C. Nieman et al. "Chia seed does not promote weight loss or alter disease risk factors in overweight adults". In: *Nutrition Research* 29.6 (2009), pp. 414–418.

1.39 Flawed reasoning. Identify the flaw(s) in reasoning in the following scenarios. Explain what the individuals in the study should have done differently if they wanted to make such strong conclusions.

(a) Students at an elementary school are given a questionnaire that they are asked to return after their parents have completed it. One of the questions asked is, "Do you find that your work schedule makes it difficult for you to spend time with your kids after school?" Of the parents who replied, 85% said "no". Based on these results, the school officials conclude that a great majority of the parents have no difficulty spending time with their kids after school.

(b) A survey is conducted on a simple random sample of 1,000 women who recently gave birth, asking them about whether or not they smoked during pregnancy. A follow-up survey asking if the children have respiratory problems is conducted 3 years later. However, only 567 of these women are reached at the same address. The researcher reports that these 567 women are representative of all mothers.

(c) An orthopedist administers a questionnaire to 30 of his patients who do not have any joint problems and finds that 20 of them regularly go running. He concludes that running decreases the risk of joint problems.

1.40 Income and education in US counties. The scatterplot below shows the relationship between per capita income (in thousands of dollars) and percent of population with a bachelor's degree in 3,143 counties in the US in 2010.

(a) What are the explanatory and response variables?

(b) Describe the relationship between the two variables. Make sure to discuss unusual observations, if any.

(c) Can we conclude that having a bachelor's degree increases one's income?

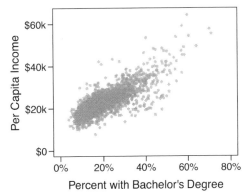

1.41 Eat better, feel better? In a public health study on the effects of consumption of fruits and vegetables on psychological well-being in young adults, participants were randomly assigned to three groups: (1) diet-as-usual, (2) an ecological momentary intervention involving text message reminders to increase their fruits and vegetable consumption plus a voucher to purchase them, or (3) a fruit and vegetable intervention in which participants were given two additional daily servings of fresh fruits and vegetables to consume on top of their normal diet. Participants were asked to take a nightly survey on their smartphones. Participants were student volunteers at the University of Otago, New Zealand. At the end of the 14-day study, only participants in the third group showed improvements to their psychological well-being across the 14-days relative to the other groups.[34]

(a) What type of study is this?

(b) Identify the explanatory and response variables.

(c) Comment on whether the results of the study can be generalized to the population.

(d) Comment on whether the results of the study can be used to establish causal relationships.

(e) A newspaper article reporting on the study states, "The results of this study provide proof that giving young adults fresh fruits and vegetables to eat can have psychological benefits, even over a brief period of time." How would you suggest revising this statement so that it can be supported by the study?

---

[34]Tamlin S Conner et al. "Let them eat fruit! The effect of fruit and vegetable consumption on psychological well-being in young adults: A randomized controlled trial". In: *PloS one* 12.2 (2017), e0171206.

1.42 Screens, teens, and psychological well-being. In a study of three nationally representative large-scale data sets from Ireland, the United States, and the United Kingdom (n = 17,247), teenagers between the ages of 12 to 15 were asked to keep a diary of their screen time and answer questions about how they felt or acted. The answers to these questions were then used to compute a psychological well-being score. Additional data were collected and included in the analysis, such as each child's sex and age, and on the mother's education, ethnicity, psychological distress, and employment. The study concluded that there is little clear-cut evidence that screen time decreases adolescent well-being.[35]

(a) What type of study is this?

(b) Identify the explanatory variables.

(c) Identify the response variable.

(d) Comment on whether the results of the study can be generalized to the population, and why.

(e) Comment on whether the results of the study can be used to establish causal relationships.

1.43 Stanford Open Policing. The Stanford Open Policing project gathers, analyzes, and releases records from traffic stops by law enforcement agencies across the United States. Their goal is to help researchers, journalists, and policymakers investigate and improve interactions between police and the public.[36] The following is an excerpt from a summary table created based off of the data collected as part of this project.

County	State	Driver's race	No. of stops per year	% of stopped cars searched	% of stopped drivers arrested
Apaice County	Arizona	Black	266	0.08	0.02
Apaice County	Arizona	Hispanic	1008	0.05	0.02
Apaice County	Arizona	White	6322	0.02	0.01
Cochise County	Arizona	Black	1169	0.05	0.01
Cochise County	Arizona	Hispanic	9453	0.04	0.01
Cochise County	Arizona	White	10826	0.02	0.01
...	...	...	...	...	...
Wood County	Wisconsin	Black	16	0.24	0.10
Wood County	Wisconsin	Hispanic	27	0.04	0.03
Wood County	Wisconsin	White	1157	0.03	0.03

(a) What variables were collected on each individual traffic stop in order to create to the summary table above?

(b) State whether each variable is numerical or categorical. If numerical, state whether it is continuous or discrete. If categorical, state whether it is ordinal or not.

(c) Suppose we wanted to evaluate whether vehicle search rates are different for drivers of different races. In this analysis, which variable would be the response variable and which variable would be the explanatory variable?

1.44 Space launches. The following summary table shows the number of space launches in the US by the type of launching agency and the outcome of the launch (success or failure).[37]

	1957 - 1999		2000 - 2018	
	Failure	Success	Failure	Success
Private	13	295	10	562
State	281	3751	33	711
Startup	-	-	5	65

(a) What variables were collected on each launch in order to create to the summary table above?

(b) State whether each variable is numerical or categorical. If numerical, state whether it is continuous or discrete. If categorical, state whether it is ordinal or not.

(c) Suppose we wanted to study how the success rate of launches vary between launching agencies and over time. In this analysis, which variable would be the response variable and which variable would be the explanatory variable?

---

[35]Amy Orben and AK Baukney-Przybylski. "Screens, Teens and Psychological Well-Being: Evidence from three time-use diary studies". In: *Psychological Science* (2018).

[36]Emma Pierson et al. "A large-scale analysis of racial disparities in police stops across the United States". In: *arXiv preprint arXiv:1706.05678* (2017).

[37]JSR Launch Vehicle Database, A comprehensive list of suborbital space launches, 2019 Feb 10 Edition.

Chapter 2

Summarizing data

This chapter focuses on the mechanics and construction of summary statistics and graphs. We use statistical software for generating the summaries and graphs presented in this chapter and book. However, since this might be your first exposure to these concepts, we take our time in this chapter to detail how to create them. Mastery of the content presented in this chapter will be crucial for understanding the methods and techniques introduced in rest of the book.

For videos, slides, and other resources, please visit
www.openintro.org/os

2.1 Examining numerical data

In this section we will explore techniques for summarizing numerical variables. For example, consider the `loan_amount` variable from the `loan50` data set, which represents the loan size for all 50 loans in the data set. This variable is numerical since we can sensibly discuss the numerical difference of the size of two loans. On the other hand, area codes and zip codes are not numerical, but rather they are categorical variables.

Throughout this section and the next, we will apply these methods using the `loan50` and `county` data sets, which were introduced in Section 1.2. If you'd like to review the variables from either data set, see Figures 1.3 and 1.5.

2.1.1 Scatterplots for paired data

A **scatterplot** provides a case-by-case view of data for two numerical variables. In Figure 1.8 on page 16, a scatterplot was used to examine the homeownership rate against the fraction of housing units that were part of multi-unit properties (e.g. apartments) in the `county` data set. Another scatterplot is shown in Figure 2.1, comparing the total income of a borrower (`total_income`) and the amount they borrowed (`loan_amount`) for the `loan50` data set. In any scatterplot, each point represents a single case. Since there are 50 cases in `loan50`, there are 50 points in Figure 2.1.

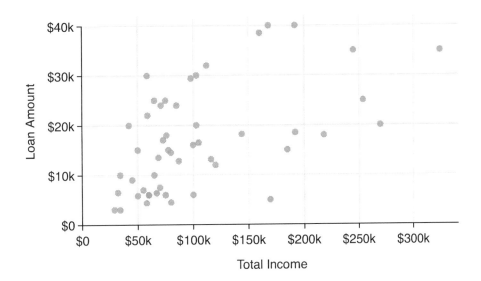

Figure 2.1: A scatterplot of `total_income` versus `loan_amount` for the `loan50` data set.

Looking at Figure 2.1, we see that there are many borrowers with an income below $100,000 on the left side of the graph, while there are a handful of borrowers with income above $250,000.

EXAMPLE 2.1

Figure 2.2 shows a plot of median household income against the poverty rate for 3,142 counties. What can be said about the relationship between these variables?

The relationship is evidently **nonlinear**, as highlighted by the dashed line. This is different from previous scatterplots we've seen, which show relationships that do not show much, if any, curvature in the trend.

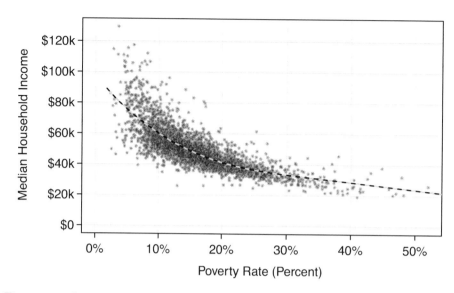

Figure 2.2: A scatterplot of the median household income against the poverty rate for the `county` data set. A statistical model has also been fit to the data and is shown as a dashed line.

GUIDED PRACTICE 2.2

What do scatterplots reveal about the data, and how are they useful?[1]

GUIDED PRACTICE 2.3

Describe two variables that would have a horseshoe-shaped association in a scatterplot ($\cap$ or $\frown$).[2]

2.1.2 Dot plots and the mean

Sometimes two variables are one too many: only one variable may be of interest. In these cases, a dot plot provides the most basic of displays. A **dot plot** is a one-variable scatterplot; an example using the interest rate of 50 loans is shown in Figure 2.3. A stacked version of this dot plot is shown in Figure 2.4.

Interest Rate

Figure 2.3: A dot plot of `interest_rate` for the `loan50` data set. The distribution's mean is shown as a red triangle.

---

[1]Answers may vary. Scatterplots are helpful in quickly spotting associations relating variables, whether those associations come in the form of simple trends or whether those relationships are more complex.

[2]Consider the case where your vertical axis represents something "good" and your horizontal axis represents something that is only good in moderation. Health and water consumption fit this description: we require some water to survive, but consume too much and it becomes toxic and can kill a person.

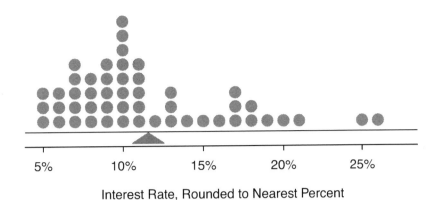

Interest Rate, Rounded to Nearest Percent

Figure 2.4: A stacked dot plot of `interest_rate` for the `loan50` data set. The rates have been rounded to the nearest percent in this plot, and the distribution's mean is shown as a red triangle.

The **mean**, often called the **average**, is a common way to measure the center of a **distribution** of data. To compute the mean interest rate, we add up all the interest rates and divide by the number of observations:

$$\bar{x} = \frac{10.90\% + 9.92\% + 26.30\% + \cdots + 6.08\%}{50} = 11.57\%$$

The sample mean is often labeled $\bar{x}$. The letter x is being used as a generic placeholder for the variable of interest, `interest_rate`, and the bar over the x communicates we're looking at the average interest rate, which for these 50 loans was 11.57%. It is useful to think of the mean as the balancing point of the distribution, and it's shown as a triangle in Figures 2.3 and 2.4.

MEAN

The sample mean can be computed as the sum of the observed values divided by the number of observations:

$$\bar{x} = \frac{x_1 + x_2 + \cdots + x_n}{n}$$

where $x_1, x_2, \ldots, x_n$ represent the n observed values.

GUIDED PRACTICE 2.4

Examine the equation for the mean. What does x_1 correspond to? And x_2? Can you infer a general meaning to what x_i might represent?[3]

GUIDED PRACTICE 2.5

What was n in this sample of loans?[4]

The `loan50` data set represents a sample from a larger population of loans made through Lending Club. We could compute a mean for this population in the same way as the sample mean. However, the population mean has a special label: μ. The symbol μ is the Greek letter *mu* and represents the average of all observations in the population. Sometimes a subscript, such as $_x$, is used to represent which variable the population mean refers to, e.g. μ_x. Often times it is too expensive to measure the population mean precisely, so we often estimate μ using the sample mean, $\bar{x}$.

---

[3]x_1 corresponds to the interest rate for the first loan in the sample (10.90%), x_2 to the second loan's interest rate (9.92%), and x_i corresponds to the interest rate for the ith loan in the data set. For example, if $i = 4$, then we're examining x_4, which refers to the fourth observation in the data set.

[4]The sample size was $n = 50$.

EXAMPLE 2.6

The average interest rate across all loans in the population can be estimated using the sample data. Based on the sample of 50 loans, what would be a reasonable estimate of μ_x, the mean interest rate for all loans in the full data set?

The sample mean, 11.57%, provides a rough estimate of μ_x. While it's not perfect, this is our single best guess of the average interest rate of all the loans in the population under study.

In Chapter 5 and beyond, we will develop tools to characterize the accuracy of *point estimates* like the sample mean. As you might have guessed, point estimates based on larger samples tend to be more accurate than those based on smaller samples.

EXAMPLE 2.7

The mean is useful because it allows us to rescale or standardize a metric into something more easily interpretable and comparable. Provide 2 examples where the mean is useful for making comparisons.

1. We would like to understand if a new drug is more effective at treating asthma attacks than the standard drug. A trial of 1500 adults is set up, where 500 receive the new drug, and 1000 receive a standard drug in the control group:

	New drug	Standard drug
Number of patients	500	1000
Total asthma attacks	200	300

Comparing the raw counts of 200 to 300 asthma attacks would make it appear that the new drug is better, but this is an artifact of the imbalanced group sizes. Instead, we should look at the average number of asthma attacks per patient in each group:

New drug: $200/500 = 0.4$ Standard drug: $300/1000 = 0.3$

The standard drug has a lower average number of asthma attacks per patient than the average in the treatment group.

2. Emilio opened a food truck last year where he sells burritos, and his business has stabilized over the last 3 months. Over that 3 month period, he has made $11,000 while working 625 hours. Emilio's average hourly earnings provides a useful statistic for evaluating whether his venture is, at least from a financial perspective, worth it:

$$\frac{\$11000}{625 \text{ hours}} = \$17.60 \text{ per hour}$$

By knowing his average hourly wage, Emilio now has put his earnings into a standard unit that is easier to compare with many other jobs that he might consider.

EXAMPLE 2.8

Suppose we want to compute the average income per person in the US. To do so, we might first think to take the mean of the per capita incomes across the 3,142 counties in the county data set. What would be a better approach?

The county data set is special in that each county actually represents many individual people. If we were to simply average across the income variable, we would be treating counties with 5,000 and 5,000,000 residents equally in the calculations. Instead, we should compute the total income for each county, add up all the counties' totals, and then divide by the number of people in all the counties. If we completed these steps with the county data, we would find that the per capita income for the US is $30,861. Had we computed the *simple* mean of per capita income across counties, the result would have been just $26,093!

This example used what is called a **weighted mean**. For more information on this topic, check out the following online supplement regarding weighted means openintro.org/d?file=stat_wtd_mean.

2.1.3 Histograms and shape

Dot plots show the exact value for each observation. This is useful for small data sets, but they can become hard to read with larger samples. Rather than showing the value of each observation, we prefer to think of the value as belonging to a *bin*. For example, in the `loan50` data set, we created a table of counts for the number of loans with interest rates between 5.0% and 7.5%, then the number of loans with rates between 7.5% and 10.0%, and so on. Observations that fall on the boundary of a bin (e.g. 10.00%) are allocated to the lower bin. This tabulation is shown in Figure 2.5. These binned counts are plotted as bars in Figure 2.6 into what is called a **histogram**, which resembles a more heavily binned version of the stacked dot plot shown in Figure 2.4.

Interest Rate	5.0% - 7.5%	7.5% - 10.0%	10.0% - 12.5%	12.5% - 15.0%	⋯	25.0% - 27.5%
Count	11	15	8	4	⋯	1

Figure 2.5: Counts for the binned `interest_rate` data.

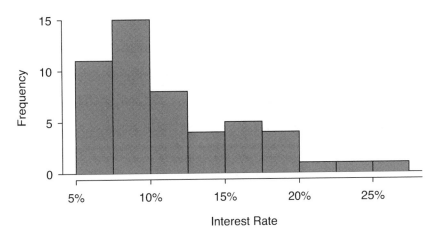

Figure 2.6: A histogram of `interest_rate`. This distribution is strongly skewed to the right.

Histograms provide a view of the **data density**. Higher bars represent where the data are relatively more common. For instance, there are many more loans with rates between 5% and 10% than loans with rates between 20% and 25% in the data set. The bars make it easy to see how the density of the data changes relative to the interest rate.

Histograms are especially convenient for understanding the shape of the data distribution. Figure 2.6 suggests that most loans have rates under 15%, while only a handful of loans have rates above 20%. When data trail off to the right in this way and has a longer right tail, the shape is said to be **right skewed**.[5]

Data sets with the reverse characteristic – a long, thinner tail to the left – are said to be **left skewed**. We also say that such a distribution has a long left tail. Data sets that show roughly equal trailing off in both directions are called **symmetric**.

LONG TAILS TO IDENTIFY SKEW

When data trail off in one direction, the distribution has a **long tail**. If a distribution has a long left tail, it is left skewed. If a distribution has a long right tail, it is right skewed.

---

[5]Other ways to describe data that are right skewed: **skewed to the right**, **skewed to the high end**, or **skewed to the positive end**.

GUIDED PRACTICE 2.9

Take a look at the dot plots in Figures 2.3 and 2.4. Can you see the skew in the data? Is it easier to see the skew in this histogram or the dot plots?[6]

GUIDED PRACTICE 2.10

Besides the mean (since it was labeled), what can you see in the dot plots that you cannot see in the histogram?[7]

In addition to looking at whether a distribution is skewed or symmetric, histograms can be used to identify modes. A **mode** is represented by a prominent peak in the distribution. There is only one prominent peak in the histogram of `loan_amount`.

A definition of *mode* sometimes taught in math classes is the value with the most occurrences in the data set. However, for many real-world data sets, it is common to have *no* observations with the same value in a data set, making this definition impractical in data analysis.

Figure 2.7 shows histograms that have one, two, or three prominent peaks. Such distributions are called **unimodal**, **bimodal**, and **multimodal**, respectively. Any distribution with more than 2 prominent peaks is called multimodal. Notice that there was one prominent peak in the unimodal distribution with a second less prominent peak that was not counted since it only differs from its neighboring bins by a few observations.

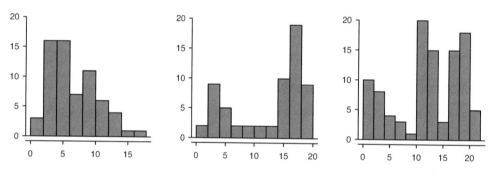

Figure 2.7: Counting only prominent peaks, the distributions are (left to right) unimodal, bimodal, and multimodal. Note that we've said the left plot is unimodal intentionally. This is because we are counting *prominent* peaks, not just any peak.

EXAMPLE 2.11

Figure 2.6 reveals only one prominent mode in the interest rate. Is the distribution unimodal, bimodal, or multimodal?

———————

Unimodal. Remember that *uni* stands for 1 (think *unicycles*). Similarly, *bi* stands for 2 (think *bicycles*). We're hoping a *multicycle* will be invented to complete this analogy.

GUIDED PRACTICE 2.12

Height measurements of young students and adult teachers at a K-3 elementary school were taken. How many modes would you expect in this height data set?[8]

Looking for modes isn't about finding a clear and correct answer about the number of modes in a distribution, which is why *prominent* is not rigorously defined in this book. The most important part of this examination is to better understand your data.

————————————————————

[6]The skew is visible in all three plots, though the flat dot plot is the least useful. The stacked dot plot and histogram are helpful visualizations for identifying skew.

[7]The interest rates for individual loans.

[8]There might be two height groups visible in the data set: one of the students and one of the adults. That is, the data are probably bimodal.

2.1.4 Variance and standard deviation

The mean was introduced as a method to describe the center of a data set, and variability in the data is also important. Here, we introduce two measures of variability: the variance and the standard deviation. Both of these are very useful in data analysis, even though their formulas are a bit tedious to calculate by hand. The standard deviation is the easier of the two to comprehend, and it roughly describes how far away the typical observation is from the mean.

We call the distance of an observation from its mean its **deviation**. Below are the deviations for the 1st, 2nd, 3rd, and 50th observations in the `interest_rate` variable:

$$x_1 - \bar{x} = 10.90 - 11.57 = -0.67$$
$$x_2 - \bar{x} = 9.92 - 11.57 = -1.65$$
$$x_3 - \bar{x} = 26.30 - 11.57 = 14.73$$
$$\vdots$$
$$x_{50} - \bar{x} = 6.08 - 11.57 = -5.49$$

If we square these deviations and then take an average, the result is equal to the sample **variance**, denoted by s^2:

$$
\begin{aligned}
s^2 &= \frac{(-0.67)^2 + (-1.65)^2 + (14.73)^2 + \cdots + (-5.49)^2}{50 - 1} \\
&= \frac{0.45 + 2.72 + 216.97 + \cdots + 30.14}{49} \\
&= 25.52
\end{aligned}
$$

We divide by $n - 1$, rather than dividing by n, when computing a sample's variance; there's some mathematical nuance here, but the end result is that doing this makes this statistic slightly more reliable and useful.

Notice that squaring the deviations does two things. First, it makes large values relatively much larger, seen by comparing $(-0.67)^2$, $(-1.65)^2$, $(14.73)^2$, and $(-5.49)^2$. Second, it gets rid of any negative signs.

The **standard deviation** is defined as the square root of the variance:

$$s = \sqrt{25.52} = 5.05$$

While often omitted, a subscript of x may be added to the variance and standard deviation, i.e. s_x^2 and s_x, if it is useful as a reminder that these are the variance and standard deviation of the observations represented by x_1, x_2, ..., x_n.

VARIANCE AND STANDARD DEVIATION

The variance is the average squared distance from the mean. The standard deviation is the square root of the variance. The standard deviation is useful when considering how far the data are distributed from the mean.

The standard deviation represents the typical deviation of observations from the mean. Usually about 70% of the data will be within one standard deviation of the mean and about 95% will be within two standard deviations. However, as seen in Figures 2.8 and 2.9, these percentages are not strict rules.

Like the mean, the population values for variance and standard deviation have special symbols: σ^2 for the variance and σ for the standard deviation. The symbol σ is the Greek letter *sigma*.

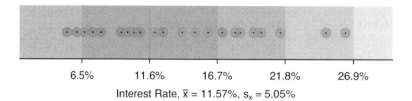

Interest Rate, $\bar{x}$ = 11.57%, s_x = 5.05%

Figure 2.8: For the `interest_rate` variable, 34 of the 50 loans (68%) had interest rates within 1 standard deviation of the mean, and 48 of the 50 loans (96%) had rates within 2 standard deviations. Usually about 70% of the data are within 1 standard deviation of the mean and 95% within 2 standard deviations, though this is far from a hard rule.

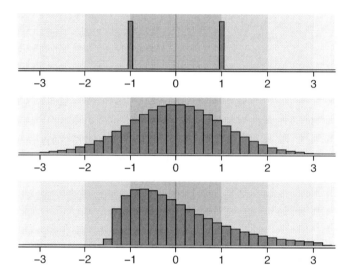

Figure 2.9: Three very different population distributions with the same mean $\mu = 0$ and standard deviation $\sigma = 1$.

GUIDED PRACTICE 2.13

On page 45, the concept of shape of a distribution was introduced. A good description of the shape of a distribution should include modality and whether the distribution is symmetric or skewed to one side. Using Figure 2.9 as an example, explain why such a description is important.[9]

EXAMPLE 2.14

Describe the distribution of the `interest_rate` variable using the histogram in Figure 2.6. The description should incorporate the center, variability, and shape of the distribution, and it should also be placed in context. Also note any especially unusual cases.

The distribution of interest rates is unimodal and skewed to the high end. Many of the rates fall near the mean at 11.57%, and most fall within one standard deviation (5.05%) of the mean. There are a few exceptionally large interest rates in the sample that are above 20%.

In practice, the variance and standard deviation are sometimes used as a means to an end, where the "end" is being able to accurately estimate the uncertainty associated with a sample statistic. For example, in Chapter 5 the standard deviation is used in calculations that help us understand how much a sample mean varies from one sample to the next.

---

[9]Figure 2.9 shows three distributions that look quite different, but all have the same mean, variance, and standard deviation. Using modality, we can distinguish between the first plot (bimodal) and the last two (unimodal). Using skewness, we can distinguish between the last plot (right skewed) and the first two. While a picture, like a histogram, tells a more complete story, we can use modality and shape (symmetry/skew) to characterize basic information about a distribution.

2.1.5 Box plots, quartiles, and the median

A **box plot** summarizes a data set using five statistics while also plotting unusual observations. Figure 2.10 provides a vertical dot plot alongside a box plot of the `interest_rate` variable from the `loan50` data set.

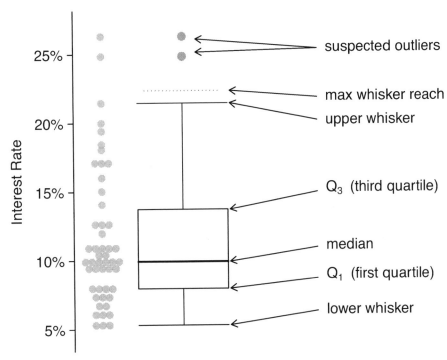

Figure 2.10: A vertical dot plot, where points have been horizontally stacked, next to a labeled box plot for the interest rates of the 50 loans.

The first step in building a box plot is drawing a dark line denoting the **median**, which splits the data in half. Figure 2.10 shows 50% of the data falling below the median and other 50% falling above the median. There are 50 loans in the data set (an even number) so the data are perfectly split into two groups of 25. We take the median in this case to be the average of the two observations closest to the 50th percentile, which happen to be the same value in this data set: $(9.93\% + 9.93\%)/2 = 9.93\%$. When there are an odd number of observations, there will be exactly one observation that splits the data into two halves, and in such a case that observation is the median (no average needed).

MEDIAN: THE NUMBER IN THE MIDDLE

If the data are ordered from smallest to largest, the **median** is the observation right in the middle. If there are an even number of observations, there will be two values in the middle, and the median is taken as their average.

The second step in building a box plot is drawing a rectangle to represent the middle 50% of the data. The total length of the box, shown vertically in Figure 2.10, is called the **interquartile range** (IQR, for short). It, like the standard deviation, is a measure of variability in data. The more variable the data, the larger the standard deviation and IQR tend to be. The two boundaries of the box are called the **first quartile** (the 25th percentile, i.e. 25% of the data fall below this value) and the **third quartile** (the 75th percentile), and these are often labeled Q_1 and Q_3, respectively.

INTERQUARTILE RANGE (IQR)

The IQR is the length of the box in a box plot. It is computed as

$$IQR = Q_3 - Q_1$$

where Q_1 and Q_3 are the 25th and 75th percentiles.

GUIDED PRACTICE 2.15

What percent of the data fall between Q_1 and the median? What percent is between the median and Q_3?[10]

Extending out from the box, the **whiskers** attempt to capture the data outside of the box. However, their reach is never allowed to be more than $1.5 \times IQR$. They capture everything within this reach. In Figure 2.10, the upper whisker does not extend to the last two points, which is beyond $Q_3 + 1.5 \times IQR$, and so it extends only to the last point below this limit. The lower whisker stops at the lowest value, 5.31%, since there is no additional data to reach; the lower whisker's limit is not shown in the figure because the plot does not extend down to $Q_1 - 1.5 \times IQR$. In a sense, the box is like the body of the box plot and the whiskers are like its arms trying to reach the rest of the data.

Any observation lying beyond the whiskers is labeled with a dot. The purpose of labeling these points – instead of extending the whiskers to the minimum and maximum observed values – is to help identify any observations that appear to be unusually distant from the rest of the data. Unusually distant observations are called **outliers**. In this case, it would be reasonable to classify the interest rates of 24.85% and 26.30% as outliers since they are numerically distant from most of the data.

OUTLIERS ARE EXTREME

An **outlier** is an observation that appears extreme relative to the rest of the data.

Examining data for outliers serves many useful purposes, including

1. Identifying strong skew in the distribution.
2. Identifying possible data collection or data entry errors.
3. Providing insight into interesting properties of the data.

GUIDED PRACTICE 2.16

Using Figure 2.10, estimate the following values for `interest_rate` in the `loan50` data set: (a) Q_1, (b) Q_3, and (c) IQR.[11]

---

[10]Since Q_1 and Q_3 capture the middle 50% of the data and the median splits the data in the middle, 25% of the data fall between Q_1 and the median, and another 25% falls between the median and Q_3.

[11]These visual estimates will vary a little from one person to the next: $Q_1 = 8\%$, $Q_3 = 14\%$, IQR $= Q_3 - Q_1 = 6\%$. (The true values: $Q_1 = 7.96\%$, $Q_3 = 13.72\%$, IQR $= 5.76\%$.)

2.1.6   Robust statistics

How are the sample statistics of the `interest_rate` data set affected by the observation, 26.3%? What would have happened if this loan had instead been only 15%? What would happen to these summary statistics if the observation at 26.3% had been even larger, say 35%? These scenarios are plotted alongside the original data in Figure 2.11, and sample statistics are computed under each scenario in Figure 2.12.

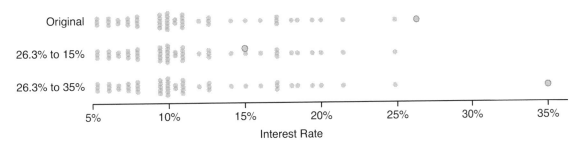

Figure 2.11: Dot plots of the original interest rate data and two modified data sets.

scenario	robust		not robust	
	median	IQR	$\bar{x}$	s
original `interest_rate` data	9.93%	5.76%	11.57%	5.05%
move 26.3% → 15%	9.93%	5.76%	11.34%	4.61%
move 26.3% → 35%	9.93%	5.76%	11.74%	5.68%

Figure 2.12: A comparison of how the median, IQR, mean ($\bar{x}$), and standard deviation (s) change had an extreme observations from the `interest_rate` variable been different.

GUIDED PRACTICE 2.17

(a) Which is more affected by extreme observations, the mean or median? Figure 2.12 may be helpful. (b) Is the standard deviation or IQR more affected by extreme observations?[12]

The median and IQR are called **robust statistics** because extreme observations have little effect on their values: moving the most extreme value generally has little influence on these statistics. On the other hand, the mean and standard deviation are more heavily influenced by changes in extreme observations, which can be important in some situations.

EXAMPLE 2.18

The median and IQR did not change under the three scenarios in Figure 2.12. Why might this be the case?

The median and IQR are only sensitive to numbers near Q_1, the median, and Q_3. Since values in these regions are stable in the three data sets, the median and IQR estimates are also stable.

GUIDED PRACTICE 2.19

The distribution of loan amounts in the `loan50` data set is right skewed, with a few large loans lingering out into the right tail. If you were wanting to understand the typical loan size, should you be more interested in the mean or median?[13]

---

[12](a) Mean is affected more. (b) Standard deviation is affected more. Complete explanations are provided in the material following Guided Practice 2.17.

[13]Answers will vary! If we're looking to simply understand what a typical individual loan looks like, the median is probably more useful. However, if the goal is to understand something that scales well, such as the total amount of money we might need to have on hand if we were to offer 1,000 loans, then the mean would be more useful.

2.1.7 Transforming data (special topic)

When data are very strongly skewed, we sometimes transform them so they are easier to model.

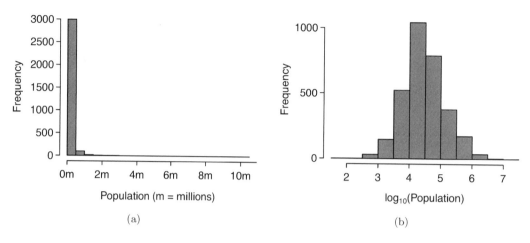

Figure 2.13: (a) A histogram of the populations of all US counties. (b) A histogram of $\log_{10}$-transformed county populations. For this plot, the x-value corresponds to the power of 10, e.g. "4" on the x-axis corresponds to $10^4 = 10,000$.

EXAMPLE 2.20

Consider the histogram of county populations shown in Figure 2.13(a), which shows extreme skew. What isn't useful about this plot?

Nearly all of the data fall into the left-most bin, and the extreme skew obscures many of the potentially interesting details in the data.

There are some standard transformations that may be useful for strongly right skewed data where much of the data is positive but clustered near zero. A **transformation** is a rescaling of the data using a function. For instance, a plot of the logarithm (base 10) of county populations results in the new histogram in Figure 2.13(b). This data is symmetric, and any potential outliers appear much less extreme than in the original data set. By reigning in the outliers and extreme skew, transformations like this often make it easier to build statistical models against the data.

Transformations can also be applied to one or both variables in a scatterplot. A scatterplot of the population change from 2010 to 2017 against the population in 2010 is shown in Figure 2.14(a). In this first scatterplot, it's hard to decipher any interesting patterns because the population variable is so strongly skewed. However, if we apply a $\log_{10}$ transformation to the population variable, as shown in Figure 2.14(b), a positive association between the variables is revealed. In fact, we may be interested in fitting a trend line to the data when we explore methods around fitting regression lines in Chapter 8.

Transformations other than the logarithm can be useful, too. For instance, the square root ($\sqrt{\text{original observation}}$) and inverse ($\frac{1}{\text{original observation}}$) are commonly used by data scientists. Common goals in transforming data are to see the data structure differently, reduce skew, assist in modeling, or straighten a nonlinear relationship in a scatterplot.

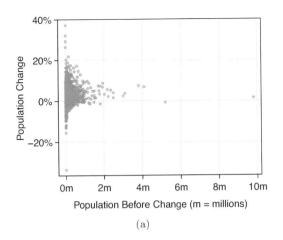
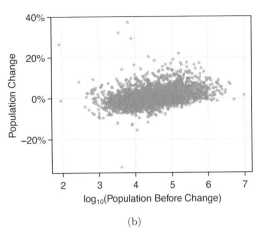

(a)   (b)

Figure 2.14: (a) Scatterplot of population change against the population before the change. (b) A scatterplot of the same data but where the population size has been log-transformed.

2.1.8 Mapping data (special topic)

The `county` data set offers many numerical variables that we could plot using dot plots, scatterplots, or box plots, but these miss the true nature of the data. Rather, when we encounter geographic data, we should create an **intensity map**, where colors are used to show higher and lower values of a variable. Figures 2.15 and 2.16 shows intensity maps for poverty rate in percent (`poverty`), unemployment rate (`unemployment_rate`), homeownership rate in percent (`homeownership`), and median household income (`median_hh_income`). The color key indicates which colors correspond to which values. The intensity maps are not generally very helpful for getting precise values in any given county, but they are very helpful for seeing geographic trends and generating interesting research questions or hypotheses.

EXAMPLE 2.21

What interesting features are evident in the `poverty` and `unemployment_rate` intensity maps?

Poverty rates are evidently higher in a few locations. Notably, the deep south shows higher poverty rates, as does much of Arizona and New Mexico. High poverty rates are evident in the Mississippi flood plains a little north of New Orleans and also in a large section of Kentucky.

The unemployment rate follows similar trends, and we can see correspondence between the two variables. In fact, it makes sense for higher rates of unemployment to be closely related to poverty rates. One observation that stand out when comparing the two maps: the poverty rate is much higher than the unemployment rate, meaning while many people may be working, they are not making enough to break out of poverty.

GUIDED PRACTICE 2.22

What interesting features are evident in the `median_hh_income` intensity map in Figure 2.16(b)?[14]

---

[14]Note: answers will vary. There is some correspondence between high earning and metropolitan areas, where we can see darker spots (higher median household income), though there are several exceptions. You might look for large cities you are familiar with and try to spot them on the map as dark spots.

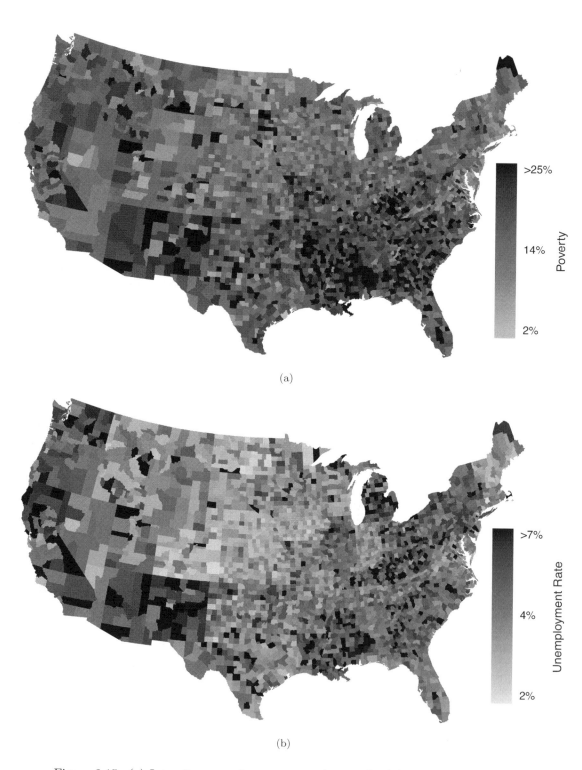

Figure 2.15: (a) Intensity map of poverty rate (percent). (b) Map of the unemployment rate (percent).

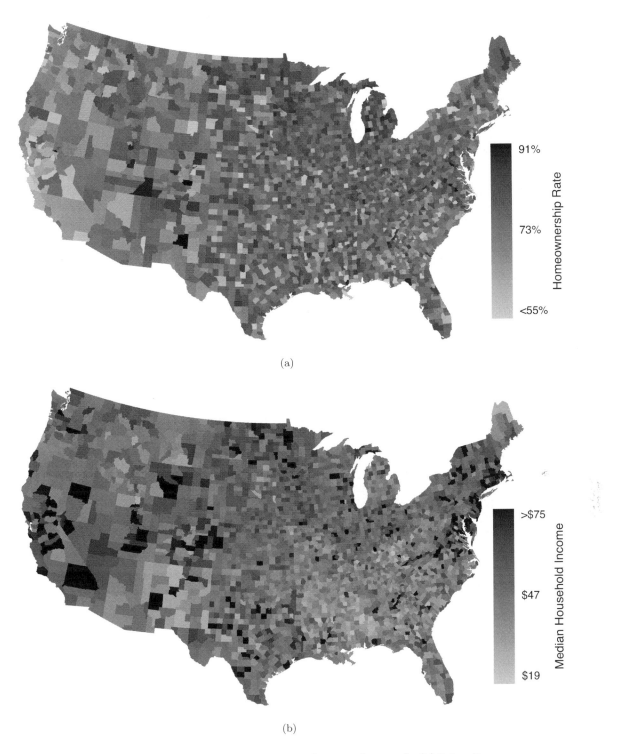

Figure 2.16: (a) Intensity map of homeownership rate (percent). (b) Intensity map of median household income ($1000s).

Exercises

2.1 Mammal life spans. Data were collected on life spans (in years) and gestation lengths (in days) for 62 mammals. A scatterplot of life span versus length of gestation is shown below.[15]

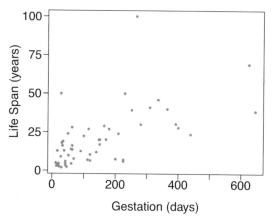

(a) What type of an association is apparent between life span and length of gestation?

(b) What type of an association would you expect to see if the axes of the plot were reversed, i.e. if we plotted length of gestation versus life span?

(c) Are life span and length of gestation independent? Explain your reasoning.

2.2 Associations. Indicate which of the plots show (a) a positive association, (b) a negative association, or (c) no association. Also determine if the positive and negative associations are linear or nonlinear. Each part may refer to more than one plot.

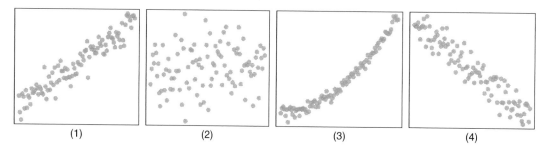

(1)   (2)   (3)   (4)

2.3 Reproducing bacteria. Suppose that there is only sufficient space and nutrients to support one million bacterial cells in a petri dish. You place a few bacterial cells in this petri dish, allow them to reproduce freely, and record the number of bacterial cells in the dish over time. Sketch a plot representing the relationship between number of bacterial cells and time.

2.4 Office productivity. Office productivity is relatively low when the employees feel no stress about their work or job security. However, high levels of stress can also lead to reduced employee productivity. Sketch a plot to represent the relationship between stress and productivity.

2.5 Parameters and statistics. Identify which value represents the sample mean and which value represents the claimed population mean.

(a) American households spent an average of about $52 in 2007 on Halloween merchandise such as costumes, decorations and candy. To see if this number had changed, researchers conducted a new survey in 2008 before industry numbers were reported. The survey included 1,500 households and found that average Halloween spending was $58 per household.

(b) The average GPA of students in 2001 at a private university was 3.37. A survey on a sample of 203 students from this university yielded an average GPA of 3.59 a decade later.

2.6 Sleeping in college. A recent article in a college newspaper stated that college students get an average of 5.5 hrs of sleep each night. A student who was skeptical about this value decided to conduct a survey by randomly sampling 25 students. On average, the sampled students slept 6.25 hours per night. Identify which value represents the sample mean and which value represents the claimed population mean.

---

[15]T. Allison and D.V. Cicchetti. "Sleep in mammals: ecological and constitutional correlates". In: *Arch. Hydrobiol* 75 (1975), p. 442.

2.7 Days off at a mining plant. Workers at a particular mining site receive an average of 35 days paid vacation, which is lower than the national average. The manager of this plant is under pressure from a local union to increase the amount of paid time off. However, he does not want to give more days off to the workers because that would be costly. Instead he decides he should fire 10 employees in such a way as to raise the average number of days off that are reported by his employees. In order to achieve this goal, should he fire employees who have the most number of days off, least number of days off, or those who have about the average number of days off?

2.8 Medians and IQRs. For each part, compare distributions (1) and (2) based on their medians and IQRs. You do not need to calculate these statistics; simply state how the medians and IQRs compare. Make sure to explain your reasoning.

(a) (1) 3, 5, 6, 7, 9
     (2) 3, 5, 6, 7, 20

(b) (1) 3, 5, 6, 7, 9
     (2) 3, 5, 7, 8, 9

(c) (1) 1, 2, 3, 4, 5
     (2) 6, 7, 8, 9, 10

(d) (1) 0, 10, 50, 60, 100
     (2) 0, 100, 500, 600, 1000

2.9 Means and SDs. For each part, compare distributions (1) and (2) based on their means and standard deviations. You do not need to calculate these statistics; simply state how the means and the standard deviations compare. Make sure to explain your reasoning. *Hint:* It may be useful to sketch dot plots of the distributions.

(a) (1) 3, 5, 5, 5, 8, 11, 11, 11, 13
     (2) 3, 5, 5, 5, 8, 11, 11, 11, 20

(b) (1) -20, 0, 0, 0, 15, 25, 30, 30
     (2) -40, 0, 0, 0, 15, 25, 30, 30

(c) (1) 0, 2, 4, 6, 8, 10
     (2) 20, 22, 24, 26, 28, 30

(d) (1) 100, 200, 300, 400, 500
     (2) 0, 50, 300, 550, 600

2.10 Mix-and-match. Describe the distribution in the histograms below and match them to the box plots.

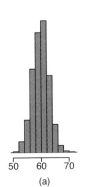

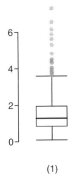
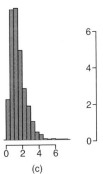
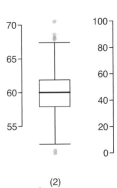

2.11 Air quality. Daily air quality is measured by the air quality index (AQI) reported by the Environmental Protection Agency. This index reports the pollution level and what associated health effects might be a concern. The index is calculated for five major air pollutants regulated by the Clean Air Act and takes values from 0 to 300, where a higher value indicates lower air quality. AQI was reported for a sample of 91 days in 2011 in Durham, NC. The relative frequency histogram below shows the distribution of the AQI values on these days.[16]

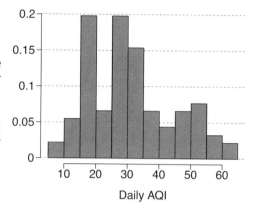

(a) Estimate the median AQI value of this sample.

(b) Would you expect the mean AQI value of this sample to be higher or lower than the median? Explain your reasoning.

(c) Estimate Q1, Q3, and IQR for the distribution.

(d) Would any of the days in this sample be considered to have an unusually low or high AQI? Explain your reasoning.

2.12 Median vs. mean. Estimate the median for the 400 observations shown in the histogram, and note whether you expect the mean to be higher or lower than the median.

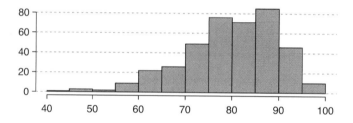

2.13 Histograms vs. box plots. Compare the two plots below. What characteristics of the distribution are apparent in the histogram and not in the box plot? What characteristics are apparent in the box plot but not in the histogram?

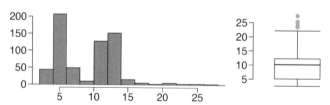

2.14 Facebook friends. Facebook data indicate that 50% of Facebook users have 100 or more friends, and that the average friend count of users is 190. What do these findings suggest about the shape of the distribution of number of friends of Facebook users?[17]

2.15 Distributions and appropriate statistics, Part I. For each of the following, state whether you expect the distribution to be symmetric, right skewed, or left skewed. Also specify whether the mean or median would best represent a typical observation in the data, and whether the variability of observations would be best represented using the standard deviation or IQR. Explain your reasoning.

(a) Number of pets per household.

(b) Distance to work, i.e. number of miles between work and home.

(c) Heights of adult males.

---

[16]US Environmental Protection Agency, AirData, 2011.

[17]Lars Backstrom. "Anatomy of Facebook". In: *Facebook Data Team's Notes* (2011).

2.16   Distributions and appropriate statistics, Part II. For each of the following, state whether you expect the distribution to be symmetric, right skewed, or left skewed. Also specify whether the mean or median would best represent a typical observation in the data, and whether the variability of observations would be best represented using the standard deviation or IQR. Explain your reasoning.

(a) Housing prices in a country where 25% of the houses cost below $350,000, 50% of the houses cost below $450,000, 75% of the houses cost below $1,000,000 and there are a meaningful number of houses that cost more than $6,000,000.

(b) Housing prices in a country where 25% of the houses cost below $300,000, 50% of the houses cost below $600,000, 75% of the houses cost below $900,000 and very few houses that cost more than $1,200,000.

(c) Number of alcoholic drinks consumed by college students in a given week. Assume that most of these students don't drink since they are under 21 years old, and only a few drink excessively.

(d) Annual salaries of the employees at a Fortune 500 company where only a few high level executives earn much higher salaries than all the other employees.

2.17   Income at the coffee shop. The first histogram below shows the distribution of the yearly incomes of 40 patrons at a college coffee shop. Suppose two new people walk into the coffee shop: one making $225,000 and the other $250,000. The second histogram shows the new income distribution. Summary statistics are also provided.

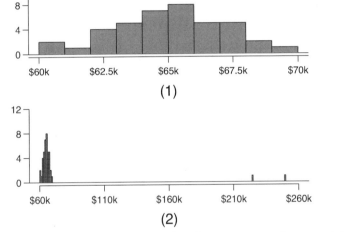

	(1)	(2)
n	40	42
Min.	60,680	60,680
1st Qu.	63,620	63,710
Median	65,240	65,350
Mean	65,090	73,300
3rd Qu.	66,160	66,540
Max.	69,890	250,000
SD	2,122	37,321

(a) Would the mean or the median best represent what we might think of as a typical income for the 42 patrons at this coffee shop? What does this say about the robustness of the two measures?

(b) Would the standard deviation or the IQR best represent the amount of variability in the incomes of the 42 patrons at this coffee shop? What does this say about the robustness of the two measures?

2.18   Midrange. The *midrange* of a distribution is defined as the average of the maximum and the minimum of that distribution. Is this statistic robust to outliers and extreme skew? Explain your reasoning

2.19 Commute times. The US census collects data on time it takes Americans to commute to work, among many other variables. The histogram below shows the distribution of average commute times in 3,142 US counties in 2010. Also shown below is a spatial intensity map of the same data.

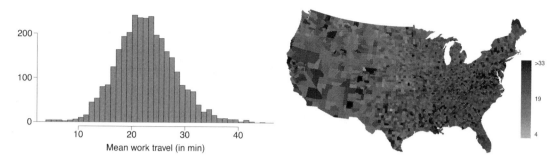

(a) Describe the numerical distribution and comment on whether or not a log transformation may be advisable for these data.

(b) Describe the spatial distribution of commuting times using the map below.

2.20 Hispanic population. The US census collects data on race and ethnicity of Americans, among many other variables. The histogram below shows the distribution of the percentage of the population that is Hispanic in 3,142 counties in the US in 2010. Also shown is a histogram of logs of these values.

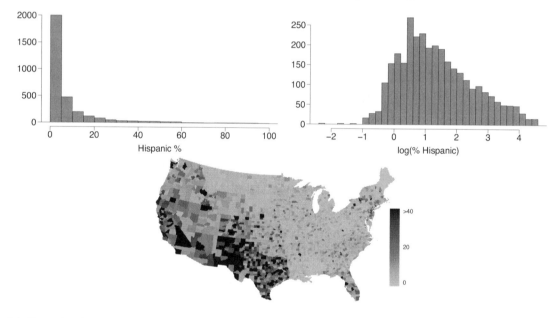

(a) Describe the numerical distribution and comment on why we might want to use log-transformed values in analyzing or modeling these data.

(b) What features of the distribution of the Hispanic population in US counties are apparent in the map but not in the histogram? What features are apparent in the histogram but not the map?

(c) Is one visualization more appropriate or helpful than the other? Explain your reasoning.

2.2 Considering categorical data

In this section, we will introduce tables and other basic tools for categorical data that are used throughout this book. The `loan50` data set represents a sample from a larger loan data set called `loans`. This larger data set contains information on 10,000 loans made through Lending Club. We will examine the relationship between `homeownership`, which for the `loans` data can take a value of `rent`, `mortgage` (owns but has a mortgage), or `own`, and `app_type`, which indicates whether the loan application was made with a partner or whether it was an individual application.

2.2.1 Contingency tables and bar plots

Figure 2.17 summarizes two variables: `app_type` and `homeownership`. A table that summarizes data for two categorical variables in this way is called a **contingency table**. Each value in the table represents the number of times a particular combination of variable outcomes occurred. For example, the value 3496 corresponds to the number of loans in the data set where the borrower rents their home and the application type was by an individual. Row and column totals are also included. The **row totals** provide the total counts across each row (e.g. $3496 + 3839 + 1170 = 8505$), and **column totals** are total counts down each column. We can also create a table that shows only the overall percentages or proportions for each combination of categories, or we can create a table for a single variable, such as the one shown in Figure 2.18 for the `homeownership` variable.

		homeownership			
		rent	mortgage	own	Total
app_type	individual	3496	3839	1170	8505
	joint	362	950	183	1495
	Total	3858	4789	1353	10000

Figure 2.17: A contingency table for `app_type` and `homeownership`.

homeownership	Count
rent	3858
mortgage	4789
own	1353
Total	10000

Figure 2.18: A table summarizing the frequencies of each value for the `homeownership` variable.

A bar plot is a common way to display a single categorical variable. The left panel of Figure 2.19 shows a **bar plot** for the `homeownership` variable. In the right panel, the counts are converted into proportions, showing the proportion of observations that are in each level (e.g. $3858/10000 = 0.3858$ for `rent`).

Figure 2.19: Two bar plots of `number`. The left panel shows the counts, and the right panel shows the proportions in each group.

2.2.2 Row and column proportions

Sometimes it is useful to understand the fractional breakdown of one variable in another, and we can modify our contingency table to provide such a view. Figure 2.20 shows the **row proportions** for Figure 2.17, which are computed as the counts divided by their row totals. The value 3496 at the intersection of `individual` and `rent` is replaced by 3496/8505 = 0.411, i.e. 3496 divided by its row total, 8505. So what does 0.411 represent? It corresponds to the proportion of individual applicants who rent.

	rent	mortgage	own	Total
individual	0.411	0.451	0.138	1.000
joint	0.242	0.635	0.122	1.000
Total	0.386	0.479	0.135	1.000

Figure 2.20: A contingency table with row proportions for the `app_type` and `homeownership` variables. The row total is off by 0.001 for the `joint` row due to a rounding error.

A contingency table of the column proportions is computed in a similar way, where each **column proportion** is computed as the count divided by the corresponding column total. Figure 2.21 shows such a table, and here the value 0.906 indicates that 90.6% of renters applied as individuals for the loan. This rate is higher compared to loans from people with mortgages (80.2%) or who own their home (86.5%). Because these rates vary between the three levels of `homeownership` (`rent`, `mortgage`, `own`), this provides evidence that the `app_type` and `homeownership` variables are associated.

	rent	mortgage	own	Total
individual	0.906	0.802	0.865	0.851
joint	0.094	0.198	0.135	0.150
Total	1.000	1.000	1.000	1.000

Figure 2.21: A contingency table with column proportions for the `app_type` and `homeownership` variables. The total for the last column is off by 0.001 due to a rounding error.

We could also have checked for an association between `app_type` and `homeownership` in Figure 2.20 using row proportions. When comparing these row proportions, we would look down columns to see if the fraction of loans where the borrower rents, has a mortgage, or owns varied across the `individual` to `joint` application types.

GUIDED PRACTICE 2.23

(a) What does 0.451 represent in Figure 2.20?
(b) What does 0.802 represent in Figure 2.21?[18]

GUIDED PRACTICE 2.24

(a) What does 0.122 at the intersection of `joint` and `own` represent in Figure 2.20?
(b) What does 0.135 represent in the Figure 2.21?[19]

EXAMPLE 2.25

Data scientists use statistics to filter spam from incoming email messages. By noting specific characteristics of an email, a data scientist may be able to classify some emails as spam or not spam with high accuracy. One such characteristic is whether the email contains no numbers, small numbers, or big numbers. Another characteristic is the email format, which indicates whether or not an email has any HTML content, such as bolded text. We'll focus on email format and spam status using the `email` data set, and these variables are summarized in a contingency table in Figure 2.22. Which would be more helpful to someone hoping to classify email as spam or regular email for this table: row or column proportions?

A data scientist would be interested in how the proportion of spam changes within each email format. This corresponds to column proportions: the proportion of spam in plain text emails and the proportion of spam in HTML emails.

If we generate the column proportions, we can see that a higher fraction of plain text emails are spam ($209/1195 = 17.5\%$) than compared to HTML emails ($158/2726 = 5.8\%$). This information on its own is insufficient to classify an email as spam or not spam, as over 80% of plain text emails are not spam. Yet, when we carefully combine this information with many other characteristics, we stand a reasonable chance of being able to classify some emails as spam or not spam with confidence.

	text	HTML	Total
spam	209	158	367
not spam	986	2568	3554
Total	1195	2726	3921

Figure 2.22: A contingency table for `spam` and `format`.

Example 2.25 points out that row and column proportions are not equivalent. Before settling on one form for a table, it is important to consider each to ensure that the most useful table is constructed. However, sometimes it simply isn't clear which, if either, is more useful.

EXAMPLE 2.26

Look back to Tables 2.20 and 2.21. Are there any obvious scenarios where one might be more useful than the other?

None that we thought were obvious! What is distinct about `app_type` and `homeownership` vs the email example is that these two variables don't have a clear explanatory-response variable relationship that we might hypothesize (see Section 1.2.4 for these terms). Usually it is most useful to "condition" on the explanatory variable. For instance, in the email example, the email format was seen as a possible explanatory variable of whether the message was spam, so we would find it more interesting to compute the relative frequencies (proportions) for each email format.

---

[18](a) 0.451 represents the proportion of individual applicants who have a mortgage. (b) 0.802 represents the fraction of applicants with mortgages who applied as individuals.
[19](a) 0.122 represents the fraction of joint borrowers who own their home. (b) 0.135 represents the home-owning borrowers who had a joint application for the loan.

2.2.3   Using a bar plot with two variables

Contingency tables using row or column proportions are especially useful for examining how two categorical variables are related. Stacked bar plots provide a way to visualize the information in these tables.

A **stacked bar plot** is a graphical display of contingency table information. For example, a stacked bar plot representing Figure 2.21 is shown in Figure 2.23(a), where we have first created a bar plot using the `homeownership` variable and then divided each group by the levels of `app_type`.

One related visualization to the stacked bar plot is the **side-by-side bar plot**, where an example is shown in Figure 2.23(b).

For the last type of bar plot we introduce, the column proportions for the `app_type` and `homeownership` contingency table have been translated into a standardized stacked bar plot in Figure 2.23(c). This type of visualization is helpful in understanding the fraction of individual or joint loan applications for borrowers in each level of `homeownership`. Additionally, since the proportions of `joint` and `individual` vary across the groups, we can conclude that the two variables are associated.

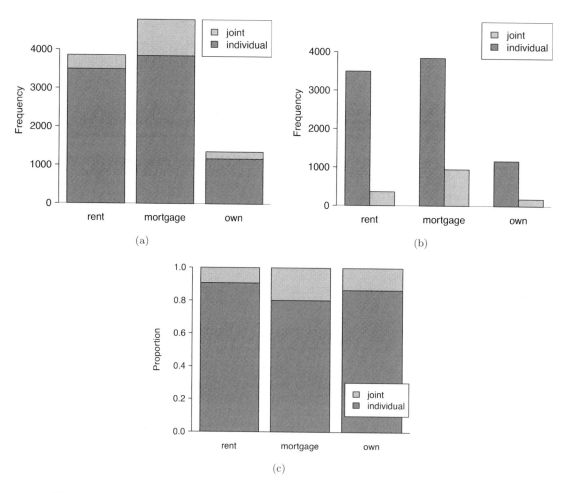

Figure 2.23: (a) Stacked bar plot for `homeownership`, where the counts have been further broken down by `app_type`. (b) Side-by-side bar plot for `homeownership` and `app_type`. (c) Standardized version of the stacked bar plot.

EXAMPLE 2.27

Examine the three bar plots in Figure 2.23. When is the stacked, side-by-side, or standardized stacked bar plot the most useful?

The stacked bar plot is most useful when it's reasonable to assign one variable as the explanatory variable and the other variable as the response, since we are effectively grouping by one variable first and then breaking it down by the others.

Side-by-side bar plots are more agnostic in their display about which variable, if any, represents the explanatory and which the response variable. It is also easy to discern the number of cases in of the six different group combinations. However, one downside is that it tends to require more horizontal space; the narrowness of Figure 2.23(b) makes the plot feel a bit cramped. Additionally, when two groups are of very different sizes, as we see in the `own` group relative to either of the other two groups, it is difficult to discern if there is an association between the variables.

The standardized stacked bar plot is helpful if the primary variable in the stacked bar plot is relatively imbalanced, e.g. the `own` category has only a third of the observations in the `mortgage` category, making the simple stacked bar plot less useful for checking for an association. The major downside of the standardized version is that we lose all sense of how many cases each of the bars represents.

2.2.4 Mosaic plots

A **mosaic plot** is a visualization technique suitable for contingency tables that resembles a standardized stacked bar plot with the benefit that we still see the relative group sizes of the primary variable as well.

To get started in creating our first mosaic plot, we'll break a square into columns for each category of the `homeownership` variable, with the result shown in Figure 2.24(a). Each column represents a level of `homeownership`, and the column widths correspond to the proportion of loans in each of those categories. For instance, there are fewer loans where the borrower is an owner than where the borrower has a mortgage. In general, mosaic plots use box *areas* to represent the number of cases in each category.

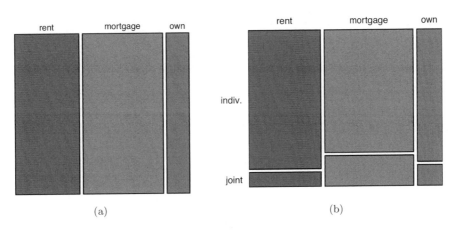

Figure 2.24: (a) The one-variable mosaic plot for `homeownership`. (b) Two-variable mosaic plot for both `homeownership` and `app_type`.

To create a completed mosaic plot, the single-variable mosaic plot is further divided into pieces in Figure 2.24(b) using the `app_type` variable. Each column is split proportional to the number of loans from individual and joint borrowers. For example, the second column represents loans where the borrower has a mortgage, and it was divided into individual loans (upper) and joint loans (lower). As another example, the bottom segment of the third column represents loans where the borrower owns their home and applied jointly, while the upper segment of this column represents borrowers who are homeowners and filed individually. We can again use this plot to see that the `homeownership` and `app_type` variables are associated, since some columns are divided in different vertical locations than others, which was the same technique used for checking an association in the standardized stacked bar plot.

In Figure 2.24, we chose to first split by the homeowner status of the borrower. However, we could have instead first split by the application type, as in Figure 2.25. Like with the bar plots, it's common to use the explanatory variable to represent the first split in a mosaic plot, and then for the response to break up each level of the explanatory variable, if these labels are reasonable to attach to the variables under consideration.

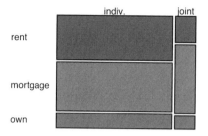

Figure 2.25: Mosaic plot where loans are grouped by the `homeownership` variable after they've been divided into the `individual` and `joint` application types.

2.2.5   The only pie chart you will see in this book

A **pie chart** is shown in Figure 2.26 alongside a bar plot representing the same information. Pie charts can be useful for giving a high-level overview to show how a set of cases break down. However, it is also difficult to decipher details in a pie chart. For example, it takes a couple seconds longer to recognize that there are more loans where the borrower has a mortgage than rent when looking at the pie chart, while this detail is very obvious in the bar plot. While pie charts can be useful, we prefer bar plots for their ease in comparing groups.

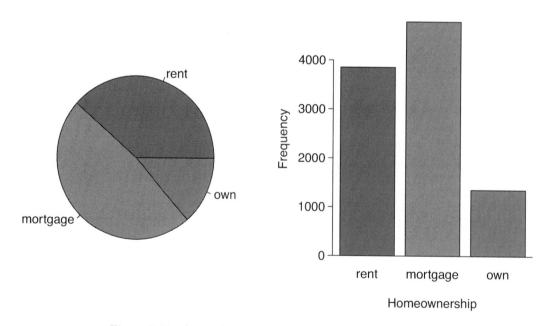

Figure 2.26: A pie chart and bar plot of `homeownership`.

2.2.6 Comparing numerical data across groups

Some of the more interesting investigations can be considered by examining numerical data across groups. The methods required here aren't really new: all that's required is to make a numerical plot for each group in the same graph. Here two convenient methods are introduced: side-by-side box plots and hollow histograms.

We will take a look again at the `county` data set and compare the median household income for counties that gained population from 2010 to 2017 versus counties that had no gain. While we might like to make a causal connection here, remember that these are observational data and so such an interpretation would be, at best, half-baked.

There were 1,454 counties where the population increased from 2010 to 2017, and there were 1,672 counties with no gain (all but one were a loss). A random sample of 100 counties from the first group and 50 from the second group are shown in Figure 2.27 to give a better sense of some of the raw median income data.

Median Income for 150 Counties, in $1000s

Population Gain						No Population Gain		
38.2	43.6	42.2	61.5	51.1	45.7	48.3	60.3	50.7
44.6	51.8	40.7	48.1	56.4	41.9	39.3	40.4	40.3
40.6	63.3	52.1	60.3	49.8	51.7	57	47.2	45.9
51.1	34.1	45.5	52.8	49.1	51	42.3	41.5	46.1
80.8	46.3	82.2	43.6	39.7	49.4	44.9	51.7	46.4
75.2	40.6	46.3	62.4	44.1	51.3	29.1	51.8	50.5
51.9	34.7	54	42.9	52.2	45.1	27	30.9	34.9
61	51.4	56.5	62	46	46.4	40.7	51.8	61.1
53.8	57.6	69.2	48.4	40.5	48.6	43.4	34.7	45.7
53.1	54.6	55	46.4	39.9	56.7	33.1	21	37
63	49.1	57.2	44.1	50	38.9	52	31.9	45.7
46.6	46.5	38.9	50.9	56	34.6	56.3	38.7	45.7
74.2	63	49.6	53.7	77.5	60	56.2	43	21.7
63.2	47.6	55.9	39.1	57.8	42.6	44.5	34.5	48.9
50.4	49	45.6	39	38.8	37.1	50.9	42.1	43.2
57.2	44.7	71.7	35.3	100.2		35.4	41.3	33.6
42.6	55.5	38.6	52.7	63		43.4	56.5	

Figure 2.27: In this table, median household income (in $1000s) from a random sample of 100 counties that had population gains are shown on the left. Median incomes from a random sample of 50 counties that had no population gain are shown on the right.

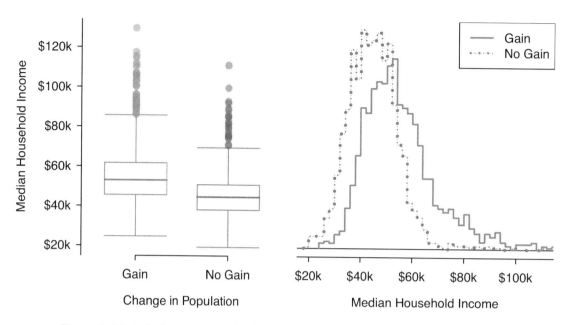

Figure 2.28: Side-by-side box plot (left panel) and hollow histograms (right panel) for med_hh_income, where the counties are split by whether there was a population gain or loss.

The **side-by-side box plot** is a traditional tool for comparing across groups. An example is shown in the left panel of Figure 2.28, where there are two box plots, one for each group, placed into one plotting window and drawn on the same scale.

Another useful plotting method uses **hollow histograms** to compare numerical data across groups. These are just the outlines of histograms of each group put on the same plot, as shown in the right panel of Figure 2.28.

GUIDED PRACTICE 2.28

Use the plots in Figure 2.28 to compare the incomes for counties across the two groups. What do you notice about the approximate center of each group? What do you notice about the variability between groups? Is the shape relatively consistent between groups? How many *prominent* modes are there for each group?[20]

GUIDED PRACTICE 2.29

What components of each plot in Figure 2.28 do you find most useful?[21]

---

[20]Answers may vary a little. The counties with population gains tend to have higher income (median of about $45,000) versus counties without a gain (median of about $40,000). The variability is also slightly larger for the population gain group. This is evident in the IQR, which is about 50% bigger in the *gain* group. Both distributions show slight to moderate right skew and are unimodal. The box plots indicate there are many observations far above the median in each group, though we should anticipate that many observations will fall beyond the whiskers when examining any data set that contain more than a couple hundred data points.

[21]Answers will vary. The side-by-side box plots are especially useful for comparing centers and spreads, while the hollow histograms are more useful for seeing distribution shape, skew, and potential anomalies.

Exercises

2.21 Antibiotic use in children. The bar plot and the pie chart below show the distribution of pre-existing medical conditions of children involved in a study on the optimal duration of antibiotic use in treatment of tracheitis, which is an upper respiratory infection.

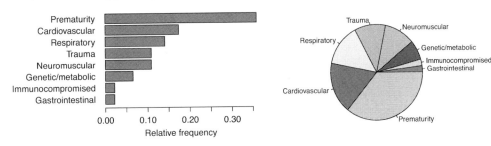

(a) What features are apparent in the bar plot but not in the pie chart?

(b) What features are apparent in the pie chart but not in the bar plot?

(c) Which graph would you prefer to use for displaying these categorical data?

2.22 Views on immigration. 910 randomly sampled registered voters from Tampa, FL were asked if they thought workers who have illegally entered the US should be (i) allowed to keep their jobs and apply for US citizenship, (ii) allowed to keep their jobs as temporary guest workers but not allowed to apply for US citizenship, or (iii) lose their jobs and have to leave the country. The results of the survey by political ideology are shown below.[22]

		Political ideology			
		Conservative	Moderate	Liberal	Total
Response	(i) Apply for citizenship	57	120	101	278
	(ii) Guest worker	121	113	28	262
	(iii) Leave the country	179	126	45	350
	(iv) Not sure	15	4	1	20
	Total	372	363	175	910

(a) What percent of these Tampa, FL voters identify themselves as conservatives?

(b) What percent of these Tampa, FL voters are in favor of the citizenship option?

(c) What percent of these Tampa, FL voters identify themselves as conservatives and are in favor of the citizenship option?

(d) What percent of these Tampa, FL voters who identify themselves as conservatives are also in favor of the citizenship option? What percent of moderates share this view? What percent of liberals share this view?

(e) Do political ideology and views on immigration appear to be independent? Explain your reasoning.

---

[22]SurveyUSA, News Poll #18927, data collected Jan 27-29, 2012.

2.23 Views on the DREAM Act. A random sample of registered voters from Tampa, FL were asked if they support the DREAM Act, a proposed law which would provide a path to citizenship for people brought illegally to the US as children. The survey also collected information on the political ideology of the respondents. Based on the mosaic plot shown below, do views on the DREAM Act and political ideology appear to be independent? Explain your reasoning.[23]

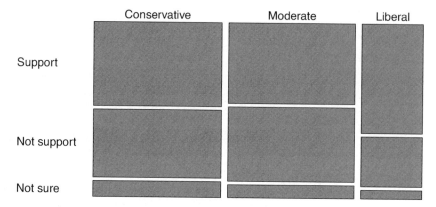

2.24 Raise taxes. A random sample of registered voters nationally were asked whether they think it's better to raise taxes on the rich or raise taxes on the poor. The survey also collected information on the political party affiliation of the respondents. Based on the mosaic plot shown below, do views on raising taxes and political affiliation appear to be independent? Explain your reasoning.[24]

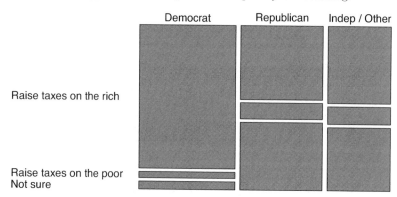

[23]SurveyUSA, News Poll #18927, data collected Jan 27-29, 2012.
[24]Public Policy Polling, Americans on College Degrees, Classic Literature, the Seasons, and More, data collected Feb 20-22, 2015.

2.3 Case study: malaria vaccine

EXAMPLE 2.30

Suppose your professor splits the students in class into two groups: students on the left and students on the right. If $\hat{p}_L$ and $\hat{p}_R$ represent the proportion of students who own an Apple product on the left and right, respectively, would you be surprised if $\hat{p}_L$ did not exactly equal $\hat{p}_R$?

While the proportions would probably be close to each other, it would be unusual for them to be exactly the same. We would probably observe a small difference due to chance.

GUIDED PRACTICE 2.31

If we don't think the side of the room a person sits on in class is related to whether the person owns an Apple product, what assumption are we making about the relationship between these two variables?[25]

2.3.1 Variability within data

We consider a study on a new malaria vaccine called PfSPZ. In this study, volunteer patients were randomized into one of two experiment groups: 14 patients received an experimental vaccine or 6 patients received a placebo vaccine. Nineteen weeks later, all 20 patients were exposed to a drug-sensitive malaria virus strain; the motivation of using a drug-sensitive strain of virus here is for ethical considerations, allowing any infections to be treated effectively. The results are summarized in Figure 2.29, where 9 of the 14 treatment patients remained free of signs of infection while all of the 6 patients in the control group patients showed some baseline signs of infection.

		outcome		
		infection	no infection	Total
treatment	vaccine	5	9	14
	placebo	6	0	6
	Total	11	9	20

Figure 2.29: Summary results for the malaria vaccine experiment.

GUIDED PRACTICE 2.32

Is this an observational study or an experiment? What implications does the study type have on what can be inferred from the results?[26]

In this study, a smaller proportion of patients who received the vaccine showed signs of an infection (35.7% versus 100%). However, the sample is very small, and it is unclear whether the difference provides *convincing evidence* that the vaccine is effective.

---

[25]We would be assuming that these two variables are independent.

[26]The study is an experiment, as patients were randomly assigned an experiment group. Since this is an experiment, the results can be used to evaluate a causal relationship between the malaria vaccine and whether patients showed signs of an infection.

EXAMPLE 2.33

Data scientists are sometimes called upon to evaluate the strength of evidence. When looking at the rates of infection for patients in the two groups in this study, what comes to mind as we try to determine whether the data show convincing evidence of a real difference?

The observed infection rates (35.7% for the treatment group versus 100% for the control group) suggest the vaccine may be effective. However, we cannot be sure if the observed difference represents the vaccine's efficacy or is just from random chance. Generally there is a little bit of fluctuation in sample data, and we wouldn't expect the sample proportions to be *exactly* equal, even if the truth was that the infection rates were independent of getting the vaccine. Additionally, with such small samples, perhaps it's common to observe such large differences when we randomly split a group due to chance alone!

Example 2.33 is a reminder that the observed outcomes in the data sample may not perfectly reflect the true relationships between variables since there is **random noise**. While the observed difference in rates of infection is large, the sample size for the study is small, making it unclear if this observed difference represents efficacy of the vaccine or whether it is simply due to chance. We label these two competing claims, H_0 and H_A, which are spoken as "H-nought" and "H-A":

H_0: **Independence model.** The variables `treatment` and `outcome` are independent. They have no relationship, and the observed difference between the proportion of patients who developed an infection in the two groups, 64.3%, was due to chance.

H_A: **Alternative model.** The variables are *not* independent. The difference in infection rates of 64.3% was not due to chance, and vaccine affected the rate of infection.

What would it mean if the independence model, which says the vaccine had no influence on the rate of infection, is true? It would mean 11 patients were going to develop an infection *no matter which group they were randomized into*, and 9 patients would not develop an infection *no matter which group they were randomized into*. That is, if the vaccine did not affect the rate of infection, the difference in the infection rates was due to chance alone in how the patients were randomized.

Now consider the alternative model: infection rates were influenced by whether a patient received the vaccine or not. If this was true, and especially if this influence was substantial, we would expect to see some difference in the infection rates of patients in the groups.

We choose between these two competing claims by assessing if the data conflict so much with H_0 that the independence model cannot be deemed reasonable. If this is the case, and the data support H_A, then we will reject the notion of independence and conclude the vaccine was effective.

2.3.2 Simulating the study

We're going to implement **simulations**, where we will pretend we know that the malaria vaccine being tested does *not* work. Ultimately, we want to understand if the large difference we observed is common in these simulations. If it is common, then maybe the difference we observed was purely due to chance. If it is very uncommon, then the possibility that the vaccine was helpful seems more plausible.

Figure 2.29 shows that 11 patients developed infections and 9 did not. For our simulation, we will suppose the infections were independent of the vaccine and we were able to *rewind* back to when the researchers randomized the patients in the study. If we happened to randomize the patients differently, we may get a different result in this hypothetical world where the vaccine doesn't influence the infection. Let's complete another **randomization** using a simulation.

In this **simulation**, we take 20 notecards to represent the 20 patients, where we write down "infection" on 11 cards and "no infection" on 9 cards. In this hypothetical world, we believe each patient that got an infection was going to get it regardless of which group they were in, so let's see what happens if we randomly assign the patients to the treatment and control groups again. We thoroughly shuffle the notecards and deal 14 into a `vaccine` pile and 6 into a `placebo` pile. Finally, we tabulate the results, which are shown in Figure 2.30.

		outcome		Total
		infection	no infection	
treatment	vaccine	7	7	14
(simulated)	placebo	4	2	6
	Total	11	9	20

Figure 2.30: Simulation results, where any difference in infection rates is purely due to chance.

GUIDED PRACTICE 2.34

What is the difference in infection rates between the two simulated groups in Figure 2.30? How does this compare to the observed 64.3% difference in the actual data?[27]

2.3.3 Checking for independence

We computed one possible difference under the independence model in Guided Practice 2.34, which represents one difference due to chance. While in this first simulation, we physically dealt out notecards to represent the patients, it is more efficient to perform this simulation using a computer. Repeating the simulation on a computer, we get another difference due to chance:

$$\frac{2}{6} - \frac{9}{14} = -0.310$$

And another:

$$\frac{3}{6} - \frac{8}{14} = -0.071$$

And so on until we repeat the simulation enough times that we have a good idea of what represents the *distribution of differences from chance alone*. Figure 2.31 shows a stacked plot of the differences found from 100 simulations, where each dot represents a simulated difference between the infection rates (control rate minus treatment rate).

Note that the distribution of these simulated differences is centered around 0. We simulated these differences assuming that the independence model was true, and under this condition, we expect the difference to be near zero with some random fluctuation, where *near* is pretty generous in this case since the sample sizes are so small in this study.

EXAMPLE 2.35

How often would you observe a difference of at least 64.3% (0.643) according to Figure 2.31? Often, sometimes, rarely, or never?

It appears that a difference of at least 64.3% due to chance alone would only happen about 2% of the time according to Figure 2.31. Such a low probability indicates a rare event.

---

[27]$4/6 - 7/14 = 0.167$ or about 16.7% in favor of the vaccine. This difference due to chance is much smaller than the difference observed in the actual groups.

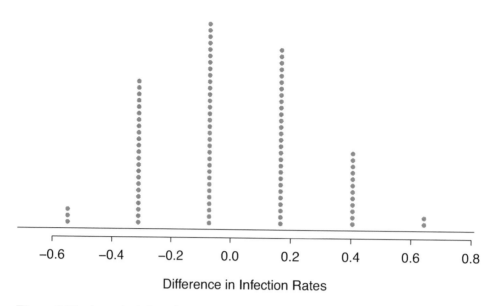

Figure 2.31: A stacked dot plot of differences from 100 simulations produced under the independence model, H_0, where in these simulations infections are unaffected by the vaccine. Two of the 100 simulations had a difference of at least 64.3%, the difference observed in the study.

The difference of 64.3% being a rare event suggests two possible interpretations of the results of the study:

H_0 **Independence model.** The vaccine has no effect on infection rate, and we just happened to observe a difference that would only occur on a rare occasion.

H_A **Alternative model.** The vaccine has an effect on infection rate, and the difference we observed was actually due to the vaccine being effective at combatting malaria, which explains the large difference of 64.3%.

Based on the simulations, we have two options. (1) We conclude that the study results do not provide strong evidence against the independence model. That is, we do not have sufficiently strong evidence to conclude the vaccine had an effect in this clinical setting. (2) We conclude the evidence is sufficiently strong to reject H_0 and assert that the vaccine was useful. When we conduct formal studies, usually we reject the notion that we just happened to observe a rare event.[28] So in this case, we reject the independence model in favor of the alternative. That is, we are concluding the data provide strong evidence that the vaccine provides some protection against malaria in this clinical setting.

One field of statistics, statistical inference, is built on evaluating whether such differences are due to chance. In statistical inference, data scientists evaluate which model is most reasonable given the data. Errors do occur, just like rare events, and we might choose the wrong model. While we do not always choose correctly, statistical inference gives us tools to control and evaluate how often these errors occur. In Chapter 5, we give a formal introduction to the problem of model selection. We spend the next two chapters building a foundation of probability and theory necessary to make that discussion rigorous.

---

[28]This reasoning does not generally extend to anecdotal observations. Each of us observes incredibly rare events every day, events we could not possibly hope to predict. However, in the non-rigorous setting of anecdotal evidence, almost anything may appear to be a rare event, so the idea of looking for rare events in day-to-day activities is treacherous. For example, we might look at the lottery: there was only a 1 in 292 million chance that the Powerball numbers for the largest jackpot in history (January 13th, 2016) would be (04, 08, 19, 27, 34) with a Powerball of (10), but nonetheless those numbers came up! However, no matter what numbers had turned up, they would have had the same incredibly rare odds. That is, *any set of numbers we could have observed would ultimately be incredibly rare*. This type of situation is typical of our daily lives: each possible event in itself seems incredibly rare, but if we consider every alternative, those outcomes are also incredibly rare. We should be cautious not to misinterpret such anecdotal evidence.

Exercises

2.25 Side effects of Avandia. Rosiglitazone is the active ingredient in the controversial type 2 diabetes medicine Avandia and has been linked to an increased risk of serious cardiovascular problems such as stroke, heart failure, and death. A common alternative treatment is pioglitazone, the active ingredient in a diabetes medicine called Actos. In a nationwide retrospective observational study of 227,571 Medicare beneficiaries aged 65 years or older, it was found that 2,593 of the 67,593 patients using rosiglitazone and 5,386 of the 159,978 using pioglitazone had serious cardiovascular problems. These data are summarized in the contingency table below.[29]

		Cardiovascular problems		Total
		Yes	No	
Treatment	Rosiglitazone	2,593	65,000	67,593
	Pioglitazone	5,386	154,592	159,978
	Total	7,979	219,592	227,571

(a) Determine if each of the following statements is true or false. If false, explain why. *Be careful:* The reasoning may be wrong even if the statement's conclusion is correct. In such cases, the statement should be considered false.

  i. Since more patients on pioglitazone had cardiovascular problems (5,386 vs. 2,593), we can conclude that the rate of cardiovascular problems for those on a pioglitazone treatment is higher.

  ii. The data suggest that diabetic patients who are taking rosiglitazone are more likely to have cardiovascular problems since the rate of incidence was (2,593 / 67,593 = 0.038) 3.8% for patients on this treatment, while it was only (5,386 / 159,978 = 0.034) 3.4% for patients on pioglitazone.

  iii. The fact that the rate of incidence is higher for the rosiglitazone group proves that rosiglitazone causes serious cardiovascular problems.

  iv. Based on the information provided so far, we cannot tell if the difference between the rates of incidences is due to a relationship between the two variables or due to chance.

(b) What proportion of all patients had cardiovascular problems?

(c) If the type of treatment and having cardiovascular problems were independent, about how many patients in the rosiglitazone group would we expect to have had cardiovascular problems?

(d) We can investigate the relationship between outcome and treatment in this study using a randomization technique. While in reality we would carry out the simulations required for randomization using statistical software, suppose we actually simulate using index cards. In order to simulate from the independence model, which states that the outcomes were independent of the treatment, we write whether or not each patient had a cardiovascular problem on cards, shuffled all the cards together, then deal them into two groups of size 67,593 and 159,978. We repeat this simulation 1,000 times and each time record the number of people in the rosiglitazone group who had cardiovascular problems. Use the relative frequency histogram of these counts to answer (i)-(iii).

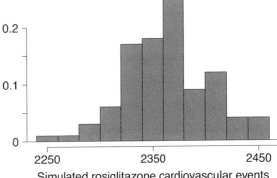

  i. What are the claims being tested?

  ii. Compared to the number calculated in part (b), which would provide more support for the alternative hypothesis, *more* or *fewer* patients with cardiovascular problems in the rosiglitazone group?

  iii. What do the simulation results suggest about the relationship between taking rosiglitazone and having cardiovascular problems in diabetic patients?

[29]D.J. Graham et al. "Risk of acute myocardial infarction, stroke, heart failure, and death in elderly Medicare patients treated with rosiglitazone or pioglitazone". In: *JAMA* 304.4 (2010), p. 411. ISSN: 0098-7484.

2.26 Heart transplants. The Stanford University Heart Transplant Study was conducted to determine whether an experimental heart transplant program increased lifespan. Each patient entering the program was designated an official heart transplant candidate, meaning that he was gravely ill and would most likely benefit from a new heart. Some patients got a transplant and some did not. The variable `transplant` indicates which group the patients were in; patients in the treatment group got a transplant and those in the control group did not. Of the 34 patients in the control group, 30 died. Of the 69 people in the treatment group, 45 died. Another variable called `survived` was used to indicate whether or not the patient was alive at the end of the study.[30]

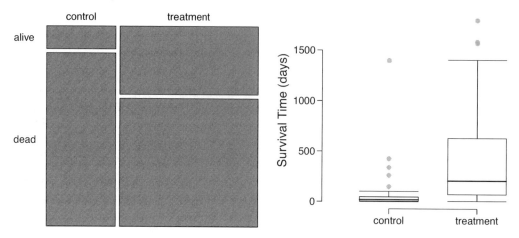

(a) Based on the mosaic plot, is survival independent of whether or not the patient got a transplant? Explain your reasoning.

(b) What do the box plots below suggest about the efficacy (effectiveness) of the heart transplant treatment.

(c) What proportion of patients in the treatment group and what proportion of patients in the control group died?

(d) One approach for investigating whether or not the treatment is effective is to use a randomization technique.

   i. What are the claims being tested?

   ii. The paragraph below describes the set up for such approach, if we were to do it without using statistical software. Fill in the blanks with a number or phrase, whichever is appropriate.

   > We write *alive* on _____ cards representing patients who were alive at the end of the study, and *dead* on _____ cards representing patients who were not. Then, we shuffle these cards and split them into two groups: one group of size _____ representing treatment, and another group of size _____ representing control. We calculate the difference between the proportion of *dead* cards in the treatment and control groups (treatment - control) and record this value. We repeat this 100 times to build a distribution centered at _____. Lastly, we calculate the fraction of simulations where the simulated differences in proportions are _____. If this fraction is low, we conclude that it is unlikely to have observed such an outcome by chance and that the null hypothesis should be rejected in favor of the alternative.

   iii. What do the simulation results shown below suggest about the effectiveness of the transplant program?

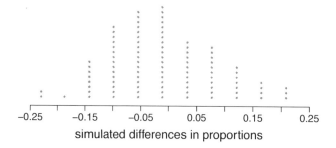

simulated differences in proportions

[30]B. Turnbull et al. "Survivorship of Heart Transplant Data". In: *Journal of the American Statistical Association* 69 (1974), pp. 74–80.

Chapter exercises

2.27 Make-up exam. In a class of 25 students, 24 of them took an exam in class and 1 student took a make-up exam the following day. The professor graded the first batch of 24 exams and found an average score of 74 points with a standard deviation of 8.9 points. The student who took the make-up the following day scored 64 points on the exam.

(a) Does the new student's score increase or decrease the average score?

(b) What is the new average?

(c) Does the new student's score increase or decrease the standard deviation of the scores?

2.28 Infant mortality. The infant mortality rate is defined as the number of infant deaths per 1,000 live births. This rate is often used as an indicator of the level of health in a country. The relative frequency histogram below shows the distribution of estimated infant death rates for 224 countries for which such data were available in 2014.[31]

(a) Estimate Q1, the median, and Q3 from the histogram.

(b) Would you expect the mean of this data set to be smaller or larger than the median? Explain your reasoning.

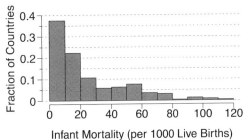

2.29 TV watchers. Students in an AP Statistics class were asked how many hours of television they watch per week (including online streaming). This sample yielded an average of 4.71 hours, with a standard deviation of 4.18 hours. Is the distribution of number of hours students watch television weekly symmetric? If not, what shape would you expect this distribution to have? Explain your reasoning.

2.30 A new statistic. The statistic $\frac{\bar{x}}{median}$ can be used as a measure of skewness. Suppose we have a distribution where all observations are greater than 0, $x_i > 0$. What is the expected shape of the distribution under the following conditions? Explain your reasoning.

(a) $\frac{\bar{x}}{median} = 1$

(b) $\frac{\bar{x}}{median} < 1$

(c) $\frac{\bar{x}}{median} > 1$

2.31 Oscar winners. The first Oscar awards for best actor and best actress were given out in 1929. The histograms below show the age distribution for all of the best actor and best actress winners from 1929 to 2018. Summary statistics for these distributions are also provided. Compare the distributions of ages of best actor and actress winners.[32]

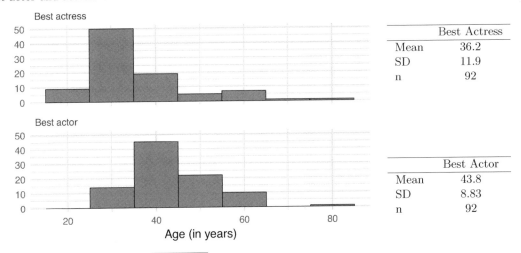

	Best Actress
Mean	36.2
SD	11.9
n	92

	Best Actor
Mean	43.8
SD	8.83
n	92

2.32 Exam scores. The average on a history exam (scored out of 100 points) was 85, with a standard deviation of 15. Is the distribution of the scores on this exam symmetric? If not, what shape would you expect this distribution to have? Explain your reasoning.

2.33 Stats scores. Below are the final exam scores of twenty introductory statistics students.

57, 66, 69, 71, 72, 73, 74, 77, 78, 78, 79, 79, 81, 81, 82, 83, 83, 88, 89, 94

Create a box plot of the distribution of these scores. The five number summary provided below may be useful.

Min	Q1	Q2 (Median)	Q3	Max
57	72.5	78.5	82.5	94

2.34 Marathon winners. The histogram and box plots below show the distribution of finishing times for male and female winners of the New York Marathon between 1970 and 1999.

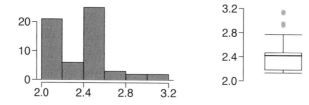

(a) What features of the distribution are apparent in the histogram and not the box plot? What features are apparent in the box plot but not in the histogram?

(b) What may be the reason for the bimodal distribution? Explain.

(c) Compare the distribution of marathon times for men and women based on the box plot shown below.

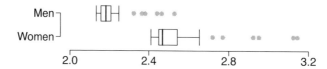

(d) The time series plot shown below is another way to look at these data. Describe what is visible in this plot but not in the others.

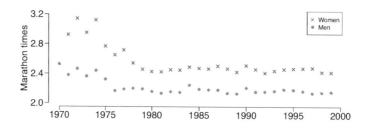

Chapter 3

Probability

Probability forms the foundation of statistics, and you're probably already aware of many of the ideas presented in this chapter. However, formalization of probability concepts is likely new for most readers.

While this chapter provides a theoretical foundation for the ideas in later chapters and provides a path to a deeper understanding, mastery of the concepts introduced in this chapter is not required for applying the methods introduced in the rest of this book.

For videos, slides, and other resources, please visit
www.openintro.org/os

3.1 Defining probability

Statistics is based on probability, and while probability is not required for the applied techniques in this book, it may help you gain a deeper understanding of the methods and set a better foundation for future courses.

3.1.1 Introductory examples

Before we get into technical ideas, let's walk through some basic examples that may feel more familiar.

EXAMPLE 3.1

A "die", the singular of dice, is a cube with six faces numbered 1, 2, 3, 4, 5, and 6. What is the chance of getting 1 when rolling a die?

If the die is fair, then the chance of a 1 is as good as the chance of any other number. Since there are six outcomes, the chance must be 1-in-6 or, equivalently, 1/6.

EXAMPLE 3.2

What is the chance of getting a 1 or 2 in the next roll?

1 and 2 constitute two of the six equally likely possible outcomes, so the chance of getting one of these two outcomes must be $2/6 = 1/3$.

EXAMPLE 3.3

What is the chance of getting either 1, 2, 3, 4, 5, or 6 on the next roll?

100%. The outcome must be one of these numbers.

EXAMPLE 3.4

What is the chance of not rolling a 2?

Since the chance of rolling a 2 is 1/6 or $16.\bar{6}\%$, the chance of not rolling a 2 must be $100\% - 16.\bar{6}\% = 83.\bar{3}\%$ or 5/6.

Alternatively, we could have noticed that not rolling a 2 is the same as getting a 1, 3, 4, 5, or 6, which makes up five of the six equally likely outcomes and has probability 5/6.

EXAMPLE 3.5

Consider rolling two dice. If 1/6 of the time the first die is a 1 and 1/6 of those times the second die is a 1, what is the chance of getting two 1s?

If $16.\bar{6}\%$ of the time the first die is a 1 and 1/6 of *those* times the second die is also a 1, then the chance that both dice are 1 is $(1/6) \times (1/6)$ or 1/36.

3.1.2 Probability

We use probability to build tools to describe and understand apparent randomness. We often frame probability in terms of a **random process** giving rise to an **outcome**.

$$\begin{array}{lcl} \text{Roll a die} & \rightarrow & \text{1, 2, 3, 4, 5, or 6} \\ \text{Flip a coin} & \rightarrow & \text{H or T} \end{array}$$

Rolling a die or flipping a coin is a seemingly random process and each gives rise to an outcome.

PROBABILITY

The **probability** of an outcome is the proportion of times the outcome would occur if we observed the random process an infinite number of times.

Probability is defined as a proportion, and it always takes values between 0 and 1 (inclusively). It may also be displayed as a percentage between 0% and 100%.

Probability can be illustrated by rolling a die many times. Let $\hat{p}_n$ be the proportion of outcomes that are 1 after the first n rolls. As the number of rolls increases, $\hat{p}_n$ will converge to the probability of rolling a 1, $p = 1/6$. Figure 3.1 shows this convergence for 100,000 die rolls. The tendency of $\hat{p}_n$ to stabilize around p is described by the **Law of Large Numbers**.

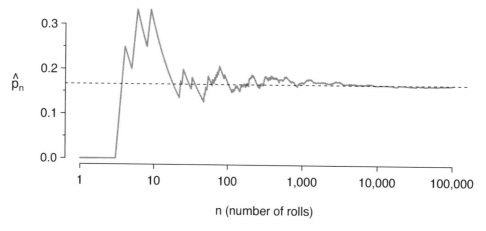

Figure 3.1: The fraction of die rolls that are 1 at each stage in a simulation. The proportion tends to get closer to the probability $1/6 \approx 0.167$ as the number of rolls increases.

LAW OF LARGE NUMBERS

As more observations are collected, the proportion $\hat{p}_n$ of occurrences with a particular outcome converges to the probability p of that outcome.

Occasionally the proportion will veer off from the probability and appear to defy the Law of Large Numbers, as $\hat{p}_n$ does many times in Figure 3.1. However, these deviations become smaller as the number of rolls increases.

Above we write p as the probability of rolling a 1. We can also write this probability as

$$P(\text{rolling a 1})$$

As we become more comfortable with this notation, we will abbreviate it further. For instance, if it is clear that the process is "rolling a die", we could abbreviate $P(\text{rolling a 1})$ as $P(1)$.

GUIDED PRACTICE 3.6

Random processes include rolling a die and flipping a coin. (a) Think of another random process. (b) Describe all the possible outcomes of that process. For instance, rolling a die is a random process with possible outcomes 1, 2, ..., 6.[1]

What we think of as random processes are not necessarily random, but they may just be too difficult to understand exactly. The fourth example in the footnote solution to Guided Practice 3.6 suggests a roommate's behavior is a random process. However, even if a roommate's behavior is not truly random, modeling her behavior as a random process can still be useful.

3.1.3   Disjoint or mutually exclusive outcomes

Two outcomes are called **disjoint** or **mutually exclusive** if they cannot both happen. For instance, if we roll a die, the outcomes 1 and 2 are disjoint since they cannot both occur. On the other hand, the outcomes 1 and "rolling an odd number" are not disjoint since both occur if the outcome of the roll is a 1. The terms *disjoint* and *mutually exclusive* are equivalent and interchangeable.

Calculating the probability of disjoint outcomes is easy. When rolling a die, the outcomes 1 and 2 are disjoint, and we compute the probability that one of these outcomes will occur by adding their separate probabilities:

$$P(1 \text{ or } 2) = P(1) + P(2) = 1/6 + 1/6 = 1/3$$

What about the probability of rolling a 1, 2, 3, 4, 5, or 6? Here again, all of the outcomes are disjoint so we add the probabilities:

$$P(1 \text{ or } 2 \text{ or } 3 \text{ or } 4 \text{ or } 5 \text{ or } 6)$$
$$= P(1) + P(2) + P(3) + P(4) + P(5) + P(6)$$
$$= 1/6 + 1/6 + 1/6 + 1/6 + 1/6 + 1/6 = 1$$

The **Addition Rule** guarantees the accuracy of this approach when the outcomes are disjoint.

ADDITION RULE OF DISJOINT OUTCOMES

If A_1 and A_2 represent two disjoint outcomes, then the probability that one of them occurs is given by

$$P(A_1 \text{ or } A_2) = P(A_1) + P(A_2)$$

If there are many disjoint outcomes A_1, ..., A_k, then the probability that one of these outcomes will occur is

$$P(A_1) + P(A_2) + \cdots + P(A_k)$$

---

[1] Here are four examples. (i) Whether someone gets sick in the next month or not is an apparently random process with outcomes `sick` and `not`. (ii) We can *generate* a random process by randomly picking a person and measuring that person's height. The outcome of this process will be a positive number. (iii) Whether the stock market goes up or down next week is a seemingly random process with possible outcomes `up`, `down`, and `no_change`. Alternatively, we could have used the percent change in the stock market as a numerical outcome. (iv) Whether your roommate cleans her dishes tonight probably seems like a random process with possible outcomes `cleans_dishes` and `leaves_dishes`.

GUIDED PRACTICE 3.7

We are interested in the probability of rolling a 1, 4, or 5. (a) Explain why the outcomes 1, 4, and 5 are disjoint. (b) Apply the Addition Rule for disjoint outcomes to determine $P(1 \text{ or } 4 \text{ or } 5)$.[2]

GUIDED PRACTICE 3.8

In the `loans` data set in Chapter 2, the `homeownership` variable described whether the borrower rents, has a mortgage, or owns her property. Of the 10,000 borrowers, 3858 rented, 4789 had a mortgage, and 1353 owned their home.[3]

(a) Are the outcomes `rent`, `mortgage`, and `own` disjoint?

(b) Determine the proportion of loans with value `mortgage` and `own` separately.

(c) Use the Addition Rule for disjoint outcomes to compute the probability a randomly selected loan from the data set is for someone who has a mortgage or owns her home.

Data scientists rarely work with individual outcomes and instead consider *sets* or *collections* of outcomes. Let A represent the event where a die roll results in 1 or 2 and B represent the event that the die roll is a 4 or a 6. We write A as the set of outcomes $\{1, 2\}$ and $B = \{4, 6\}$. These sets are commonly called **events**. Because A and B have no elements in common, they are disjoint events. A and B are represented in Figure 3.2.

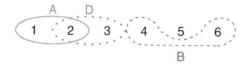

Figure 3.2: Three events, A, B, and D, consist of outcomes from rolling a die. A and B are disjoint since they do not have any outcomes in common.

The Addition Rule applies to both disjoint outcomes and disjoint events. The probability that one of the disjoint events A or B occurs is the sum of the separate probabilities:

$$P(A \text{ or } B) = P(A) + P(B) = 1/3 + 1/3 = 2/3$$

GUIDED PRACTICE 3.9

(a) Verify the probability of event A, $P(A)$, is 1/3 using the Addition Rule. (b) Do the same for event B.[4]

GUIDED PRACTICE 3.10

(a) Using Figure 3.2 as a reference, what outcomes are represented by event D? (b) Are events B and D disjoint? (c) Are events A and D disjoint?[5]

GUIDED PRACTICE 3.11

In Guided Practice 3.10, you confirmed B and D from Figure 3.2 are disjoint. Compute the probability that event B or event D occurs.[6]

---

[2](a) The random process is a die roll, and at most one of these outcomes can come up. This means they are disjoint outcomes. (b) $P(1 \text{ or } 4 \text{ or } 5) = P(1) + P(4) + P(5) = \frac{1}{6} + \frac{1}{6} + \frac{1}{6} = \frac{3}{6} = \frac{1}{2}$

[3](a) Yes. Each loan is categorized in only one level of `homeownership`. (b) Mortgage: $\frac{4789}{10000} = 0.479$. Own: $\frac{1353}{10000} = 0.135$. (c) $P(\text{mortgage or own}) = P(\text{mortgage}) + P(\text{own}) = 0.479 + 0.135 = 0.614$.

[4](a) $P(A) = P(1 \text{ or } 2) = P(1) + P(2) = \frac{1}{6} + \frac{1}{6} = \frac{2}{6} = \frac{1}{3}$. (b) Similarly, $P(B) = 1/3$.

[5](a) Outcomes 2 and 3. (b) Yes, events B and D are disjoint because they share no outcomes. (c) The events A and D share an outcome in common, 2, and so are not disjoint.

[6]Since B and D are disjoint events, use the Addition Rule: $P(B \text{ or } D) = P(B) + P(D) = \frac{1}{3} + \frac{1}{3} = \frac{2}{3}$.

3.1.4 Probabilities when events are not disjoint

Let's consider calculations for two events that are not disjoint in the context of a regular deck of 52 cards, represented in Figure 3.3. If you are unfamiliar with the cards in a regular deck, please see the footnote.[7]

2♣	3♣	4♣	5♣	6♣	7♣	8♣	9♣	10♣	J♣	Q♣	K♣	A♣
2♢	3♢	4♢	5♢	6♢	7♢	8♢	9♢	10♢	J♢	Q♢	K♢	A♢
2♡	3♡	4♡	5♡	6♡	7♡	8♡	9♡	10♡	J♡	Q♡	K♡	A♡
2♠	3♠	4♠	5♠	6♠	7♠	8♠	9♠	10♠	J♠	Q♠	K♠	A♠

Figure 3.3: Representations of the 52 unique cards in a deck.

GUIDED PRACTICE 3.12

(a) What is the probability that a randomly selected card is a diamond? (b) What is the probability that a randomly selected card is a face card?[8]

Venn diagrams are useful when outcomes can be categorized as "in" or "out" for two or three variables, attributes, or random processes. The Venn diagram in Figure 3.4 uses a circle to represent diamonds and another to represent face cards. If a card is both a diamond and a face card, it falls into the intersection of the circles. If it is a diamond but not a face card, it will be in part of the left circle that is not in the right circle (and so on). The total number of cards that are diamonds is given by the total number of cards in the diamonds circle: $10 + 3 = 13$. The probabilities are also shown (e.g. $10/52 = 0.1923$).

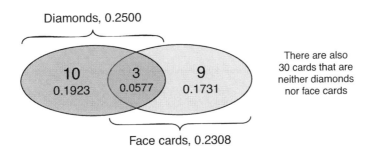

Figure 3.4: A Venn diagram for diamonds and face cards.

Let A represent the event that a randomly selected card is a diamond and B represent the event that it is a face card. How do we compute $P(A \text{ or } B)$? Events A and B are not disjoint – the cards $J\diamondsuit$, $Q\diamondsuit$, and $K\diamondsuit$ fall into both categories – so we cannot use the Addition Rule for disjoint events. Instead we use the Venn diagram. We start by adding the probabilities of the two events:

$$P(A) + P(B) = P(\diamondsuit) + P(\text{face card}) = 13/52 + 12/52$$

---

[7]The 52 cards are split into four **suits**: ♣ (club), ♢ (diamond), ♡ (heart), ♠ (spade). Each suit has its 13 cards labeled: 2, 3, ..., 10, J (jack), Q (queen), K (king), and A (ace). Thus, each card is a unique combination of a suit and a label, e.g. 4♡ and J♣. The 12 cards represented by the jacks, queens, and kings are called face cards. The cards that are ♢ or ♡ are typically colored red while the other two suits are typically colored black.

[8](a) There are 52 cards and 13 diamonds. If the cards are thoroughly shuffled, each card has an equal chance of being drawn, so the probability that a randomly selected card is a diamond is $P(\diamondsuit) = \frac{13}{52} = 0.250$. (b) Likewise, there are 12 face cards, so $P(\text{face card}) = \frac{12}{52} = \frac{3}{13} = 0.231$.

However, the three cards that are in both events were counted twice, once in each probability. We must correct this double counting:

$$P(A \text{ or } B) = P(\diamond \text{ or face card})$$
$$= P(\diamond) + P(\text{face card}) - P(\diamond \text{ and face card})$$
$$= 13/52 + 12/52 - 3/52$$
$$= 22/52 = 11/26$$

This equation is an example of the **General Addition Rule**.

GENERAL ADDITION RULE

If A and B are any two events, disjoint or not, then the probability that at least one of them will occur is

$$P(A \text{ or } B) = P(A) + P(B) - P(A \text{ and } B)$$

where $P(A \text{ and } B)$ is the probability that both events occur.

TIP: "or" is inclusive

When we write "or" in statistics, we mean "and/or" unless we explicitly state otherwise. Thus, A or B occurs means A, B, or both A and B occur.

GUIDED PRACTICE 3.13

(a) If A and B are disjoint, describe why this implies $P(A \text{ and } B) = 0$. (b) Using part (a), verify that the General Addition Rule simplifies to the simpler Addition Rule for disjoint events if A and B are disjoint.[9]

GUIDED PRACTICE 3.14

In the `loans` data set describing 10,000 loans, 1495 loans were from joint applications (e.g. a couple applied together), 4789 applicants had a mortgage, and 950 had both of these characteristics. Create a Venn diagram for this setup.[10]

GUIDED PRACTICE 3.15

(a) Use your Venn diagram from Guided Practice 3.14 to determine the probability a randomly drawn loan from the `loans` data set is from a joint application where the couple had a mortgage. (b) What is the probability that the loan had either of these attributes?[11]

---

[9](a) If A and B are disjoint, A and B can never occur simultaneously. (b) If A and B are disjoint, then the last $P(A \text{ and } B)$ term of in the General Addition Rule formula is 0 (see part (a)) and we are left with the Addition Rule for disjoint events.

[10]Both the counts and corresponding probabilities (e.g. $3839/10000 = 0.384$) are shown. Notice that the number of loans represented in the left circle corresponds to $3839 + 950 = 4789$, and the number represented in the right circle is $950 + 545 = 1495$.

[11](a) The solution is represented by the intersection of the two circles: 0.095. (b) This is the sum of the three disjoint probabilities shown in the circles: $0.384 + 0.095 + 0.055 = 0.534$ (off by 0.001 due to a rounding error).

3.1.5 Probability distributions

A **probability distribution** is a table of all disjoint outcomes and their associated probabilities. Figure 3.5 shows the probability distribution for the sum of two dice.

Dice sum	2	3	4	5	6	7	8	9	10	11	12
Probability	$\frac{1}{36}$	$\frac{2}{36}$	$\frac{3}{36}$	$\frac{4}{36}$	$\frac{5}{36}$	$\frac{6}{36}$	$\frac{5}{36}$	$\frac{4}{36}$	$\frac{3}{36}$	$\frac{2}{36}$	$\frac{1}{36}$

Figure 3.5: Probability distribution for the sum of two dice.

RULES FOR PROBABILITY DISTRIBUTIONS

A probability distribution is a list of the possible outcomes with corresponding probabilities that satisfies three rules:

1. The outcomes listed must be disjoint.

2. Each probability must be between 0 and 1.

3. The probabilities must total 1.

GUIDED PRACTICE 3.16

Figure 3.6 suggests three distributions for household income in the United States. Only one is correct. Which one must it be? What is wrong with the other two?[12]

Income Range	$0-25k	$25k-50k	$50k-100k	$100k+
(a)	0.18	0.39	0.33	0.16
(b)	0.38	-0.27	0.52	0.37
(c)	0.28	0.27	0.29	0.16

Figure 3.6: Proposed distributions of US household incomes (Guided Practice 3.16).

Chapter 1 emphasized the importance of plotting data to provide quick summaries. Probability distributions can also be summarized in a bar plot. For instance, the distribution of US household incomes is shown in Figure 3.7 as a bar plot. The probability distribution for the sum of two dice is shown in Figure 3.5 and plotted in Figure 3.8.

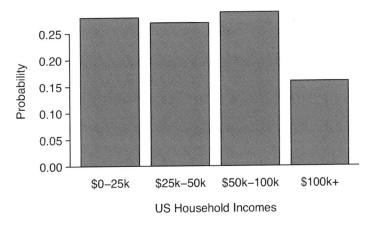

Figure 3.7: The probability distribution of US household income.

---

[12]The probabilities of (a) do not sum to 1. The second probability in (b) is negative. This leaves (c), which sure enough satisfies the requirements of a distribution. One of the three was said to be the actual distribution of US household incomes, so it must be (c).

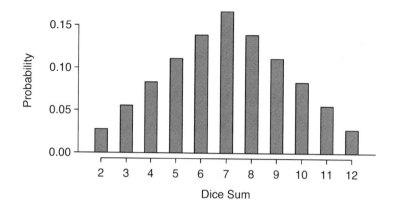

Figure 3.8: The probability distribution of the sum of two dice.

In these bar plots, the bar heights represent the probabilities of outcomes. If the outcomes are numerical and discrete, it is usually (visually) convenient to make a bar plot that resembles a histogram, as in the case of the sum of two dice. Another example of plotting the bars at their respective locations is shown in Figure 3.18 on page 115.

3.1.6 Complement of an event

Rolling a die produces a value in the set {1, 2, 3, 4, 5, 6}. This set of all possible outcomes is called the **sample space** (S) for rolling a die. We often use the sample space to examine the scenario where an event does not occur.

Let $D = \{2, 3\}$ represent the event that the outcome of a die roll is 2 or 3. Then the **complement** of D represents all outcomes in our sample space that are not in D, which is denoted by $D^c = \{1, 4, 5, 6\}$. That is, D^c is the set of all possible outcomes not already included in D. Figure 3.9 shows the relationship between D, D^c, and the sample space S.

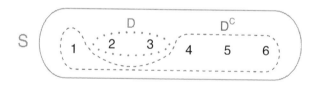

Figure 3.9: Event $D = \{2, 3\}$ and its complement, $D^c = \{1, 4, 5, 6\}$. S represents the sample space, which is the set of all possible outcomes.

GUIDED PRACTICE 3.17

(a) Compute $P(D^c) = P(\text{rolling a 1, 4, 5, or 6})$. (b) What is $P(D) + P(D^c)$?[13]

GUIDED PRACTICE 3.18

Events $A = \{1, 2\}$ and $B = \{4, 6\}$ are shown in Figure 3.2 on page 84. (a) Write out what A^c and B^c represent. (b) Compute $P(A^c)$ and $P(B^c)$. (c) Compute $P(A) + P(A^c)$ and $P(B) + P(B^c)$.[14]

---

[13](a) The outcomes are disjoint and each has probability 1/6, so the total probability is 4/6 = 2/3. (b) We can also see that $P(D) = \frac{1}{6} + \frac{1}{6} = 1/3$. Since D and D^c are disjoint, $P(D) + P(D^c) = 1$.

[14]Brief solutions: (a) $A^c = \{3, 4, 5, 6\}$ and $B^c = \{1, 2, 3, 5\}$. (b) Noting that each outcome is disjoint, add the individual outcome probabilities to get $P(A^c) = 2/3$ and $P(B^c) = 2/3$. (c) A and A^c are disjoint, and the same is true of B and B^c. Therefore, $P(A) + P(A^c) = 1$ and $P(B) + P(B^c) = 1$.

A complement of an event A is constructed to have two very important properties: (i) every possible outcome not in A is in A^c, and (ii) A and A^c are disjoint. Property (i) implies

$$P(A \text{ or } A^c) = 1$$

That is, if the outcome is not in A, it must be represented in A^c. We use the Addition Rule for disjoint events to apply Property (ii):

$$P(A \text{ or } A^c) = P(A) + P(A^c)$$

Combining the last two equations yields a very useful relationship between the probability of an event and its complement.

COMPLEMENT

The complement of event A is denoted A^c, and A^c represents all outcomes not in A. A and A^c are mathematically related:

$$P(A) + P(A^c) = 1, \quad \text{i.e.} \quad P(A) = 1 - P(A^c)$$

In simple examples, computing A or A^c is feasible in a few steps. However, using the complement can save a lot of time as problems grow in complexity.

GUIDED PRACTICE 3.19

Let A represent the event where we roll two dice and their total is less than 12. (a) What does the event A^c represent? (b) Determine $P(A^c)$ from Figure 3.5 on page 87. (c) Determine $P(A)$.[15]

GUIDED PRACTICE 3.20

Find the following probabilities for rolling two dice:[16]

(a) The sum of the dice is *not* 6.

(b) The sum is at least 4. That is, determine the probability of the event $B = \{4, 5, ..., 12\}$.

(c) The sum is no more than 10. That is, determine the probability of the event $D = \{2, 3, ..., 10\}$.

3.1.7 Independence

Just as variables and observations can be independent, random processes can be independent, too. Two processes are **independent** if knowing the outcome of one provides no useful information about the outcome of the other. For instance, flipping a coin and rolling a die are two independent processes – knowing the coin was heads does not help determine the outcome of a die roll. On the other hand, stock prices usually move up or down together, so they are not independent.

Example 3.5 provides a basic example of two independent processes: rolling two dice. We want to determine the probability that both will be 1. Suppose one of the dice is red and the other white. If the outcome of the red die is a 1, it provides no information about the outcome of the white die. We first encountered this same question in Example 3.5 (page 81), where we calculated the probability using the following reasoning: 1/6 of the time the red die is a 1, and 1/6 of *those* times the white die

---

[15](a) The complement of A: when the total is equal to 12. (b) $P(A^c) = 1/36$. (c) Use the probability of the complement from part (b), $P(A^c) = 1/36$, and the equation for the complement: $P(\text{less than } 12) = 1 - P(12) = 1 - 1/36 = 35/36$.

[16](a) First find $P(6) = 5/36$, then use the complement: $P(\text{not } 6) = 1 - P(6) = 31/36$.

(b) First find the complement, which requires much less effort: $P(2 \text{ or } 3) = 1/36 + 2/36 = 1/12$. Then calculate $P(B) = 1 - P(B^c) = 1 - 1/12 = 11/12$.

(c) As before, finding the complement is the clever way to determine $P(D)$. First find $P(D^c) = P(11 \text{ or } 12) = 2/36 + 1/36 = 1/12$. Then calculate $P(D) = 1 - P(D^c) = 11/12$.

will also be 1. This is illustrated in Figure 3.10. Because the rolls are independent, the probabilities of the corresponding outcomes can be multiplied to get the final answer: $(1/6) \times (1/6) = 1/36$. This can be generalized to many independent processes.

All rolls

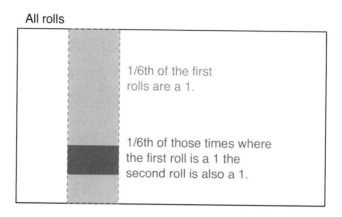

Figure 3.10: 1/6 of the time, the first roll is a **1**. Then 1/6 of *those* times, the second roll will also be a **1**.

EXAMPLE 3.21

What if there was also a blue die independent of the other two? What is the probability of rolling the three dice and getting all **1**s?

The same logic applies from Example 3.5. If 1/36 of the time the white and red dice are both **1**, then 1/6 of *those* times the blue die will also be **1**, so multiply:

$$P(white = \mathbf{1} \text{ and } red = \mathbf{1} \text{ and } blue = \mathbf{1}) = P(white = \mathbf{1}) \times P(red = \mathbf{1}) \times P(blue = \mathbf{1})$$
$$= (1/6) \times (1/6) \times (1/6) = 1/216$$

Example 3.21 illustrates what is called the Multiplication Rule for independent processes.

MULTIPLICATION RULE FOR INDEPENDENT PROCESSES

If A and B represent events from two different and independent processes, then the probability that both A and B occur can be calculated as the product of their separate probabilities:

$$P(A \text{ and } B) = P(A) \times P(B)$$

Similarly, if there are k events A_1, ..., A_k from k independent processes, then the probability they all occur is

$$P(A_1) \times P(A_2) \times \cdots \times P(A_k)$$

GUIDED PRACTICE 3.22

About 9% of people are left-handed. Suppose 2 people are selected at random from the U.S. population. Because the sample size of 2 is very small relative to the population, it is reasonable to assume these two people are independent. (a) What is the probability that both are left-handed? (b) What is the probability that both are right-handed?[17]

---

[17](a) The probability the first person is left-handed is 0.09, which is the same for the second person. We apply the Multiplication Rule for independent processes to determine the probability that both will be left-handed: $0.09 \times 0.09 = 0.0081$.

(b) It is reasonable to assume the proportion of people who are ambidextrous (both right- and left-handed) is nearly 0, which results in $P(\text{right-handed}) = 1 - 0.09 = 0.91$. Using the same reasoning as in part (a), the probability that both will be right-handed is $0.91 \times 0.91 = 0.8281$.

GUIDED PRACTICE 3.23

Suppose 5 people are selected at random.[18]

(a) What is the probability that all are right-handed?

(b) What is the probability that all are left-handed?

(c) What is the probability that not all of the people are right-handed?

Suppose the variables handedness and sex are independent, i.e. knowing someone's sex provides no useful information about their handedness and vice-versa. Then we can compute whether a randomly selected person is right-handed and female[19] using the Multiplication Rule:

$$P(\text{right-handed and female}) = P(\text{right-handed}) \times P(\text{female})$$
$$= 0.91 \times 0.50 = 0.455$$

GUIDED PRACTICE 3.24

Three people are selected at random.[20]

(a) What is the probability that the first person is male and right-handed?

(b) What is the probability that the first two people are male and right-handed?.

(c) What is the probability that the third person is female and left-handed?

(d) What is the probability that the first two people are male and right-handed and the third person is female and left-handed?

Sometimes we wonder if one outcome provides useful information about another outcome. The question we are asking is, are the occurrences of the two events independent? We say that two events A and B are independent if they satisfy $P(A \text{ and } B) = P(A) \times P(B)$.

EXAMPLE 3.25

If we shuffle up a deck of cards and draw one, is the event that the card is a heart independent of the event that the card is an ace?

The probability the card is a heart is 1/4 and the probability that it is an ace is 1/13. The probability the card is the ace of hearts is 1/52. We check whether $P(A \text{ and } B) = P(A) \times P(B)$ is satisfied:

$$P(\heartsuit) \times P(\text{ace}) = \frac{1}{4} \times \frac{1}{13} = \frac{1}{52} = P(\heartsuit \text{ and ace})$$

Because the equation holds, the event that the card is a heart and the event that the card is an ace are independent events.

---

[18](a) The abbreviations RH and LH are used for right-handed and left-handed, respectively. Since each are independent, we apply the Multiplication Rule for independent processes:

$$P(\text{all five are RH}) = P(\text{first} = \text{RH}, \text{second} = \text{RH}, ..., \text{fifth} = \text{RH})$$
$$= P(\text{first} = \text{RH}) \times P(\text{second} = \text{RH}) \times \cdots \times P(\text{fifth} = \text{RH})$$
$$= 0.91 \times 0.91 \times 0.91 \times 0.91 \times 0.91 = 0.624$$

(b) Using the same reasoning as in (a), $0.09 \times 0.09 \times 0.09 \times 0.09 \times 0.09 = 0.0000059$

(c) Use the complement, $P(\text{all five are RH})$, to answer this question:

$$P(\text{not all RH}) = 1 - P(\text{all RH}) = 1 - 0.624 = 0.376$$

[19]The actual proportion of the U.S. population that is female is about 50%, and so we use 0.5 for the probability of sampling a woman. However, this probability does differ in other countries.

[20]Brief answers are provided. (a) This can be written in probability notation as $P(\text{a randomly selected person is male and right-handed}) = 0.455$. (b) 0.207. (c) 0.045. (d) 0.0093.

Exercises

3.1 True or false. Determine if the statements below are true or false, and explain your reasoning.

(a) If a fair coin is tossed many times and the last eight tosses are all heads, then the chance that the next toss will be heads is somewhat less than 50%.

(b) Drawing a face card (jack, queen, or king) and drawing a red card from a full deck of playing cards are mutually exclusive events.

(c) Drawing a face card and drawing an ace from a full deck of playing cards are mutually exclusive events.

3.2 Roulette wheel. The game of roulette involves spinning a wheel with 38 slots: 18 red, 18 black, and 2 green. A ball is spun onto the wheel and will eventually land in a slot, where each slot has an equal chance of capturing the ball.

(a) You watch a roulette wheel spin 3 consecutive times and the ball lands on a red slot each time. What is the probability that the ball will land on a red slot on the next spin?

(b) You watch a roulette wheel spin 300 consecutive times and the ball lands on a red slot each time. What is the probability that the ball will land on a red slot on the next spin?

(c) Are you equally confident of your answers to parts (a) and (b)? Why or why not?

Photo by Håkan Dahlström
(http://flic.kr/p/93fEzp)
CC BY 2.0 license

3.3 Four games, one winner. Below are four versions of the same game. Your archnemesis gets to pick the version of the game, and then you get to choose how many times to flip a coin: 10 times or 100 times. Identify how many coin flips you should choose for each version of the game. It costs $1 to play each game. Explain your reasoning.

(a) If the proportion of heads is larger than 0.60, you win $1.

(b) If the proportion of heads is larger than 0.40, you win $1.

(c) If the proportion of heads is between 0.40 and 0.60, you win $1.

(d) If the proportion of heads is smaller than 0.30, you win $1.

3.4 Backgammon. Backgammon is a board game for two players in which the playing pieces are moved according to the roll of two dice. Players win by removing all of their pieces from the board, so it is usually good to roll high numbers. You are playing backgammon with a friend and you roll two 6s in your first roll and two 6s in your second roll. Your friend rolls two 3s in his first roll and again in his second row. Your friend claims that you are cheating, because rolling double 6s twice in a row is very unlikely. Using probability, show that your rolls were just as likely as his.

3.5 Coin flips. If you flip a fair coin 10 times, what is the probability of

(a) getting all tails?

(b) getting all heads?

(c) getting at least one tails?

3.6 Dice rolls. If you roll a pair of fair dice, what is the probability of

(a) getting a sum of 1?

(b) getting a sum of 5?

(c) getting a sum of 12?

3.7 Swing voters. A Pew Research survey asked 2,373 randomly sampled registered voters their political affiliation (Republican, Democrat, or Independent) and whether or not they identify as swing voters. 35% of respondents identified as Independent, 23% identified as swing voters, and 11% identified as both.[21]

(a) Are being Independent and being a swing voter disjoint, i.e. mutually exclusive?

(b) Draw a Venn diagram summarizing the variables and their associated probabilities.

(c) What percent of voters are Independent but not swing voters?

(d) What percent of voters are Independent or swing voters?

(e) What percent of voters are neither Independent nor swing voters?

(f) Is the event that someone is a swing voter independent of the event that someone is a political Independent?

3.8 Poverty and language. The American Community Survey is an ongoing survey that provides data every year to give communities the current information they need to plan investments and services. The 2010 American Community Survey estimates that 14.6% of Americans live below the poverty line, 20.7% speak a language other than English (foreign language) at home, and 4.2% fall into both categories.[22]

(a) Are living below the poverty line and speaking a foreign language at home disjoint?

(b) Draw a Venn diagram summarizing the variables and their associated probabilities.

(c) What percent of Americans live below the poverty line and only speak English at home?

(d) What percent of Americans live below the poverty line or speak a foreign language at home?

(e) What percent of Americans live above the poverty line and only speak English at home?

(f) Is the event that someone lives below the poverty line independent of the event that the person speaks a foreign language at home?

3.9 Disjoint vs. independent. In parts (a) and (b), identify whether the events are disjoint, independent, or neither (events cannot be both disjoint and independent).

(a) You and a randomly selected student from your class both earn A's in this course.

(b) You and your class study partner both earn A's in this course.

(c) If two events can occur at the same time, must they be dependent?

3.10 Guessing on an exam. In a multiple choice exam, there are 5 questions and 4 choices for each question (a, b, c, d). Nancy has not studied for the exam at all and decides to randomly guess the answers. What is the probability that:

(a) the first question she gets right is the 5th question?

(b) she gets all of the questions right?

(c) she gets at least one question right?

---

[21] Pew Research Center, With Voters Focused on Economy, Obama Lead Narrows, data collected between April 4-15, 2012.

[22] U.S. Census Bureau, 2010 American Community Survey 1-Year Estimates, Characteristics of People by Language Spoken at Home.

3.11 Educational attainment of couples. The table below shows the distribution of education level attained by US residents by gender based on data collected in the 2010 American Community Survey.[23]

		Gender	
		Male	Female
	Less than 9th grade	0.07	0.13
	9th to 12th grade, no diploma	0.10	0.09
Highest	HS graduate (or equivalent)	0.30	0.20
education	Some college, no degree	0.22	0.24
attained	Associate's degree	0.06	0.08
	Bachelor's degree	0.16	0.17
	Graduate or professional degree	0.09	0.09
	Total	1.00	1.00

(a) What is the probability that a randomly chosen man has at least a Bachelor's degree?

(b) What is the probability that a randomly chosen woman has at least a Bachelor's degree?

(c) What is the probability that a man and a woman getting married both have at least a Bachelor's degree? Note any assumptions you must make to answer this question.

(d) If you made an assumption in part (c), do you think it was reasonable? If you didn't make an assumption, double check your earlier answer and then return to this part.

3.12 School absences. Data collected at elementary schools in DeKalb County, GA suggest that each year roughly 25% of students miss exactly one day of school, 15% miss 2 days, and 28% miss 3 or more days due to sickness.[24]

(a) What is the probability that a student chosen at random doesn't miss any days of school due to sickness this year?

(b) What is the probability that a student chosen at random misses no more than one day?

(c) What is the probability that a student chosen at random misses at least one day?

(d) If a parent has two kids at a DeKalb County elementary school, what is the probability that neither kid will miss any school? Note any assumption you must make to answer this question.

(e) If a parent has two kids at a DeKalb County elementary school, what is the probability that both kids will miss some school, i.e. at least one day? Note any assumption you make.

(f) If you made an assumption in part (d) or (e), do you think it was reasonable? If you didn't make any assumptions, double check your earlier answers.

---

[23]U.S. Census Bureau, 2010 American Community Survey 1-Year Estimates, Educational Attainment.

[24]S.S. Mizan et al. "Absence, Extended Absence, and Repeat Tardiness Related to Asthma Status among Elementary School Children". In: *Journal of Asthma* 48.3 (2011), pp. 228–234.

3.2 Conditional probability

There can be rich relationships between two or more variables that are useful to understand. For example a car insurance company will consider information about a person's driving history to assess the risk that they will be responsible for an accident. These types of relationships are the realm of conditional probabilities.

3.2.1 Exploring probabilities with a contingency table

The `photo_classify` data set represents a classifier a sample of 1822 photos from a photo sharing website. Data scientists have been working to improve a classifier for whether the photo is about fashion or not, and these 1822 photos represent a test for their classifier. Each photo gets two classifications: the first is called `mach_learn` and gives a classification from a machine learning (ML) system of either `pred_fashion` or `pred_not`. Each of these 1822 photos have also been classified carefully by a team of people, which we take to be the source of truth; this variable is called `truth` and takes values `fashion` and `not`. Figure 3.11 summarizes the results.

| | | truth | | |
		fashion	not	Total
`mach_learn`	`pred_fashion`	197	22	219
	`pred_not`	112	1491	1603
	Total	309	1513	1822

Figure 3.11: Contingency table summarizing the `photo_classify` data set.

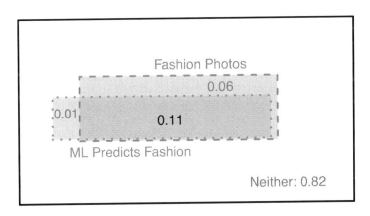

Figure 3.12: A Venn diagram using boxes for the `photo_classify` data set.

EXAMPLE 3.26

If a photo is actually about fashion, what is the chance the ML classifier correctly identified the photo as being about fashion?

We can estimate this probability using the data. Of the 309 fashion photos, the ML algorithm correctly classified 197 of the photos:

$$P(\text{mach\_learn is pred\_fashion given truth is fashion}) = \frac{197}{309} = 0.638$$

EXAMPLE 3.27

We sample a photo from the data set and learn the ML algorithm predicted this photo was not about fashion. What is the probability that it was incorrect and the photo is about fashion?

If the ML classifier suggests a photo is not about fashion, then it comes from the second row in the data set. Of these 1603 photos, 112 were actually about fashion:

$$P(\texttt{truth is fashion given mach\_learn is pred\_not}) = \frac{112}{1603} = 0.070$$

3.2.2   Marginal and joint probabilities

Figure 3.11 includes row and column totals for each variable separately in the `photo_classify` data set. These totals represent **marginal probabilities** for the sample, which are the probabilities based on a single variable without regard to any other variables. For instance, a probability based solely on the `mach_learn` variable is a marginal probability:

$$P(\texttt{mach\_learn is pred\_fashion}) = \frac{219}{1822} = 0.12$$

A probability of outcomes for two or more variables or processes is called a **joint probability**:

$$P(\texttt{mach\_learn is pred\_fashion and truth is fashion}) = \frac{197}{1822} = 0.11$$

It is common to substitute a comma for "and" in a joint probability, although using either the word "and" or a comma is acceptable:

$$P(\texttt{mach\_learn is pred\_fashion, truth is fashion})$$

means the same thing as

$$P(\texttt{mach\_learn is pred\_fashion and truth is fashion})$$

MARGINAL AND JOINT PROBABILITIES

If a probability is based on a single variable, it is a *marginal probability*. The probability of outcomes for two or more variables or processes is called a *joint probability*.

We use **table proportions** to summarize joint probabilities for the `photo_classify` sample. These proportions are computed by dividing each count in Figure 3.11 by the table's total, 1822, to obtain the proportions in Figure 3.13. The joint probability distribution of the `mach_learn` and `truth` variables is shown in Figure 3.14.

	truth: fashion	truth: not	Total
mach_learn: pred_fashion	0.1081	0.0121	0.1202
mach_learn: pred_not	0.0615	0.8183	0.8798
Total	0.1696	0.8304	1.00

Figure 3.13: Probability table summarizing the `photo_classify` data set.

Joint outcome	Probability
mach_learn is pred_fashion and truth is fashion	0.1081
mach_learn is pred_fashion and truth is not	0.0121
mach_learn is pred_not and truth is fashion	0.0615
mach_learn is pred_not and truth is not	0.8183
Total	1.0000

Figure 3.14: Joint probability distribution for the photo_classify data set.

GUIDED PRACTICE 3.28

Verify Figure 3.14 represents a probability distribution: events are disjoint, all probabilities are non-negative, and the probabilities sum to 1.[25]

We can compute marginal probabilities using joint probabilities in simple cases. For example, the probability a randomly selected photo from the data set is about fashion is found by summing the outcomes where truth takes value fashion:

$$P(\underline{\text{truth is fashion}}) = P(\text{mach\_learn is pred\_fashion and } \underline{\text{truth is fashion}})$$
$$+ P(\text{mach\_learn is pred\_not and } \underline{\text{truth is fashion}})$$
$$= 0.1081 + 0.0615$$
$$= 0.1696$$

3.2.3 Defining conditional probability

The ML classifier predicts whether a photo is about fashion, even if it is not perfect. We would like to better understand how to use information from a variable like mach_learn to improve our probability estimation of a second variable, which in this example is truth.

The probability that a random photo from the data set is about fashion is about 0.17. If we knew the machine learning classifier predicted the photo was about fashion, could we get a better estimate of the probability the photo is actually about fashion? Absolutely. To do so, we limit our view to only those 219 cases where the ML classifier predicted that the photo was about fashion and look at the fraction where the photo was actually about fashion:

$$P(\text{truth is fashion given mach\_learn is pred\_fashion}) = \frac{197}{219} = 0.900$$

We call this a **conditional probability** because we computed the probability under a condition: the ML classifier prediction said the photo was about fashion.

There are two parts to a conditional probability, the **outcome of interest** and the **condition**. It is useful to think of the condition as information we know to be true, and this information usually can be described as a known outcome or event. We generally separate the text inside our probability notation into the outcome of interest and the condition with a vertical bar:

$$P(\text{truth is fashion given mach\_learn is pred\_fashion})$$
$$= P(\text{truth is fashion} \mid \text{mach\_learn is pred\_fashion}) = \frac{197}{219} = 0.900$$

The vertical bar "|" is read as *given*.

---

[25]Each of the four outcome combination are disjoint, all probabilities are indeed non-negative, and the sum of the probabilities is $0.1081 + 0.0121 + 0.0615 + 0.8183 = 1.00$.

In the last equation, we computed the probability a photo was about fashion based on the condition that the ML algorithm predicted it was about fashion as a fraction:

$$P(\texttt{truth is fashion} \mid \texttt{mach\_learn is pred\_fashion})$$

$$= \frac{\#\text{ cases where } \texttt{truth is fashion} \text{ and } \texttt{mach\_learn is pred\_fashion}}{\#\text{ cases where } \texttt{mach\_learn is pred\_fashion}}$$

$$= \frac{197}{219} = 0.900$$

We considered only those cases that met the condition, mach_learn is pred_fashion, and then we computed the ratio of those cases that satisfied our outcome of interest, photo was actually about fashion.

Frequently, marginal and joint probabilities are provided instead of count data. For example, disease rates are commonly listed in percentages rather than in a count format. We would like to be able to compute conditional probabilities even when no counts are available, and we use the last equation as a template to understand this technique.

We considered only those cases that satisfied the condition, where the ML algorithm predicted fashion. Of these cases, the conditional probability was the fraction representing the outcome of interest, that the photo was about fashion. Suppose we were provided only the information in Figure 3.13, i.e. only probability data. Then if we took a sample of 1000 photos, we would anticipate about 12.0% or $0.120 \times 1000 = 120$ would be predicted to be about fashion (mach_learn is pred_fashion). Similarly, we would expect about 10.8% or $0.108 \times 1000 = 108$ to meet both the information criteria and represent our outcome of interest. Then the conditional probability can be computed as

$$P(\texttt{truth is fashion} \mid \texttt{mach\_learn is pred\_fashion})$$

$$= \frac{\#\ (\texttt{truth is fashion} \text{ and } \texttt{mach\_learn is pred\_fashion})}{\#\ (\texttt{mach\_learn is pred\_fashion})}$$

$$= \frac{108}{120} = \frac{0.108}{0.120} = 0.90$$

Here we are examining exactly the fraction of two probabilities, 0.108 and 0.120, which we can write as

$$P(\texttt{truth is fashion and mach\_learn is pred\_fashion}) \quad \text{and} \quad P(\texttt{mach\_learn is pred\_fashion}).$$

The fraction of these probabilities is an example of the general formula for conditional probability.

CONDITIONAL PROBABILITY

The conditional probability of outcome A given condition B is computed as the following:

$$P(A|B) = \frac{P(A \text{ and } B)}{P(B)}$$

GUIDED PRACTICE 3.29

(a) Write out the following statement in conditional probability notation: *"The probability that the ML prediction was correct, if the photo was about fashion"*. Here the condition is now based on the photo's truth status, not the ML algorithm.

(b) Determine the probability from part (a). Table 3.13 on page 96 may be helpful.[26]

---

[26](a) If the photo is about fashion and the ML algorithm prediction was correct, then the ML algorithm my have a value of pred_fashion:

$$P(\texttt{mach\_learn is pred\_fashion} \mid \texttt{truth is fashion})$$

(b) The equation for conditional probability indicates we should first find
$P(\texttt{mach\_learn is pred\_fashion and truth is fashion}) = 0.1081$ and $P(\texttt{truth is fashion}) = 0.1696$.
Then the ratio represents the conditional probability: $0.1081/0.1696 = 0.6374$.

GUIDED PRACTICE 3.30

(a) Determine the probability that the algorithm is incorrect if it is known the photo is about fashion.

(b) Using the answers from part (a) and Guided Practice 3.29(b), compute

$$P(\texttt{mach\_learn} \text{ is } \texttt{pred\_fashion} \mid \texttt{truth} \text{ is } \texttt{fashion})$$
$$+ \; P(\texttt{mach\_learn} \text{ is } \texttt{pred\_not} \mid \texttt{truth} \text{ is } \texttt{fashion})$$

(c) Provide an intuitive argument to explain why the sum in (b) is 1.[27]

3.2.4 Smallpox in Boston, 1721

The `smallpox` data set provides a sample of 6,224 individuals from the year 1721 who were exposed to smallpox in Boston. Doctors at the time believed that inoculation, which involves exposing a person to the disease in a controlled form, could reduce the likelihood of death.

Each case represents one person with two variables: `inoculated` and `result`. The variable `inoculated` takes two levels: `yes` or `no`, indicating whether the person was inoculated or not. The variable `result` has outcomes `lived` or `died`. These data are summarized in Tables 3.15 and 3.16.

| | | inoculated | | |
		yes	no	Total
result	lived	238	5136	5374
	died	6	844	850
	Total	244	5980	6224

Figure 3.15: Contingency table for the `smallpox` data set.

| | | inoculated | | |
		yes	no	Total
result	lived	0.0382	0.8252	0.8634
	died	0.0010	0.1356	0.1366
	Total	0.0392	0.9608	1.0000

Figure 3.16: Table proportions for the `smallpox` data, computed by dividing each count by the table total, 6224.

GUIDED PRACTICE 3.31

Write out, in formal notation, the probability a randomly selected person who was not inoculated died from smallpox, and find this probability.[28]

GUIDED PRACTICE 3.32

Determine the probability that an inoculated person died from smallpox. How does this result compare with the result of Guided Practice 3.31?[29]

---

[27](a) This probability is $\frac{P(\texttt{mach\_learn} \text{ is } \texttt{pred\_not}, \; \texttt{truth} \text{ is } \texttt{fashion})}{P(\texttt{truth} \text{ is } \texttt{fashion})} = \frac{0.0615}{0.1696} = 0.3626$. (b) The total equals 1. (c) Under the condition the photo is about fashion, the ML algorithm must have either predicted it was about fashion or predicted it was not about fashion. The complement still works for conditional probabilities, provided the probabilities are conditioned on the same information.

[28]$P(\texttt{result} = \texttt{died} \mid \texttt{inoculated} = \texttt{no}) = \frac{P(\texttt{result} = \texttt{died and inoculated} = \texttt{no})}{P(\texttt{inoculated} = \texttt{no})} = \frac{0.1356}{0.9608} = 0.1411$.

[29]$P(\texttt{result} = \texttt{died} \mid \texttt{inoculated} = \texttt{yes}) = \frac{P(\texttt{result} = \texttt{died and inoculated} = \texttt{yes})}{P(\texttt{inoculated} = \texttt{yes})} = \frac{0.0010}{0.0392} = 0.0255$ (if we avoided rounding errors, we'd get 6/244 = 0.0246). The death rate for individuals who were inoculated is only about 1 in 40 while the death rate is about 1 in 7 for those who were not inoculated.

GUIDED PRACTICE 3.33

The people of Boston self-selected whether or not to be inoculated. (a) Is this study observational or was this an experiment? (b) Can we infer any causal connection using these data? (c) What are some potential confounding variables that might influence whether someone lived or died and also affect whether that person was inoculated?[30]

3.2.5 General multiplication rule

Section 3.1.7 introduced the Multiplication Rule for independent processes. Here we provide the **General Multiplication Rule** for events that might not be independent.

GENERAL MULTIPLICATION RULE

If A and B represent two outcomes or events, then

$$P(A \text{ and } B) = P(A|B) \times P(B)$$

It is useful to think of A as the outcome of interest and B as the condition.

This General Multiplication Rule is simply a rearrangement of the conditional probability equation.

EXAMPLE 3.34

Consider the smallpox data set. Suppose we are given only two pieces of information: 96.08% of residents were not inoculated, and 85.88% of the residents who were not inoculated ended up surviving. How could we compute the probability that a resident was not inoculated and lived?

We will compute our answer using the General Multiplication Rule and then verify it using Figure 3.16. We want to determine

$$P(\text{result} = \text{lived and inoculated} = \text{no})$$

and we are given that

$$P(\text{result} = \text{lived} \mid \text{inoculated} = \text{no}) = 0.8588 \qquad P(\text{inoculated} = \text{no}) = 0.9608$$

Among the 96.08% of people who were not inoculated, 85.88% survived:

$$P(\text{result} = \text{lived and inoculated} = \text{no}) = 0.8588 \times 0.9608 = 0.8251$$

This is equivalent to the General Multiplication Rule. We can confirm this probability in Figure 3.16 at the intersection of no and lived (with a small rounding error).

GUIDED PRACTICE 3.35

Use $P(\text{inoculated} = \text{yes}) = 0.0392$ and $P(\text{result} = \text{lived} \mid \text{inoculated} = \text{yes}) = 0.9754$ to determine the probability that a person was both inoculated and lived.[31]

GUIDED PRACTICE 3.36

If 97.54% of the inoculated people lived, what proportion of inoculated people must have died?[32]

---

[30]Brief answers: (a) Observational. (b) No, we cannot infer causation from this observational study. (c) Accessibility to the latest and best medical care. There are other valid answers for part (c).
[31]The answer is 0.0382, which can be verified using Figure 3.16.
[32]There were only two possible outcomes: lived or died. This means that 100% - 97.54% = 2.46% of the people who were inoculated died.

SUM OF CONDITIONAL PROBABILITIES

Let $A_1, ..., A_k$ represent all the disjoint outcomes for a variable or process. Then if B is an event, possibly for another variable or process, we have:

$$P(A_1|B) + \cdots + P(A_k|B) = 1$$

The rule for complements also holds when an event and its complement are conditioned on the same information:

$$P(A|B) = 1 - P(A^c|B)$$

GUIDED PRACTICE 3.37

Based on the probabilities computed above, does it appear that inoculation is effective at reducing the risk of death from smallpox?[33]

3.2.6   Independence considerations in conditional probability

If two events are independent, then knowing the outcome of one should provide no information about the other. We can show this is mathematically true using conditional probabilities.

GUIDED PRACTICE 3.38

Let X and Y represent the outcomes of rolling two dice.[34]

(a) What is the probability that the first die, X, is 1?

(b) What is the probability that both X and Y are 1?

(c) Use the formula for conditional probability to compute $P(Y = 1 \mid X = 1)$.

(d) What is $P(Y = 1)$? Is this different from the answer from part (c)? Explain.

We can show in Guided Practice 3.38(c) that the conditioning information has no influence by using the Multiplication Rule for independence processes:

$$\begin{aligned} P(Y = 1 \mid X = 1) &= \frac{P(Y = 1 \text{ and } X = 1)}{P(X = 1)} \\ &= \frac{P(Y = 1) \times P(X = 1)}{P(X = 1)} \\ &= P(Y = 1) \end{aligned}$$

GUIDED PRACTICE 3.39

Ron is watching a roulette table in a casino and notices that the last five outcomes were black. He figures that the chances of getting black six times in a row is very small (about 1/64) and puts his paycheck on red. What is wrong with his reasoning?[35]

---

[33] The samples are large relative to the difference in death rates for the "inoculated" and "not inoculated" groups, so it seems there is an association between inoculated and outcome. However, as noted in the solution to Guided Practice 3.33, this is an observational study and we cannot be sure if there is a causal connection. (Further research has shown that inoculation is effective at reducing death rates.)

[34] Brief solutions: (a) 1/6. (b) 1/36. (c) $\frac{P(Y=1 \text{ and } X=1)}{P(X=1)} = \frac{1/36}{1/6} = 1/6$. (d) The probability is the same as in part (c): $P(Y = 1) = 1/6$. The probability that $Y = 1$ was unchanged by knowledge about X, which makes sense as X and Y are independent.

[35] He has forgotten that the next roulette spin is independent of the previous spins. Casinos do employ this practice, posting the last several outcomes of many betting games to trick unsuspecting gamblers into believing the odds are in their favor. This is called the **gambler's fallacy**.

3.2.7 Tree diagrams

Tree diagrams are a tool to organize outcomes and probabilities around the structure of the data. They are most useful when two or more processes occur in a sequence and each process is conditioned on its predecessors.

The `smallpox` data fit this description. We see the population as split by `inoculation`: yes and no. Following this split, survival rates were observed for each group. This structure is reflected in the **tree diagram** shown in Figure 3.17. The first branch for `inoculation` is said to be the **primary** branch while the other branches are **secondary**.

Figure 3.17: A tree diagram of the `smallpox` data set.

Tree diagrams are annotated with marginal and conditional probabilities, as shown in Figure 3.17. This tree diagram splits the smallpox data by `inoculation` into the yes and no groups with respective marginal probabilities 0.0392 and 0.9608. The secondary branches are conditioned on the first, so we assign conditional probabilities to these branches. For example, the top branch in Figure 3.17 is the probability that `result` = `lived` conditioned on the information that `inoculated` = yes. We may (and usually do) construct joint probabilities at the end of each branch in our tree by multiplying the numbers we come across as we move from left to right. These joint probabilities are computed using the General Multiplication Rule:

$$P(\texttt{inoculated} = \texttt{yes and result} = \texttt{lived})$$
$$= P(\texttt{inoculated} = \texttt{yes}) \times P(\texttt{result} = \texttt{lived}|\texttt{inoculated} = \texttt{yes})$$
$$= 0.0392 \times 0.9754 = 0.0382$$

EXAMPLE 3.40

Consider the midterm and final for a statistics class. Suppose 13% of students earned an A on the midterm. Of those students who earned an A on the midterm, 47% received an A on the final, and 11% of the students who earned lower than an A on the midterm received an A on the final. You randomly pick up a final exam and notice the student received an A. What is the probability that this student earned an A on the midterm?

The end-goal is to find $P(\texttt{midterm} = \texttt{A} | \texttt{final} = \texttt{A})$. To calculate this conditional probability, we need the following probabilities:

$$P(\texttt{midterm} = \texttt{A} \text{ and } \texttt{final} = \texttt{A}) \quad \text{and} \quad P(\texttt{final} = \texttt{A})$$

However, this information is not provided, and it is not obvious how to calculate these probabilities. Since we aren't sure how to proceed, it is useful to organize the information into a tree diagram:

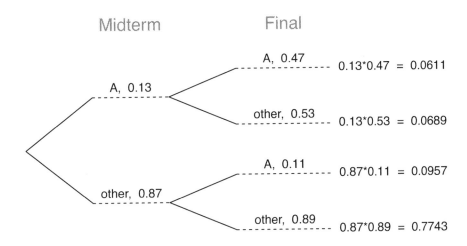

When constructing a tree diagram, variables provided with marginal probabilities are often used to create the tree's primary branches; in this case, the marginal probabilities are provided for midterm grades. The final grades, which correspond to the conditional probabilities provided, will be shown on the secondary branches.

With the tree diagram constructed, we may compute the required probabilities:

$$P(\texttt{midterm} = \texttt{A} \text{ and } \texttt{final} = \texttt{A}) = 0.0611$$

$$P(\underline{\texttt{final} = \texttt{A}})$$
$$= P(\texttt{midterm} = \texttt{other} \text{ and } \underline{\texttt{final} = \texttt{A}}) + P(\texttt{midterm} = \texttt{A} \text{ and } \underline{\texttt{final} = \texttt{A}})$$
$$= 0.0957 + 0.0611 = 0.1568$$

The marginal probability, $P(\texttt{final} = \texttt{A})$, was calculated by adding up all the joint probabilities on the right side of the tree that correspond to $\texttt{final} = \texttt{A}$. We may now finally take the ratio of the two probabilities:

$$P(\texttt{midterm} = \texttt{A} | \texttt{final} = \texttt{A}) = \frac{P(\texttt{midterm} = \texttt{A} \text{ and } \texttt{final} = \texttt{A})}{P(\texttt{final} = \texttt{A})}$$
$$= \frac{0.0611}{0.1568} = 0.3897$$

The probability the student also earned an A on the midterm is about 0.39.

GUIDED PRACTICE 3.41

After an introductory statistics course, 78% of students can successfully construct tree diagrams. Of those who can construct tree diagrams, 97% passed, while only 57% of those students who could not construct tree diagrams passed. (a) Organize this information into a tree diagram. (b) What is the probability that a randomly selected student passed? (c) Compute the probability a student is able to construct a tree diagram if it is known that she passed.[36]

3.2.8 Bayes' Theorem

In many instances, we are given a conditional probability of the form

$$P(\text{statement about variable 1} \mid \text{statement about variable 2})$$

but we would really like to know the inverted conditional probability:

$$P(\text{statement about variable 2} \mid \text{statement about variable 1})$$

Tree diagrams can be used to find the second conditional probability when given the first. However, sometimes it is not possible to draw the scenario in a tree diagram. In these cases, we can apply a very useful and general formula: Bayes' Theorem.

We first take a critical look at an example of inverting conditional probabilities where we still apply a tree diagram.

---

[36](a) The tree diagram is shown to the right.
(b) Identify which two joint probabilities represent students who passed, and add them: $P(\text{passed}) = 0.7566 + 0.1254 = 0.8820$.
(c) $P(\text{construct tree diagram} \mid \text{passed}) = \frac{0.7566}{0.8820} = 0.8578$.

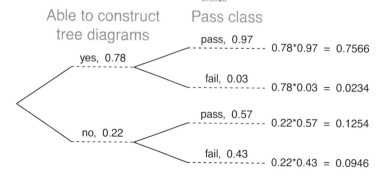

EXAMPLE 3.42

In Canada, about 0.35% of women over 40 will develop breast cancer in any given year. A common screening test for cancer is the mammogram, but this test is not perfect. In about 11% of patients with breast cancer, the test gives a **false negative**: it indicates a woman does not have breast cancer when she does have breast cancer. Similarly, the test gives a **false positive** in 7% of patients who do not have breast cancer: it indicates these patients have breast cancer when they actually do not. If we tested a random woman over 40 for breast cancer using a mammogram and the test came back positive – that is, the test suggested the patient has cancer – what is the probability that the patient actually has breast cancer?

Notice that we are given sufficient information to quickly compute the probability of testing positive if a woman has breast cancer $(1.00 - 0.11 = 0.89)$. However, we seek the inverted probability of cancer given a positive test result. (Watch out for the non-intuitive medical language: a *positive* test result suggests the possible presence of cancer in a mammogram screening.) This inverted probability may be broken into two pieces:

$$P(\text{has BC} \mid \text{mammogram}^+) = \frac{P(\text{has BC and mammogram}^+)}{P(\text{mammogram}^+)}$$

where "has BC" is an abbreviation for the patient having breast cancer and "mammogram$^+$" means the mammogram screening was positive. We can construct a tree diagram for these probabilities:

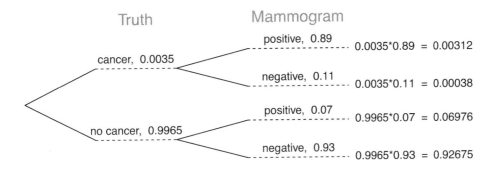

The probability the patient has breast cancer and the mammogram is positive is

$$P(\text{has BC and mammogram}^+) = P(\text{mammogram}^+ \mid \text{has BC})P(\text{has BC})$$
$$= 0.89 \times 0.0035 = 0.00312$$

The probability of a positive test result is the sum of the two corresponding scenarios:

$$P(\text{mammogram}^+) = P(\underline{\text{mammogram}^+ \text{ and has BC}})$$
$$+ P(\underline{\text{mammogram}^+ \text{ and no BC}})$$
$$= P(\text{has BC})P(\text{mammogram}^+ \mid \text{has BC})$$
$$+ P(\text{no BC})P(\text{mammogram}^+ \mid \text{no BC})$$
$$= 0.0035 \times 0.89 + 0.9965 \times 0.07 = 0.07288$$

Then if the mammogram screening is positive for a patient, the probability the patient has breast cancer is

$$P(\text{has BC} \mid \text{mammogram}^+) = \frac{P(\text{has BC and mammogram}^+)}{P(\text{mammogram}^+)}$$
$$= \frac{0.00312}{0.07288} \approx 0.0428$$

That is, even if a patient has a positive mammogram screening, there is still only a 4% chance that she has breast cancer.

Example 3.42 highlights why doctors often run more tests regardless of a first positive test result. When a medical condition is rare, a single positive test isn't generally definitive.

Consider again the last equation of Example 3.42. Using the tree diagram, we can see that the numerator (the top of the fraction) is equal to the following product:

$$P(\text{has BC and mammogram}^+) = P(\text{mammogram}^+ \mid \text{has BC})P(\text{has BC})$$

The denominator – the probability the screening was positive – is equal to the sum of probabilities for each positive screening scenario:

$$P(\underline{\text{mammogram}^+}) = P(\underline{\text{mammogram}^+} \text{ and no BC}) + P(\underline{\text{mammogram}^+} \text{ and has BC})$$

In the example, each of the probabilities on the right side was broken down into a product of a conditional probability and marginal probability using the tree diagram.

$$P(\text{mammogram}^+) = P(\text{mammogram}^+ \text{ and no BC}) + P(\text{mammogram}^+ \text{ and has BC})$$
$$= P(\text{mammogram}^+ \mid \text{no BC})P(\text{no BC})$$
$$+ P(\text{mammogram}^+ \mid \text{has BC})P(\text{has BC})$$

We can see an application of Bayes' Theorem by substituting the resulting probability expressions into the numerator and denominator of the original conditional probability.

$$P(\text{has BC} \mid \text{mammogram}^+)$$

$$= \frac{P(\text{mammogram}^+ \mid \text{has BC})P(\text{has BC})}{P(\text{mammogram}^+ \mid \text{no BC})P(\text{no BC}) + P(\text{mammogram}^+ \mid \text{has BC})P(\text{has BC})}$$

BAYES' THEOREM: INVERTING PROBABILITIES

Consider the following conditional probability for variable 1 and variable 2:

$$P(\text{outcome } A_1 \text{ of variable 1} \mid \text{outcome } B \text{ of variable 2})$$

Bayes' Theorem states that this conditional probability can be identified as the following fraction:

$$\frac{P(B|A_1)P(A_1)}{P(B|A_1)P(A_1) + P(B|A_2)P(A_2) + \cdots + P(B|A_k)P(A_k)}$$

where A_2, A_3, ..., and A_k represent all other possible outcomes of the first variable.

Bayes' Theorem is a generalization of what we have done using tree diagrams. The numerator identifies the probability of getting both A_1 and B. The denominator is the marginal probability of getting B. This bottom component of the fraction appears long and complicated since we have to add up probabilities from all of the different ways to get B. We always completed this step when using tree diagrams. However, we usually did it in a separate step so it didn't seem as complex. To apply Bayes' Theorem correctly, there are two preparatory steps:

(1) First identify the marginal probabilities of each possible outcome of the first variable: $P(A_1)$, $P(A_2)$, ..., $P(A_k)$.

(2) Then identify the probability of the outcome B, conditioned on each possible scenario for the first variable: $P(B|A_1)$, $P(B|A_2)$, ..., $P(B|A_k)$.

Once each of these probabilities are identified, they can be applied directly within the formula. Bayes' Theorem tends to be a good option when there are so many scenarios that drawing a tree diagram would be complex.

GUIDED PRACTICE 3.43

Jose visits campus every Thursday evening. However, some days the parking garage is full, often due to college events. There are academic events on 35% of evenings, sporting events on 20% of evenings, and no events on 45% of evenings. When there is an academic event, the garage fills up about 25% of the time, and it fills up 70% of evenings with sporting events. On evenings when there are no events, it only fills up about 5% of the time. If Jose comes to campus and finds the garage full, what is the probability that there is a sporting event? Use a tree diagram to solve this problem.[37]

EXAMPLE 3.44

Here we solve the same problem presented in Guided Practice 3.43, except this time we use Bayes' Theorem.

The outcome of interest is whether there is a sporting event (call this A_1), and the condition is that the lot is full (B). Let A_2 represent an academic event and A_3 represent there being no event on campus. Then the given probabilities can be written as

$$P(A_1) = 0.2 \qquad P(A_2) = 0.35 \qquad P(A_3) = 0.45$$
$$P(B|A_1) = 0.7 \qquad P(B|A_2) = 0.25 \qquad P(B|A_3) = 0.05$$

Bayes' Theorem can be used to compute the probability of a sporting event (A_1) under the condition that the parking lot is full (B):

$$
\begin{aligned}
P(A_1|B) &= \frac{P(B|A_1)P(A_1)}{P(B|A_1)P(A_1) + P(B|A_2)P(A_2) + P(B|A_3)P(A_3)} \\
&= \frac{(0.7)(0.2)}{(0.7)(0.2) + (0.25)(0.35) + (0.05)(0.45)} \\
&= 0.56
\end{aligned}
$$

Based on the information that the garage is full, there is a 56% probability that a sporting event is being held on campus that evening.

---

[37]The tree diagram, with three primary branches, is shown to the right. Next, we identify two probabilities from the tree diagram. (1) The probability that there is a sporting event and the garage is full: 0.14. (2) The probability the garage is full: $0.0875 + 0.14 + 0.0225 = 0.25$. Then the solution is the ratio of these probabilities: $\frac{0.14}{0.25} = 0.56$. If the garage is full, there is a 56% probability that there is a sporting event.

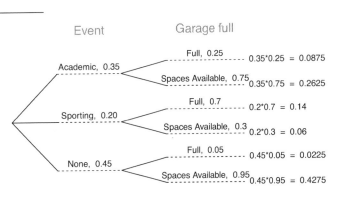

GUIDED PRACTICE 3.45

Use the information in the previous exercise and example to verify the probability that there is an academic event conditioned on the parking lot being full is 0.35.[38]

GUIDED PRACTICE 3.46

In Guided Practice 3.43 and 3.45, you found that if the parking lot is full, the probability there is a sporting event is 0.56 and the probability there is an academic event is 0.35. Using this information, compute $P(\text{no event} \mid \text{the lot is full})$.[39]

The last several exercises offered a way to update our belief about whether there is a sporting event, academic event, or no event going on at the school based on the information that the parking lot was full. This strategy of *updating beliefs* using Bayes' Theorem is actually the foundation of an entire section of statistics called **Bayesian statistics**. While Bayesian statistics is very important and useful, we will not have time to cover much more of it in this book.

---

[38]Short answer:

$$P(A_2|B) = \frac{P(B|A_2)P(A_2)}{P(B|A_1)P(A_1) + P(B|A_2)P(A_2) + P(B|A_3)P(A_3)}$$

$$= \frac{(0.25)(0.35)}{(0.7)(0.2) + (0.25)(0.35) + (0.05)(0.45)}$$

$$= 0.35$$

[39]Each probability is conditioned on the same information that the garage is full, so the complement may be used: $1.00 - 0.56 - 0.35 = 0.09$.

Exercises

3.13 Joint and conditional probabilities. $P(A) = 0.3$, $P(B) = 0.7$

(a) Can you compute P(A and B) if you only know P(A) and P(B)?

(b) Assuming that events A and B arise from independent random processes,

     i. what is P(A and B)?

     ii. what is P(A or B)?

     iii. what is P(A|B)?

(c) If we are given that P(A and B) = 0.1, are the random variables giving rise to events A and B independent?

(d) If we are given that P(A and B) = 0.1, what is P(A|B)?

3.14 PB & J. Suppose 80% of people like peanut butter, 89% like jelly, and 78% like both. Given that a randomly sampled person likes peanut butter, what's the probability that he also likes jelly?

3.15 Global warming. A Pew Research poll asked 1,306 Americans "From what you've read and heard, is there solid evidence that the average temperature on earth has been getting warmer over the past few decades, or not?". The table below shows the distribution of responses by party and ideology, where the counts have been replaced with relative frequencies.[40]

		Earth is warming	Not warming	Don't Know Refuse	Total
Party and Ideology	Conservative Republican	0.11	0.20	0.02	0.33
	Mod/Lib Republican	0.06	0.06	0.01	0.13
	Mod/Cons Democrat	0.25	0.07	0.02	0.34
	Liberal Democrat	0.18	0.01	0.01	0.20
	Total	0.60	0.34	0.06	1.00

Response

(a) Are believing that the earth is warming and being a liberal Democrat mutually exclusive?

(b) What is the probability that a randomly chosen respondent believes the earth is warming or is a liberal Democrat?

(c) What is the probability that a randomly chosen respondent believes the earth is warming given that he is a liberal Democrat?

(d) What is the probability that a randomly chosen respondent believes the earth is warming given that he is a conservative Republican?

(e) Does it appear that whether or not a respondent believes the earth is warming is independent of their party and ideology? Explain your reasoning.

(f) What is the probability that a randomly chosen respondent is a moderate/liberal Republican given that he does not believe that the earth is warming?

---

[40]Pew Research Center, Majority of Republicans No Longer See Evidence of Global Warming, data collected on October 27, 2010.

3.16 Health coverage, relative frequencies. The Behavioral Risk Factor Surveillance System (BRFSS) is an annual telephone survey designed to identify risk factors in the adult population and report emerging health trends. The following table displays the distribution of health status of respondents to this survey (excellent, very good, good, fair, poor) and whether or not they have health insurance.

		Excellent	Very good	Good	Fair	Poor	Total
Health	No	0.0230	0.0364	0.0427	0.0192	0.0050	0.1262
Coverage	Yes	0.2099	0.3123	0.2410	0.0817	0.0289	0.8738
	Total	0.2329	0.3486	0.2838	0.1009	0.0338	1.0000

Health Status spans the Excellent, Very good, Good, Fair, Poor columns.

(a) Are being in excellent health and having health coverage mutually exclusive?

(b) What is the probability that a randomly chosen individual has excellent health?

(c) What is the probability that a randomly chosen individual has excellent health given that he has health coverage?

(d) What is the probability that a randomly chosen individual has excellent health given that he doesn't have health coverage?

(e) Do having excellent health and having health coverage appear to be independent?

3.17 Burger preferences. A 2010 SurveyUSA poll asked 500 Los Angeles residents, "What is the best hamburger place in Southern California? Five Guys Burgers? In-N-Out Burger? Fat Burger? Tommy's Hamburgers? Umami Burger? Or somewhere else?" The distribution of responses by gender is shown below.[41]

		Male	Female	Total
	Five Guys Burgers	5	6	11
	In-N-Out Burger	162	181	343
Best	Fat Burger	10	12	22
hamburger	Tommy's Hamburgers	27	27	54
place	Umami Burger	5	1	6
	Other	26	20	46
	Not Sure	13	5	18
	Total	248	252	500

Gender spans the Male, Female columns.

(a) Are being female and liking Five Guys Burgers mutually exclusive?

(b) What is the probability that a randomly chosen male likes In-N-Out the best?

(c) What is the probability that a randomly chosen female likes In-N-Out the best?

(d) What is the probability that a man and a woman who are dating both like In-N-Out the best? Note any assumption you make and evaluate whether you think that assumption is reasonable.

(e) What is the probability that a randomly chosen person likes Umami best or that person is female?

---

[41]SurveyUSA, Results of SurveyUSA News Poll #17718, data collected on December 2, 2010.

3.18 Assortative mating. Assortative mating is a nonrandom mating pattern where individuals with similar genotypes and/or phenotypes mate with one another more frequently than what would be expected under a random mating pattern. Researchers studying this topic collected data on eye colors of 204 Scandinavian men and their female partners. The table below summarizes the results. For simplicity, we only include heterosexual relationships in this exercise.[42]

		Partner (female)			Total
		Blue	Brown	Green	
Self (male)	Blue	78	23	13	114
	Brown	19	23	12	54
	Green	11	9	16	36
	Total	108	55	41	204

(a) What is the probability that a randomly chosen male respondent or his partner has blue eyes?

(b) What is the probability that a randomly chosen male respondent with blue eyes has a partner with blue eyes?

(c) What is the probability that a randomly chosen male respondent with brown eyes has a partner with blue eyes? What about the probability of a randomly chosen male respondent with green eyes having a partner with blue eyes?

(d) Does it appear that the eye colors of male respondents and their partners are independent? Explain your reasoning.

3.19 Drawing box plots. After an introductory statistics course, 80% of students can successfully construct box plots. Of those who can construct box plots, 86% passed, while only 65% of those students who could not construct box plots passed.

(a) Construct a tree diagram of this scenario.

(b) Calculate the probability that a student is able to construct a box plot if it is known that he passed.

3.20 Predisposition for thrombosis. A genetic test is used to determine if people have a predisposition for *thrombosis*, which is the formation of a blood clot inside a blood vessel that obstructs the flow of blood through the circulatory system. It is believed that 3% of people actually have this predisposition. The genetic test is 99% accurate if a person actually has the predisposition, meaning that the probability of a positive test result when a person actually has the predisposition is 0.99. The test is 98% accurate if a person does not have the predisposition. What is the probability that a randomly selected person who tests positive for the predisposition by the test actually has the predisposition?

3.21 It's never lupus. Lupus is a medical phenomenon where antibodies that are supposed to attack foreign cells to prevent infections instead see plasma proteins as foreign bodies, leading to a high risk of blood clotting. It is believed that 2% of the population suffer from this disease. The test is 98% accurate if a person actually has the disease. The test is 74% accurate if a person does not have the disease. There is a line from the Fox television show *House* that is often used after a patient tests positive for lupus: "It's never lupus." Do you think there is truth to this statement? Use appropriate probabilities to support your answer.

3.22 Exit poll. Edison Research gathered exit poll results from several sources for the Wisconsin recall election of Scott Walker. They found that 53% of the respondents voted in favor of Scott Walker. Additionally, they estimated that of those who did vote in favor for Scott Walker, 37% had a college degree, while 44% of those who voted against Scott Walker had a college degree. Suppose we randomly sampled a person who participated in the exit poll and found that he had a college degree. What is the probability that he voted in favor of Scott Walker?[43]

---

[42]B. Laeng et al. "Why do blue-eyed men prefer women with the same eye color?" In: *Behavioral Ecology and Sociobiology* 61.3 (2007), pp. 371–384.
[43]New York Times, Wisconsin recall exit polls.

3.3   Sampling from a small population

When we sample observations from a population, usually we're only sampling a small fraction of the possible individuals or cases. However, sometimes our sample size is large enough or the population is small enough that we sample more than 10% of a population[44] *without replacement* (meaning we do not have a chance of sampling the same cases twice). Sampling such a notable fraction of a population can be important for how we analyze the sample.

EXAMPLE 3.47

Professors sometimes select a student at random to answer a question. If each student has an equal chance of being selected and there are 15 people in your class, what is the chance that she will pick you for the next question?

If there are 15 people to ask and none are skipping class, then the probability is 1/15, or about 0.067.

EXAMPLE 3.48

If the professor asks 3 questions, what is the probability that you will not be selected? Assume that she will not pick the same person twice in a given lecture.

For the first question, she will pick someone else with probability 14/15. When she asks the second question, she only has 14 people who have not yet been asked. Thus, if you were not picked on the first question, the probability you are again not picked is 13/14. Similarly, the probability you are again not picked on the third question is 12/13, and the probability of not being picked for any of the three questions is

$$P(\text{not picked in 3 questions})$$
$$= P(\texttt{Q1} = \texttt{not\_picked}, \texttt{Q2} = \texttt{not\_picked}, \texttt{Q3} = \texttt{not\_picked}.)$$
$$= \frac{14}{15} \times \frac{13}{14} \times \frac{12}{13} = \frac{12}{15} = 0.80$$

GUIDED PRACTICE 3.49

What rule permitted us to multiply the probabilities in Example 3.48?[45]

---

[44]The 10% guideline is a rule of thumb cutoff for when these considerations become more important.

[45]The three probabilities we computed were actually one marginal probability, $P(\texttt{Q1}=\texttt{not\_picked})$, and two conditional probabilities:

$$P(\texttt{Q2} = \texttt{not\_picked} \mid \texttt{Q1} = \texttt{not\_picked})$$
$$P(\texttt{Q3} = \texttt{not\_picked} \mid \texttt{Q1} = \texttt{not\_picked}, \texttt{Q2} = \texttt{not\_picked})$$

Using the General Multiplication Rule, the product of these three probabilities is the probability of not being picked in 3 questions.

EXAMPLE 3.50

Suppose the professor randomly picks without regard to who she already selected, i.e. students can be picked more than once. What is the probability that you will not be picked for any of the three questions?

Each pick is independent, and the probability of not being picked for any individual question is 14/15. Thus, we can use the Multiplication Rule for independent processes.

$$P(\text{not picked in 3 questions})$$
$$= P(\texttt{Q1} = \texttt{not\_picked}, \texttt{Q2} = \texttt{not\_picked}, \texttt{Q3} = \texttt{not\_picked}.)$$
$$= \frac{14}{15} \times \frac{14}{15} \times \frac{14}{15} = 0.813$$

You have a slightly higher chance of not being picked compared to when she picked a new person for each question. However, you now may be picked more than once.

GUIDED PRACTICE 3.51

Under the setup of Example 3.50, what is the probability of being picked to answer all three questions?[46]

If we sample from a small population **without replacement**, we no longer have independence between our observations. In Example 3.48, the probability of not being picked for the second question was conditioned on the event that you were not picked for the first question. In Example 3.50, the professor sampled her students **with replacement**: she repeatedly sampled the entire class without regard to who she already picked.

GUIDED PRACTICE 3.52

Your department is holding a raffle. They sell 30 tickets and offer seven prizes. (a) They place the tickets in a hat and draw one for each prize. The tickets are sampled without replacement, i.e. the selected tickets are not placed back in the hat. What is the probability of winning a prize if you buy one ticket? (b) What if the tickets are sampled with replacement?[47]

GUIDED PRACTICE 3.53

Compare your answers in Guided Practice 3.52. How much influence does the sampling method have on your chances of winning a prize?[48]

Had we repeated Guided Practice 3.52 with 300 tickets instead of 30, we would have found something interesting: the results would be nearly identical. The probability would be 0.0233 without replacement and 0.0231 with replacement. When the sample size is only a small fraction of the population (under 10%), observations are nearly independent even when sampling without replacement.

---

[46]$P(\text{being picked to answer all three questions}) = \left(\frac{1}{15}\right)^3 = 0.00030.$

[47](a) First determine the probability of not winning. The tickets are sampled without replacement, which means the probability you do not win on the first draw is 29/30, 28/29 for the second, ..., and 23/24 for the seventh. The probability you win no prize is the product of these separate probabilities: 23/30. That is, the probability of winning a prize is $1 - 23/30 = 7/30 = 0.233$. (b) When the tickets are sampled with replacement, there are seven independent draws. Again we first find the probability of not winning a prize: $(29/30)^7 = 0.789$. Thus, the probability of winning (at least) one prize when drawing with replacement is 0.211.

[48]There is about a 10% larger chance of winning a prize when using sampling without replacement. However, at most one prize may be won under this sampling procedure.

Exercises

3.23 Marbles in an urn. Imagine you have an urn containing 5 red, 3 blue, and 2 orange marbles in it.

(a) What is the probability that the first marble you draw is blue?

(b) Suppose you drew a blue marble in the first draw. If drawing with replacement, what is the probability of drawing a blue marble in the second draw?

(c) Suppose you instead drew an orange marble in the first draw. If drawing with replacement, what is the probability of drawing a blue marble in the second draw?

(d) If drawing with replacement, what is the probability of drawing two blue marbles in a row?

(e) When drawing with replacement, are the draws independent? Explain.

3.24 Socks in a drawer. In your sock drawer you have 4 blue, 5 gray, and 3 black socks. Half asleep one morning you grab 2 socks at random and put them on. Find the probability you end up wearing

(a) 2 blue socks

(b) no gray socks

(c) at least 1 black sock

(d) a green sock

(e) matching socks

3.25 Chips in a bag. Imagine you have a bag containing 5 red, 3 blue, and 2 orange chips.

(a) Suppose you draw a chip and it is blue. If drawing without replacement, what is the probability the next is also blue?

(b) Suppose you draw a chip and it is orange, and then you draw a second chip without replacement. What is the probability this second chip is blue?

(c) If drawing without replacement, what is the probability of drawing two blue chips in a row?

(d) When drawing without replacement, are the draws independent? Explain.

3.26 Books on a bookshelf. The table below shows the distribution of books on a bookcase based on whether they are nonfiction or fiction and hardcover or paperback.

		Format		
		Hardcover	Paperback	Total
Type	Fiction	13	59	72
	Nonfiction	15	8	23
	Total	28	67	95

(a) Find the probability of drawing a hardcover book first then a paperback fiction book second when drawing without replacement.

(b) Determine the probability of drawing a fiction book first and then a hardcover book second, when drawing without replacement.

(c) Calculate the probability of the scenario in part (b), except this time complete the calculations under the scenario where the first book is placed back on the bookcase before randomly drawing the second book.

(d) The final answers to parts (b) and (c) are very similar. Explain why this is the case.

3.27 Student outfits. In a classroom with 24 students, 7 students are wearing jeans, 4 are wearing shorts, 8 are wearing skirts, and the rest are wearing leggings. If we randomly select 3 students without replacement, what is the probability that one of the selected students is wearing leggings and the other two are wearing jeans? Note that these are mutually exclusive clothing options.

3.28 The birthday problem. Suppose we pick three people at random. For each of the following questions, ignore the special case where someone might be born on February 29th, and assume that births are evenly distributed throughout the year.

(a) What is the probability that the first two people share a birthday?

(b) What is the probability that at least two people share a birthday?

3.4 Random variables

It's often useful to model a process using what's called a **random variable**. Such a model allows us to apply a mathematical framework and statistical principles for better understanding and predicting outcomes in the real world.

EXAMPLE 3.54

Two books are assigned for a statistics class: a textbook and its corresponding study guide. The university bookstore determined 20% of enrolled students do not buy either book, 55% buy the textbook only, and 25% buy both books, and these percentages are relatively constant from one term to another. If there are 100 students enrolled, how many books should the bookstore expect to sell to this class?

Around 20 students will not buy either book (0 books total), about 55 will buy one book (55 books total), and approximately 25 will buy two books (totaling 50 books for these 25 students). The bookstore should expect to sell about 105 books for this class.

GUIDED PRACTICE 3.55

Would you be surprised if the bookstore sold slightly more or less than 105 books?[49]

EXAMPLE 3.56

The textbook costs $137 and the study guide $33. How much revenue should the bookstore expect from this class of 100 students?

About 55 students will just buy a textbook, providing revenue of

$$\$137 \times 55 = \$7,535$$

The roughly 25 students who buy both the textbook and the study guide would pay a total of

$$(\$137 + \$33) \times 25 = \$170 \times 25 = \$4,250$$

Thus, the bookstore should expect to generate about $7,535 + $4,250 = $11,785 from these 100 students for this one class. However, there might be some *sampling variability* so the actual amount may differ by a little bit.

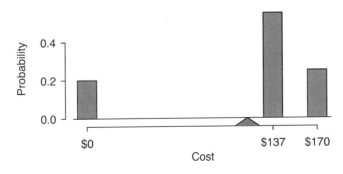

Figure 3.18: Probability distribution for the bookstore's revenue from one student. The triangle represents the average revenue per student.

---

[49]If they sell a little more or a little less, this should not be a surprise. Hopefully Chapter 1 helped make clear that there is natural variability in observed data. For example, if we would flip a coin 100 times, it will not usually come up heads exactly half the time, but it will probably be close.

EXAMPLE 3.57

What is the average revenue per student for this course?

The expected total revenue is \$11,785, and there are 100 students. Therefore the expected revenue per student is \$11,785/100 = \$117.85.

3.4.1 Expectation

We call a variable or process with a numerical outcome a **random variable**, and we usually represent this random variable with a capital letter such as X, Y, or Z. The amount of money a single student will spend on her statistics books is a random variable, and we represent it by X.

RANDOM VARIABLE

A random process or variable with a numerical outcome.

The possible outcomes of X are labeled with a corresponding lower case letter x and subscripts. For example, we write $x_1 = \$0$, $x_2 = \$137$, and $x_3 = \$170$, which occur with probabilities 0.20, 0.55, and 0.25. The distribution of X is summarized in Figure 3.18 and Figure 3.19.

i	1	2	3	Total
x_i	\$0	\$137	\$170	–
$P(X = x_i)$	0.20	0.55	0.25	1.00

Figure 3.19: The probability distribution for the random variable X, representing the bookstore's revenue from a single student.

We computed the average outcome of X as \$117.85 in Example 3.57. We call this average the **expected value** of X, denoted by $E(X)$. The expected value of a random variable is computed by adding each outcome weighted by its probability:

$$E(X) = 0 \times P(X = 0) + 137 \times P(X = 137) + 170 \times P(X = 170)$$
$$= 0 \times 0.20 + 137 \times 0.55 + 170 \times 0.25 = 117.85$$

EXPECTED VALUE OF A DISCRETE RANDOM VARIABLE

If X takes outcomes x_1, ..., x_k with probabilities $P(X = x_1)$, ..., $P(X = x_k)$, the expected value of X is the sum of each outcome multiplied by its corresponding probability:

$$E(X) = x_1 \times P(X = x_1) + \cdots + x_k \times P(X = x_k)$$
$$= \sum_{i=1}^{k} x_i P(X = x_i)$$

The Greek letter μ may be used in place of the notation $E(X)$.

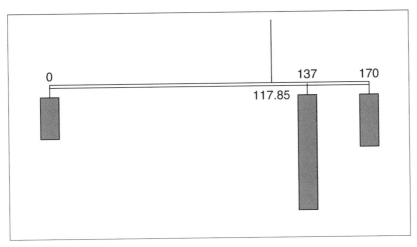

Figure 3.20: A weight system representing the probability distribution for X. The string holds the distribution at the mean to keep the system balanced.

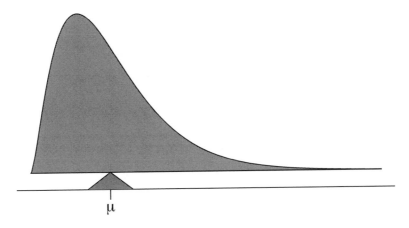

Figure 3.21: A continuous distribution can also be balanced at its mean.

The expected value for a random variable represents the average outcome. For example, $E(X) = 117.85$ represents the average amount the bookstore expects to make from a single student, which we could also write as $\mu = 117.85$.

It is also possible to compute the expected value of a continuous random variable (see Section 3.5). However, it requires a little calculus and we save it for a later class.[50]

In physics, the expectation holds the same meaning as the center of gravity. The distribution can be represented by a series of weights at each outcome, and the mean represents the balancing point. This is represented in Figures 3.18 and 3.20. The idea of a center of gravity also expands to continuous probability distributions. Figure 3.21 shows a continuous probability distribution balanced atop a wedge placed at the mean.

---

[50]$\mu = \int x f(x) dx$ where $f(x)$ represents a function for the density curve.

3.4.2   Variability in random variables

Suppose you ran the university bookstore. Besides how much revenue you expect to generate, you might also want to know the volatility (variability) in your revenue.

The variance and standard deviation can be used to describe the variability of a random variable. Section 2.1.4 introduced a method for finding the variance and standard deviation for a data set. We first computed deviations from the mean $(x_i - \mu)$, squared those deviations, and took an average to get the variance. In the case of a random variable, we again compute squared deviations. However, we take their sum weighted by their corresponding probabilities, just like we did for the expectation. This weighted sum of squared deviations equals the variance, and we calculate the standard deviation by taking the square root of the variance, just as we did in Section 2.1.4.

GENERAL VARIANCE FORMULA

If X takes outcomes $x_1, ..., x_k$ with probabilities $P(X = x_1), ..., P(X = x_k)$ and expected value $\mu = E(X)$, then the variance of X, denoted by $Var(X)$ or the symbol σ^2, is

$$\sigma^2 = (x_1 - \mu)^2 \times P(X = x_1) + \cdots$$
$$\cdots + (x_k - \mu)^2 \times P(X = x_k)$$
$$= \sum_{j=1}^{k} (x_j - \mu)^2 P(X = x_j)$$

The standard deviation of X, labeled σ, is the square root of the variance.

EXAMPLE 3.58

Compute the expected value, variance, and standard deviation of X, the revenue of a single statistics student for the bookstore.

It is useful to construct a table that holds computations for each outcome separately, then add up the results.

i	1	2	3	Total
x_i	$0	$137	$170	
$P(X = x_i)$	0.20	0.55	0.25	
$x_i \times P(X = x_i)$	0	75.35	42.50	117.85

Thus, the expected value is $\mu = 117.85$, which we computed earlier. The variance can be constructed by extending this table:

i	1	2	3	Total
x_i	$0	$137	$170	
$P(X = x_i)$	0.20	0.55	0.25	
$x_i \times P(X = x_i)$	0	75.35	42.50	117.85
$x_i - \mu$	-117.85	19.15	52.15	
$(x_i - \mu)^2$	13888.62	366.72	2719.62	
$(x_i - \mu)^2 \times P(X = x_i)$	2777.7	201.7	679.9	3659.3

The variance of X is $\sigma^2 = 3659.3$, which means the standard deviation is $\sigma = \sqrt{3659.3} = \$60.49$.

GUIDED PRACTICE 3.59

The bookstore also offers a chemistry textbook for $159 and a book supplement for $41. From past experience, they know about 25% of chemistry students just buy the textbook while 60% buy both the textbook and supplement.[51]

(a) What proportion of students don't buy either book? Assume no students buy the supplement without the textbook.

(b) Let Y represent the revenue from a single student. Write out the probability distribution of Y, i.e. a table for each outcome and its associated probability.

(c) Compute the expected revenue from a single chemistry student.

(d) Find the standard deviation to describe the variability associated with the revenue from a single student.

3.4.3 Linear combinations of random variables

So far, we have thought of each variable as being a complete story in and of itself. Sometimes it is more appropriate to use a combination of variables. For instance, the amount of time a person spends commuting to work each week can be broken down into several daily commutes. Similarly, the total gain or loss in a stock portfolio is the sum of the gains and losses in its components.

EXAMPLE 3.60

John travels to work five days a week. We will use X_1 to represent his travel time on Monday, X_2 to represent his travel time on Tuesday, and so on. Write an equation using X_1, ..., X_5 that represents his travel time for the week, denoted by W.

His total weekly travel time is the sum of the five daily values:

$$W = X_1 + X_2 + X_3 + X_4 + X_5$$

Breaking the weekly travel time W into pieces provides a framework for understanding each source of randomness and is useful for modeling W.

---

[51](a) 100% - 25% - 60% = 15% of students do not buy any books for the class. Part (b) is represented by the first two lines in the table below. The expectation for part (c) is given as the total on the line $y_i \times P(Y = y_i)$. The result of part (d) is the square-root of the variance listed on in the total on the last line: $\sigma = \sqrt{Var(Y)} = \$69.28$.

i (scenario)	1 (noBook)	2 (textbook)	3 (both)	Total
y_i	0.00	159.00	200.00	
$P(Y = y_i)$	0.15	0.25	0.60	
$y_i \times P(Y = y_i)$	0.00	39.75	120.00	$E(Y) = 159.75$
$y_i - E(Y)$	-159.75	-0.75	40.25	
$(y_i - E(Y))^2$	25520.06	0.56	1620.06	
$(y_i - E(Y))^2 \times P(Y)$	3828.0	0.1	972.0	$Var(Y) \approx 4800$

EXAMPLE 3.61

It takes John an average of 18 minutes each day to commute to work. What would you expect his average commute time to be for the week?

We were told that the average (i.e. expected value) of the commute time is 18 minutes per day: $E(X_i) = 18$. To get the expected time for the sum of the five days, we can add up the expected time for each individual day:

$$\begin{aligned} E(W) &= E(X_1 + X_2 + X_3 + X_4 + X_5) \\ &= E(X_1) + E(X_2) + E(X_3) + E(X_4) + E(X_5) \\ &= 18 + 18 + 18 + 18 + 18 = 90 \text{ minutes} \end{aligned}$$

The expectation of the total time is equal to the sum of the expected individual times. More generally, the expectation of a sum of random variables is always the sum of the expectation for each random variable.

GUIDED PRACTICE 3.62

Elena is selling a TV at a cash auction and also intends to buy a toaster oven in the auction. If X represents the profit for selling the TV and Y represents the cost of the toaster oven, write an equation that represents the net change in Elena's cash.[52]

GUIDED PRACTICE 3.63

Based on past auctions, Elena figures she should expect to make about \$175 on the TV and pay about \$23 for the toaster oven. In total, how much should she expect to make or spend?[53]

GUIDED PRACTICE 3.64

Would you be surprised if John's weekly commute wasn't exactly 90 minutes or if Elena didn't make exactly \$152? Explain.[54]

Two important concepts concerning combinations of random variables have so far been introduced. First, a final value can sometimes be described as the sum of its parts in an equation. Second, intuition suggests that putting the individual average values into this equation gives the average value we would expect in total. This second point needs clarification – it is guaranteed to be true in what are called *linear combinations of random variables*.

A **linear combination** of two random variables X and Y is a fancy phrase to describe a combination

$$aX + bY$$

where a and b are some fixed and known numbers. For John's commute time, there were five random variables – one for each work day – and each random variable could be written as having a fixed coefficient of 1:

$$1X_1 + 1X_2 + 1X_3 + 1X_4 + 1X_5$$

For Elena's net gain or loss, the X random variable had a coefficient of +1 and the Y random variable had a coefficient of -1.

---

[52]She will make X dollars on the TV but spend Y dollars on the toaster oven: $X - Y$.

[53]$E(X - Y) = E(X) - E(Y) = 175 - 23 = \$152$. She should expect to make about \$152.

[54]No, since there is probably some variability. For example, the traffic will vary from one day to next, and auction prices will vary depending on the quality of the merchandise and the interest of the attendees.

When considering the average of a linear combination of random variables, it is safe to plug in the mean of each random variable and then compute the final result. For a few examples of nonlinear combinations of random variables – cases where we cannot simply plug in the means – see the footnote.[55]

LINEAR COMBINATIONS OF RANDOM VARIABLES AND THE AVERAGE RESULT

If X and Y are random variables, then a linear combination of the random variables is given by

$$aX + bY$$

where a and b are some fixed numbers. To compute the average value of a linear combination of random variables, plug in the average of each individual random variable and compute the result:

$$a \times E(X) + b \times E(Y)$$

Recall that the expected value is the same as the mean, e.g. $E(X) = \mu_X$.

EXAMPLE 3.65

Leonard has invested \$6000 in Caterpillar Inc (stock ticker: CAT) and \$2000 in Exxon Mobil Corp (XOM). If X represents the change in Caterpillar's stock next month and Y represents the change in Exxon Mobil's stock next month, write an equation that describes how much money will be made or lost in Leonard's stocks for the month.

For simplicity, we will suppose X and Y are not in percents but are in decimal form (e.g. if Caterpillar's stock increases 1%, then $X = 0.01$; or if it loses 1%, then $X = -0.01$). Then we can write an equation for Leonard's gain as

$$\$6000 \times X + \$2000 \times Y$$

If we plug in the change in the stock value for X and Y, this equation gives the change in value of Leonard's stock portfolio for the month. A positive value represents a gain, and a negative value represents a loss.

GUIDED PRACTICE 3.66

Caterpillar stock has recently been rising at 2.0% and Exxon Mobil's at 0.2% per month, respectively. Compute the expected change in Leonard's stock portfolio for next month.[56]

GUIDED PRACTICE 3.67

You should have found that Leonard expects a positive gain in Guided Practice 3.66. However, would you be surprised if he actually had a loss this month?[57]

---

[55] If X and Y are random variables, consider the following combinations: X^{1+Y}, $X \times Y$, X/Y. In such cases, plugging in the average value for each random variable and computing the result will not generally lead to an accurate average value for the end result.

[56] $E(\$6000 \times X + \$2000 \times Y) = \$6000 \times 0.020 + \$2000 \times 0.002 = \$124$.

[57] No. While stocks tend to rise over time, they are often volatile in the short term.

3.4.4   Variability in linear combinations of random variables

Quantifying the average outcome from a linear combination of random variables is helpful, but it is also important to have some sense of the uncertainty associated with the total outcome of that combination of random variables. The expected net gain or loss of Leonard's stock portfolio was considered in Guided Practice 3.66. However, there was no quantitative discussion of the volatility of this portfolio. For instance, while the average monthly gain might be about $124 according to the data, that gain is not guaranteed. Figure 3.22 shows the monthly changes in a portfolio like Leonard's during a three year period. The gains and losses vary widely, and quantifying these fluctuations is important when investing in stocks.

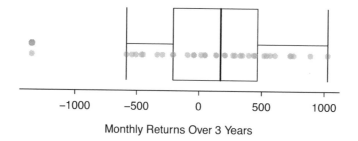

Figure 3.22: The change in a portfolio like Leonard's for 36 months, where $6000 is in Caterpillar's stock and $2000 is in Exxon Mobil's.

Just as we have done in many previous cases, we use the variance and standard deviation to describe the uncertainty associated with Leonard's monthly returns. To do so, the variances of each stock's monthly return will be useful, and these are shown in Figure 3.23. The stocks' returns are nearly independent.

Here we use an equation from probability theory to describe the uncertainty of Leonard's monthly returns; we leave the proof of this method to a dedicated probability course. The variance of a linear combination of random variables can be computed by plugging in the variances of the individual random variables and squaring the coefficients of the random variables:

$$Var(aX + bY) = a^2 \times Var(X) + b^2 \times Var(Y)$$

It is important to note that this equality assumes the random variables are independent; if independence doesn't hold, then a modification to this equation would be required that we leave as a topic for a future course to cover. This equation can be used to compute the variance of Leonard's monthly return:

$$Var(6000 \times X + 2000 \times Y) = 6000^2 \times Var(X) + 2000^2 \times Var(Y)$$
$$= 36,000,000 \times 0.0057 + 4,000,000 \times 0.0021$$
$$\approx 213,600$$

The standard deviation is computed as the square root of the variance: $\sqrt{213,600} = \$463$. While an average monthly return of $124 on an $8000 investment is nothing to scoff at, the monthly returns are so volatile that Leonard should not expect this income to be very stable.

	Mean ($\bar{x}$)	Standard deviation (s)	Variance (s^2)
CAT	0.0204	0.0757	0.0057
XOM	0.0025	0.0455	0.0021

Figure 3.23: The mean, standard deviation, and variance of the CAT and XOM stocks. These statistics were estimated from historical stock data, so notation used for sample statistics has been used.

VARIABILITY OF LINEAR COMBINATIONS OF RANDOM VARIABLES

The variance of a linear combination of random variables may be computed by squaring the constants, substituting in the variances for the random variables, and computing the result:

$$Var(aX + bY) = a^2 \times Var(X) + b^2 \times Var(Y)$$

This equation is valid as long as the random variables are independent of each other. The standard deviation of the linear combination may be found by taking the square root of the variance.

EXAMPLE 3.68

Suppose John's daily commute has a standard deviation of 4 minutes. What is the uncertainty in his total commute time for the week?

The expression for John's commute time was

$$X_1 + X_2 + X_3 + X_4 + X_5$$

Each coefficient is 1, and the variance of each day's time is $4^2 = 16$. Thus, the variance of the total weekly commute time is

$$\text{variance} = 1^2 \times 16 + 1^2 \times 16 + 1^2 \times 16 + 1^2 \times 16 + 1^2 \times 16 = 5 \times 16 = 80$$
$$\text{standard deviation} = \sqrt{\text{variance}} = \sqrt{80} = 8.94$$

The standard deviation for John's weekly work commute time is about 9 minutes.

GUIDED PRACTICE 3.69

The computation in Example 3.68 relied on an important assumption: the commute time for each day is independent of the time on other days of that week. Do you think this is valid? Explain.[58]

GUIDED PRACTICE 3.70

Consider Elena's two auctions from Guided Practice 3.62 on page 120. Suppose these auctions are approximately independent and the variability in auction prices associated with the TV and toaster oven can be described using standard deviations of $25 and $8. Compute the standard deviation of Elena's net gain.[59]

Consider again Guided Practice 3.70. The negative coefficient for Y in the linear combination was eliminated when we squared the coefficients. This generally holds true: negatives in a linear combination will have no impact on the variability computed for a linear combination, but they do impact the expected value computations.

---

[58]One concern is whether traffic patterns tend to have a weekly cycle (e.g. Fridays may be worse than other days). If that is the case, and John drives, then the assumption is probably not reasonable. However, if John walks to work, then his commute is probably not affected by any weekly traffic cycle.

[59]The equation for Elena can be written as

$$(1) \times X + (-1) \times Y$$

The variances of X and Y are 625 and 64. We square the coefficients and plug in the variances:

$$(1)^2 \times Var(X) + (-1)^2 \times Var(Y) = 1 \times 625 + 1 \times 64 = 689$$

The variance of the linear combination is 689, and the standard deviation is the square root of 689: about $26.25.

Exercises

3.29 College smokers. At a university, 13% of students smoke.

(a) Calculate the expected number of smokers in a random sample of 100 students from this university.

(b) The university gym opens at 9 am on Saturday mornings. One Saturday morning at 8:55 am there are 27 students outside the gym waiting for it to open. Should you use the same approach from part (a) to calculate the expected number of smokers among these 27 students?

3.30 Ace of clubs wins. Consider the following card game with a well-shuffled deck of cards. If you draw a red card, you win nothing. If you get a spade, you win $5. For any club, you win $10 plus an extra $20 for the ace of clubs.

(a) Create a probability model for the amount you win at this game. Also, find the expected winnings for a single game and the standard deviation of the winnings.

(b) What is the maximum amount you would be willing to pay to play this game? Explain your reasoning.

3.31 Hearts win. In a new card game, you start with a well-shuffled full deck and draw 3 cards without replacement. If you draw 3 hearts, you win $50. If you draw 3 black cards, you win $25. For any other draws, you win nothing.

(a) Create a probability model for the amount you win at this game, and find the expected winnings. Also compute the standard deviation of this distribution.

(b) If the game costs $5 to play, what would be the expected value and standard deviation of the net profit (or loss)? *(Hint: profit = winnings − cost; X − 5)*

(c) If the game costs $5 to play, should you play this game? Explain.

3.32 Is it worth it? Andy is always looking for ways to make money fast. Lately, he has been trying to make money by gambling. Here is the game he is considering playing: The game costs $2 to play. He draws a card from a deck. If he gets a number card (2-10), he wins nothing. For any face card (jack, queen or king), he wins $3. For any ace, he wins $5, and he wins an *extra* $20 if he draws the ace of clubs.

(a) Create a probability model and find Andy's expected profit per game.

(b) Would you recommend this game to Andy as a good way to make money? Explain.

3.33 Portfolio return. A portfolio's value increases by 18% during a financial boom and by 9% during normal times. It decreases by 12% during a recession. What is the expected return on this portfolio if each scenario is equally likely?

3.34 Baggage fees. An airline charges the following baggage fees: $25 for the first bag and $35 for the second. Suppose 54% of passengers have no checked luggage, 34% have one piece of checked luggage and 12% have two pieces. We suppose a negligible portion of people check more than two bags.

(a) Build a probability model, compute the average revenue per passenger, and compute the corresponding standard deviation.

(b) About how much revenue should the airline expect for a flight of 120 passengers? With what standard deviation? Note any assumptions you make and if you think they are justified.

3.35 American roulette. The game of American roulette involves spinning a wheel with 38 slots: 18 red, 18 black, and 2 green. A ball is spun onto the wheel and will eventually land in a slot, where each slot has an equal chance of capturing the ball. Gamblers can place bets on red or black. If the ball lands on their color, they double their money. If it lands on another color, they lose their money. Suppose you bet $1 on red. What's the expected value and standard deviation of your winnings?

3.36 European roulette. The game of European roulette involves spinning a wheel with 37 slots: 18 red, 18 black, and 1 green. A ball is spun onto the wheel and will eventually land in a slot, where each slot has an equal chance of capturing the ball. Gamblers can place bets on red or black. If the ball lands on their color, they double their money. If it lands on another color, they lose their money.

(a) Suppose you play roulette and bet $3 on a single round. What is the expected value and standard deviation of your total winnings?

(b) Suppose you bet $1 in three different rounds. What is the expected value and standard deviation of your total winnings?

(c) How do your answers to parts (a) and (b) compare? What does this say about the riskiness of the two games?

3.5 Continuous distributions

So far in this chapter we've discussed cases where the outcome of a variable is discrete. In this section, we consider a context where the outcome is a continuous numerical variable.

EXAMPLE 3.71

Figure 3.24 shows a few different hollow histograms for the heights of US adults. How does changing the number of bins allow you to make different interpretations of the data?

Adding more bins provides greater detail. This sample is extremely large, which is why much smaller bins still work well. Usually we do not use so many bins with smaller sample sizes since small counts per bin mean the bin heights are very volatile.

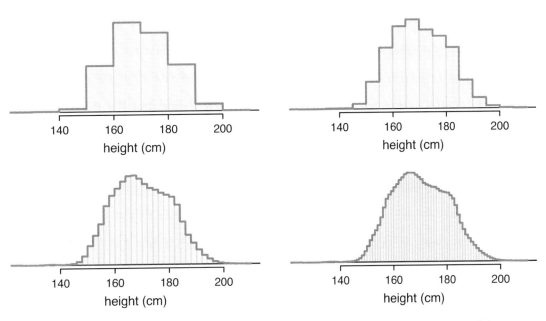

Figure 3.24: Four hollow histograms of US adults heights with varying bin widths.

EXAMPLE 3.72

What proportion of the sample is between 180 cm and 185 cm tall (about 5'11" to 6'1")?

We can add up the heights of the bins in the range 180 cm and 185 and divide by the sample size. For instance, this can be done with the two shaded bins shown in Figure 3.25. The two bins in this region have counts of 195,307 and 156,239 people, resulting in the following estimate of the probability:

$$\frac{195307 + 156239}{3,000,000} = 0.1172$$

This fraction is the same as the proportion of the histogram's area that falls in the range 180 to 185 cm.

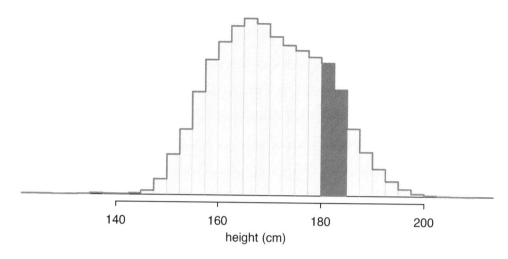

Figure 3.25: A histogram with bin sizes of 2.5 cm. The shaded region represents individuals with heights between 180 and 185 cm.

3.5.1 From histograms to continuous distributions

Examine the transition from a boxy hollow histogram in the top-left of Figure 3.24 to the much smoother plot in the lower-right. In this last plot, the bins are so slim that the hollow histogram is starting to resemble a smooth curve. This suggests the population height as a *continuous* numerical variable might best be explained by a curve that represents the outline of extremely slim bins.

This smooth curve represents a **probability density function** (also called a **density** or **distribution**), and such a curve is shown in Figure 3.26 overlaid on a histogram of the sample. A density has a special property: the total area under the density's curve is 1.

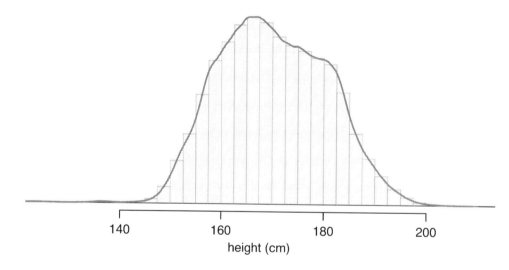

Figure 3.26: The continuous probability distribution of heights for US adults.

3.5.2 Probabilities from continuous distributions

We computed the proportion of individuals with heights 180 to 185 cm in Example 3.72 as a fraction:

$$\frac{\text{number of people between 180 and 185}}{\text{total sample size}}$$

We found the number of people with heights between 180 and 185 cm by determining the fraction of the histogram's area in this region. Similarly, we can use the area in the shaded region under the curve to find a probability (with the help of a computer):

$$P(\text{height between 180 and 185}) = \text{area between 180 and 185} = 0.1157$$

The probability that a randomly selected person is between 180 and 185 cm is 0.1157. This is very close to the estimate from Example 3.72: 0.1172.

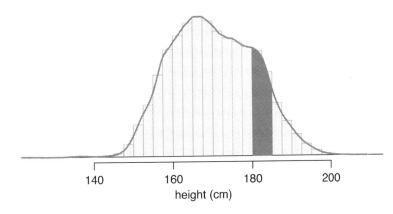

Figure 3.27: Density for heights in the US adult population with the area between 180 and 185 cm shaded. Compare this plot with Figure 3.25.

GUIDED PRACTICE 3.73

Three US adults are randomly selected. The probability a single adult is between 180 and 185 cm is 0.1157.[60]

(a) What is the probability that all three are between 180 and 185 cm tall?

(b) What is the probability that none are between 180 and 185 cm?

EXAMPLE 3.74

What is the probability that a randomly selected person is **exactly** 180 cm? Assume you can measure perfectly.

This probability is zero. A person might be close to 180 cm, but not exactly 180 cm tall. This also makes sense with the definition of probability as area; there is no area captured between 180 cm and 180 cm.

GUIDED PRACTICE 3.75

Suppose a person's height is rounded to the nearest centimeter. Is there a chance that a random person's **measured** height will be 180 cm?[61]

---

[60]Brief answers: (a) $0.1157 \times 0.1157 \times 0.1157 = 0.0015$. (b) $(1 - 0.1157)^3 = 0.692$

[61]This has positive probability. Anyone between 179.5 cm and 180.5 cm will have a *measured* height of 180 cm. This is probably a more realistic scenario to encounter in practice versus Example 3.74.

Exercises

3.37 Cat weights. The histogram shown below represents the weights (in kg) of 47 female and 97 male cats.[62]

(a) What fraction of these cats weigh less than 2.5 kg?

(b) What fraction of these cats weigh between 2.5 and 2.75 kg?

(c) What fraction of these cats weigh between 2.75 and 3.5 kg?

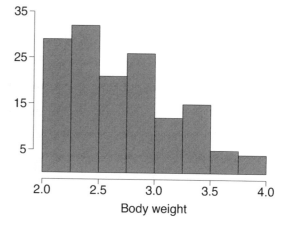

3.38 Income and gender. The relative frequency table below displays the distribution of annual total personal income (in 2009 inflation-adjusted dollars) for a representative sample of 96,420,486 Americans. These data come from the American Community Survey for 2005-2009. This sample is comprised of 59% males and 41% females.[63]

(a) Describe the distribution of total personal income.

(b) What is the probability that a randomly chosen US resident makes less than $50,000 per year?

(c) What is the probability that a randomly chosen US resident makes less than $50,000 per year and is female? Note any assumptions you make.

(d) The same data source indicates that 71.8% of females make less than $50,000 per year. Use this value to determine whether or not the assumption you made in part (c) is valid.

Income	Total
$1 to $9,999 or loss	2.2%
$10,000 to $14,999	4.7%
$15,000 to $24,999	15.8%
$25,000 to $34,999	18.3%
$35,000 to $49,999	21.2%
$50,000 to $64,999	13.9%
$65,000 to $74,999	5.8%
$75,000 to $99,999	8.4%
$100,000 or more	9.7%

[62]W. N. Venables and B. D. Ripley. *Modern Applied Statistics with S.* Fourth Edition. www.stats.ox.ac.uk/pub/MASS4. New York: Springer, 2002.

[63]U.S. Census Bureau, 2005-2009 American Community Survey.

Chapter exercises

3.39 Grade distributions. Each row in the table below is a proposed grade distribution for a class. Identify each as a valid or invalid probability distribution, and explain your reasoning.

	Grades				
	A	B	C	D	F
(a)	0.3	0.3	0.3	0.2	0.1
(b)	0	0	1	0	0
(c)	0.3	0.3	0.3	0	0
(d)	0.3	0.5	0.2	0.1	-0.1
(e)	0.2	0.4	0.2	0.1	0.1
(f)	0	-0.1	1.1	0	0

3.40 Health coverage, frequencies. The Behavioral Risk Factor Surveillance System (BRFSS) is an annual telephone survey designed to identify risk factors in the adult population and report emerging health trends. The following table summarizes two variables for the respondents: health status and health coverage, which describes whether each respondent had health insurance.[64]

		Health Status					
		Excellent	Very good	Good	Fair	Poor	Total
Health	No	459	727	854	385	99	2,524
Coverage	Yes	4,198	6,245	4,821	1,634	578	17,476
	Total	4,657	6,972	5,675	2,019	677	20,000

(a) If we draw one individual at random, what is the probability that the respondent has excellent health and doesn't have health coverage?

(b) If we draw one individual at random, what is the probability that the respondent has excellent health or doesn't have health coverage?

3.41 HIV in Swaziland. Swaziland has the highest HIV prevalence in the world: 25.9% of this country's population is infected with HIV.[65] The ELISA test is one of the first and most accurate tests for HIV. For those who carry HIV, the ELISA test is 99.7% accurate. For those who do not carry HIV, the test is 92.6% accurate. If an individual from Swaziland has tested positive, what is the probability that he carries HIV?

3.42 Twins. About 30% of human twins are identical, and the rest are fraternal. Identical twins are necessarily the same sex – half are males and the other half are females. One-quarter of fraternal twins are both male, one-quarter both female, and one-half are mixes: one male, one female. You have just become a parent of twins and are told they are both girls. Given this information, what is the probability that they are identical?

3.43 Cost of breakfast. Sally gets a cup of coffee and a muffin every day for breakfast from one of the many coffee shops in her neighborhood. She picks a coffee shop each morning at random and independently of previous days. The average price of a cup of coffee is $1.40 with a standard deviation of 30¢ ($0.30), the average price of a muffin is $2.50 with a standard deviation of 15¢, and the two prices are independent of each other.

(a) What is the mean and standard deviation of the amount she spends on breakfast daily?

(b) What is the mean and standard deviation of the amount she spends on breakfast weekly (7 days)?

---

[64]Office of Surveillance, Epidemiology, and Laboratory Services Behavioral Risk Factor Surveillance System, BRFSS 2010 Survey Data.

[65]Source: CIA Factbook, Country Comparison: HIV/AIDS - Adult Prevalence Rate.

3.44 Scooping ice cream. Ice cream usually comes in 1.5 quart boxes (48 fluid ounces), and ice cream scoops hold about 2 ounces. However, there is some variability in the amount of ice cream in a box as well as the amount of ice cream scooped out. We represent the amount of ice cream in the box as X and the amount scooped out as Y. Suppose these random variables have the following means, standard deviations, and variances:

	mean	SD	variance
X	48	1	1
Y	2	0.25	0.0625

(a) An entire box of ice cream, plus 3 scoops from a second box is served at a party. How much ice cream do you expect to have been served at this party? What is the standard deviation of the amount of ice cream served?

(b) How much ice cream would you expect to be left in the box after scooping out one scoop of ice cream? That is, find the expected value of $X - Y$. What is the standard deviation of the amount left in the box?

(c) Using the context of this exercise, explain why we add variances when we subtract one random variable from another.

3.45 Variance of a mean, Part I. Suppose we have independent observations X_1 and X_2 from a distribution with mean μ and standard deviation σ. What is the variance of the mean of the two values: $\frac{X_1+X_2}{2}$?

3.46 Variance of a mean, Part II. Suppose we have 3 independent observations X_1, X_2, X_3 from a distribution with mean μ and standard deviation σ. What is the variance of the mean of these 3 values: $\frac{X_1+X_2+X_3}{3}$?

3.47 Variance of a mean, Part III. Suppose we have n independent observations X_1, X_2, ..., X_n from a distribution with mean μ and standard deviation σ. What is the variance of the mean of these n values: $\frac{X_1+X_2+\cdots+X_n}{n}$?

Chapter 4

Distributions of random variables

In this chapter, we discuss statistical distributions that frequently arise in the context of data analysis or statistical inference. We start with the normal distribution in the first section, which is used frequently in later chapters of this book. The remaining sections will occasionally be referenced but may be considered optional for the content in this book.

For videos, slides, and other resources, please visit
www.openintro.org/os

4.1 Normal distribution

Among all the distributions we see in practice, one is overwhelmingly the most common. The symmetric, unimodal, bell curve is ubiquitous throughout statistics. Indeed it is so common, that people often know it as the **normal curve** or **normal distribution**,[1] shown in Figure 4.1. Variables such as SAT scores and heights of US adult males closely follow the normal distribution.

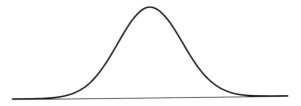

Figure 4.1: A normal curve.

NORMAL DISTRIBUTION FACTS

Many variables are nearly normal, but none are exactly normal. Thus the normal distribution, while not perfect for any single problem, is very useful for a variety of problems. We will use it in data exploration and to solve important problems in statistics.

4.1.1 Normal distribution model

The **normal distribution** always describes a symmetric, unimodal, bell-shaped curve. However, these curves can look different depending on the details of the model. Specifically, the normal distribution model can be adjusted using two parameters: mean and standard deviation. As you can probably guess, changing the mean shifts the bell curve to the left or right, while changing the standard deviation stretches or constricts the curve. Figure 4.2 shows the normal distribution with mean 0 and standard deviation 1 in the left panel and the normal distributions with mean 19 and standard deviation 4 in the right panel. Figure 4.3 shows these distributions on the same axis.

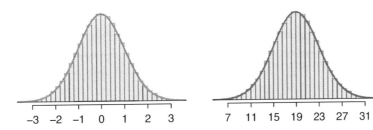

Figure 4.2: Both curves represent the normal distribution. However, they differ in their center and spread.

If a normal distribution has mean μ and standard deviation σ, we may write the distribution as $N(\mu, \sigma)$. The two distributions in Figure 4.3 may be written as

$$N(\mu = 0, \sigma = 1) \quad \text{and} \quad N(\mu = 19, \sigma = 4)$$

Because the mean and standard deviation describe a normal distribution exactly, they are called the distribution's **parameters**. The normal distribution with mean $\mu = 0$ and standard deviation $\sigma = 1$ is called the **standard normal distribution**.

---

[1]It is also introduced as the Gaussian distribution after Frederic Gauss, the first person to formalize its mathematical expression.

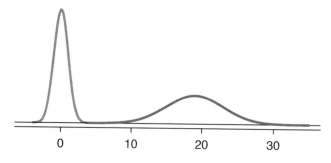

Figure 4.3: The normal distributions shown in Figure 4.2 but plotted together and on the same scale.

GUIDED PRACTICE 4.1

Ⓖ

Write down the short-hand for a normal distribution with[2]
(a) mean 5 and standard deviation 3,
(b) mean -100 and standard deviation 10, and
(c) mean 2 and standard deviation 9.

4.1.2   Standardizing with Z-scores

We often want to put data onto a standardized scale, which can make comparisons more reasonable.

EXAMPLE 4.2

Ⓔ

Table 4.4 shows the mean and standard deviation for total scores on the SAT and ACT. The distribution of SAT and ACT scores are both nearly normal. Suppose Ann scored 1300 on her SAT and Tom scored 24 on his ACT. Who performed better?

We use the standard deviation as a guide. Ann is 1 standard deviation above average on the SAT: $1100 + 200 = 1300$. Tom is 0.5 standard deviations above the mean on the ACT: $21 + 0.5 \times 6 = 24$. In Figure 4.5, we can see that Ann tends to do better with respect to everyone else than Tom did, so her score was better.

	SAT	ACT
Mean	1100	21
SD	200	6

Figure 4.4: Mean and standard deviation for the SAT and ACT.

Example 4.2 used a standardization technique called a Z-score, a method most commonly employed for nearly normal observations but that may be used with any distribution. The **Z-score** of an observation is defined as the number of standard deviations it falls above or below the mean. If the observation is one standard deviation above the mean, its Z-score is 1. If it is 1.5 standard deviations *below* the mean, then its Z-score is -1.5. If x is an observation from a distribution $N(\mu, \sigma)$, we define the Z-score mathematically as

$$Z = \frac{x - \mu}{\sigma}$$

Using $\mu_{SAT} = 1100$, $\sigma_{SAT} = 200$, and $x_{Ann} = 1300$, we find Ann's Z-score:

$$Z_{Ann} = \frac{x_{Ann} - \mu_{SAT}}{\sigma_{SAT}} = \frac{1300 - 1100}{200} = 1$$

---

[2](a) $N(\mu = 5, \sigma = 3)$. (b) $N(\mu = -100, \sigma = 10)$. (c) $N(\mu = 2, \sigma = 9)$.

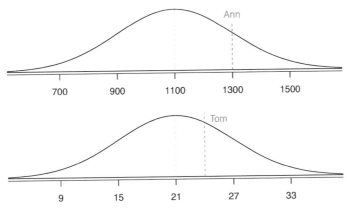

Figure 4.5: Ann's and Tom's scores shown against the SAT and ACT distributions.

THE Z-SCORE

The Z-score of an observation is the number of standard deviations it falls above or below the mean. We compute the Z-score for an observation x that follows a distribution with mean μ and standard deviation σ using

$$Z = \frac{x - \mu}{\sigma}$$

GUIDED PRACTICE 4.3

Use Tom's ACT score, 24, along with the ACT mean and standard deviation to find his Z-score.[3]

Observations above the mean always have positive Z-scores, while those below the mean always have negative Z-scores. If an observation is equal to the mean, such as an SAT score of 1100, then the Z-score is 0.

GUIDED PRACTICE 4.4

Let X represent a random variable from $N(\mu = 3, \sigma = 2)$, and suppose we observe $x = 5.19$.
(a) Find the Z-score of x.
(b) Use the Z-score to determine how many standard deviations above or below the mean x falls.[4]

GUIDED PRACTICE 4.5

Head lengths of brushtail possums follow a normal distribution with mean 92.6 mm and standard deviation 3.6 mm. Compute the Z-scores for possums with head lengths of 95.4 mm and 85.8 mm.[5]

We can use Z-scores to roughly identify which observations are more unusual than others. An observation x_1 is said to be more unusual than another observation x_2 if the absolute value of its Z-score is larger than the absolute value of the other observation's Z-score: $|Z_1| > |Z_2|$. This technique is especially insightful when a distribution is symmetric.

GUIDED PRACTICE 4.6

Which of the observations in Guided Practice 4.5 is more unusual?[6]

---

[3] $Z_{Tom} = \frac{x_{Tom} - \mu_{ACT}}{\sigma_{ACT}} = \frac{24 - 21}{6} = 0.5$

[4](a) Its Z-score is given by $Z = \frac{x - \mu}{\sigma} = \frac{5.19 - 3}{2} = 2.19/2 = 1.095$. (b) The observation x is 1.095 standard deviations *above* the mean. We know it must be above the mean since Z is positive.

[5]For $x_1 = 95.4$ mm: $Z_1 = \frac{x_1 - \mu}{\sigma} = \frac{95.4 - 92.6}{3.6} = 0.78$. For $x_2 = 85.8$ mm: $Z_2 = \frac{85.8 - 92.6}{3.6} = -1.89$.

[6]Because the *absolute value* of Z-score for the second observation is larger than that of the first, the second observation has a more unusual head length.

4.1.3   Finding tail areas

It's very useful in statistics to be able to identify tail areas of distributions. For instance, what fraction of people have an SAT score below Ann's score of 1300? This is the same as the **percentile** Ann is at, which is the percentage of cases that have lower scores than Ann. We can visualize such a tail area like the curve and shading shown in Figure 4.6.

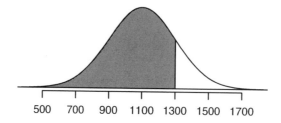

500   700   900   1100   1300   1500   1700

Figure 4.6: The area to the left of Z represents the fraction of people who scored lower than Ann.

There are many techniques for doing this, and we'll discuss three of the options.

1. The most common approach in practice is to use statistical software. For example, in the program **R**, we could find the area shown in Figure 4.6 using the following command, which takes in the Z-score and returns the lower tail area:

   ```
   > pnorm(1)
   [1] 0.8413447
   ```

   According to this calculation, the region shaded that is below 1300 represents the proportion 0.841 (84.1%) of SAT test takers who had Z-scores below $Z = 1$. More generally, we can also specify the cutoff explicitly if we also note the mean and standard deviation:

   ```
   > pnorm(1300, mean = 1100, sd = 200)
   [1] 0.8413447
   ```

   There are many other software options, such as Python or SAS; even spreadsheet programs such as Excel and Google Sheets support these calculations.

2. A common strategy in classrooms is to use a graphing calculator, such as a TI or Casio calculator. These calculators require a series of button presses that are less concisely described. You can find instructions on using these calculators for finding tail areas of a normal distribution in the OpenIntro video library:

   www.openintro.org/videos

3. The last option for finding tail areas is to use what's called a **probability table**; these are occasionally used in classrooms but rarely in practice. Appendix C.1 contains such a table and a guide for how to use it.

We will solve normal distribution problems in this section by always first finding the Z-score. The reason is that we will encounter close parallels called test statistics beginning in Chapter 5; these are, in many instances, an equivalent of a Z-score.

4.1.4 Normal probability examples

Cumulative SAT scores are approximated well by a normal model, $N(\mu = 1100, \sigma = 200)$.

EXAMPLE 4.7

Shannon is a randomly selected SAT taker, and nothing is known about Shannon's SAT aptitude. What is the probability Shannon scores at least 1190 on her SATs?

First, always draw and label a picture of the normal distribution. (Drawings need not be exact to be useful.) We are interested in the chance she scores above 1190, so we shade this upper tail:

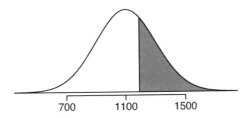

The picture shows the mean and the values at 2 standard deviations above and below the mean. The simplest way to find the shaded area under the curve makes use of the Z-score of the cutoff value. With $\mu = 1100$, $\sigma = 200$, and the cutoff value $x = 1190$, the Z-score is computed as

$$Z = \frac{x - \mu}{\sigma} = \frac{1190 - 1100}{200} = \frac{90}{200} = 0.45$$

Using statistical software (or another preferred method), we can area left of $Z = 0.45$ as 0.6736. To find the area *above* $Z = 0.45$, we compute one minus the area of the lower tail:

$$1.0000 \quad - \quad 0.6736 \quad = \quad 0.3264$$

The probability Shannon scores at least 1190 on the SAT is 0.3264.

ALWAYS DRAW A PICTURE FIRST, AND FIND THE Z-SCORE SECOND

For any normal probability situation, *always always always* draw and label the normal curve and shade the area of interest first. The picture will provide an estimate of the probability. After drawing a figure to represent the situation, identify the Z-score for the value of interest.

GUIDED PRACTICE 4.8

If the probability of Shannon scoring at least 1190 is 0.3264, then what is the probability she scores less than 1190? Draw the normal curve representing this exercise, shading the lower region instead of the upper one.[7]

---

[7]We found this probability in Example 4.7: 0.6736.

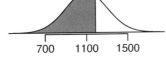

EXAMPLE 4.9

Edward earned a 1030 on his SAT. What is his percentile?

First, a picture is needed. Edward's percentile is the proportion of people who do not get as high as a 1030. These are the scores to the left of 1030.

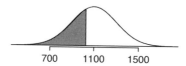

Identifying the mean $\mu = 1100$, the standard deviation $\sigma = 200$, and the cutoff for the tail area $x = 1030$ makes it easy to compute the Z-score:

$$Z = \frac{x - \mu}{\sigma} = \frac{1030 - 1100}{200} = -0.35$$

Using statistical software, we get a tail area of 0.3632. Edward is at the 36th percentile.

GUIDED PRACTICE 4.10

Use the results of Example 4.9 to compute the proportion of SAT takers who did better than Edward. Also draw a new picture.[8]

FINDING AREAS TO THE RIGHT

Many software programs return the area to the left when given a Z-score. If you would like the area to the right, first find the area to the left and then subtract this amount from one.

GUIDED PRACTICE 4.11

Stuart earned an SAT score of 1500. Draw a picture for each part.
(a) What is his percentile?
(b) What percent of SAT takers did better than Stuart?[9]

Based on a sample of 100 men, the heights of male adults in the US is nearly normal with mean 70.0" and standard deviation 3.3".

GUIDED PRACTICE 4.12

Mike is 5'7" and Jose is 6'4", and they both live in the US.
(a) What is Mike's height percentile?
(b) What is Jose's height percentile?
Also draw one picture for each part.[10]

---

[8]If Edward did better than 36% of SAT takers, then about 64% must have done better than him.

[9]We leave the drawings to you. (a) $Z = \frac{1500-1100}{200} = 2 \to 0.9772$. (b) $1 - 0.9772 = 0.0228$.
[10]First put the heights into inches: 67 and 76 inches. Figures are shown below.
(a) $Z_{\text{Mike}} = \frac{67-70}{3.3} = -0.91 \to 0.1814$. (b) $Z_{\text{Jose}} = \frac{76-70}{3.3} = 1.82 \to 0.9656$.

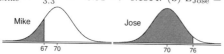

The last several problems have focused on finding the percentile (or upper tail) for a particular observation. What if you would like to know the observation corresponding to a particular percentile?

EXAMPLE 4.13

Erik's height is at the 40th percentile. How tall is he?

As always, first draw the picture.

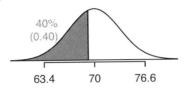

(E)

In this case, the lower tail probability is known (0.40), which can be shaded on the diagram. We want to find the observation that corresponds to this value. As a first step in this direction, we determine the Z-score associated with the 40th percentile. Using software, we can obtain the corresponding Z-score of about -0.25.

Knowing $Z_{Erik} = -0.25$ and the population parameters $\mu = 70$ and $\sigma = 3.3$ inches, the Z-score formula can be set up to determine Erik's unknown height, labeled x_{Erik}:

$$-0.25 = Z_{Erik} = \frac{x_{Erik} - \mu}{\sigma} = \frac{x_{Erik} - 70}{3.3}$$

Solving for x_{Erik} yields a height of 69.18 inches. That is, Erik is about 5'9".

EXAMPLE 4.14

What is the adult male height at the 82nd percentile?

Again, we draw the figure first.

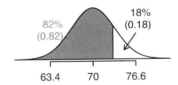

(E)

Next, we want to find the Z-score at the 82nd percentile, which will be a positive value and can be found using software as $Z = 0.92$. Finally, the height x is found using the Z-score formula with the known mean μ, standard deviation σ, and Z-score $Z = 0.92$:

$$0.92 = Z = \frac{x - \mu}{\sigma} = \frac{x - 70}{3.3}$$

This yields 73.04 inches or about 6'1" as the height at the 82nd percentile.

GUIDED PRACTICE 4.15

(G)

The SAT scores follow $N(1100, 200)$.[11]
(a) What is the 95th percentile for SAT scores?
(b) What is the 97.5^{th} percentile for SAT scores?

---

[11]Short answers: (a) $Z_{95} = 1.65 \rightarrow 1430$ SAT score. (b) $Z_{97.5} = 1.96 \rightarrow 1492$ SAT score.

GUIDED PRACTICE 4.16

Adult male heights follow $N(70.0", 3.3")$.[12]
(a) What is the probability that a randomly selected male adult is at least 6'2" (74 inches)?
(b) What is the probability that a male adult is shorter than 5'9" (69 inches)?

EXAMPLE 4.17

What is the probability that a random adult male is between 5'9" and 6'2"?

These heights correspond to 69 inches and 74 inches. First, draw the figure. The area of interest is no longer an upper or lower tail.

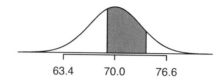

63.4	70.0	76.6

The total area under the curve is 1. If we find the area of the two tails that are not shaded (from Guided Practice 4.16, these areas are 0.3821 and 0.1131), then we can find the middle area:

$$1.0000 \ - \ 0.3821 \ - \ 0.1131 \ = \ 0.5048$$

That is, the probability of being between 5'9" and 6'2" is 0.5048.

GUIDED PRACTICE 4.18

SAT scores follow $N(1100, 200)$. What percent of SAT takers get between 1100 and 1400?[13]

GUIDED PRACTICE 4.19

Adult male heights follow $N(70.0", 3.3")$. What percent of adult males are between 5'5" and 5'7"?[14]

---

[12]Short answers: (a) $Z = 1.21 \to 0.8869$, then subtract this value from 1 to get 0.1131. (b) $Z = -0.30 \to 0.3821$.
[13]This is an abbreviated solution. (Be sure to draw a figure!) First find the percent who get below 1100 and the percent that get above 1400: $Z_{1100} = 0.00 \to 0.5000$ (area below), $Z_{1400} = 1.5 \to 0.0668$ (area above). Final answer: $1.0000 - 0.5000 - 0.0668 = 0.4332$.
[14]5'5" is 65 inches ($Z = -1.52$). 5'7" is 67 inches ($Z = -0.91$). Numerical solution: $1.000 - 0.0643 - 0.8186 = 0.1171$, i.e. 11.71%.

4.1.5 68-95-99.7 rule

Here, we present a useful rule of thumb for the probability of falling within 1, 2, and 3 standard deviations of the mean in the normal distribution. This will be useful in a wide range of practical settings, especially when trying to make a quick estimate without a calculator or Z-table.

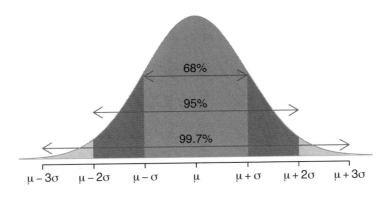

Figure 4.7: Probabilities for falling within 1, 2, and 3 standard deviations of the mean in a normal distribution.

GUIDED PRACTICE 4.20

Use software, a calculator, or a probability table to confirm that about 68%, 95%, and 99.7% of observations fall within 1, 2, and 3, standard deviations of the mean in the normal distribution, respectively. For instance, first find the area that falls between $Z = -1$ and $Z = 1$, which should have an area of about 0.68. Similarly there should be an area of about 0.95 between $Z = -2$ and $Z = 2$.[15]

It is possible for a normal random variable to fall 4, 5, or even more standard deviations from the mean. However, these occurrences are very rare if the data are nearly normal. The probability of being further than 4 standard deviations from the mean is about 1-in-15,000. For 5 and 6 standard deviations, it is about 1-in-2 million and 1-in-500 million, respectively.

GUIDED PRACTICE 4.21

SAT scores closely follow the normal model with mean $\mu = 1100$ and standard deviation $\sigma = 200$.[16]
(a) About what percent of test takers score 700 to 1500?
(b) What percent score between 1100 and 1500?

---

[15]First draw the pictures. Using software, we get 0.6827 within 1 standard deviation, 0.9545 within 2 standard deviations, and 0.9973 within 3 standard deviations.

[16](a) 700 and 1500 represent two standard deviations below and above the mean, which means about 95% of test takers will score between 700 and 1500. (b) We found that 700 to 1500 represents about 95% of test takers. These test takers would be evenly split by the center of the distribution, 1100, so $\frac{95\%}{2} = 47.5\%$ of all test takers score between 1100 and 1500.

Exercises

4.1 Area under the curve, Part I. What percent of a standard normal distribution $N(\mu = 0, \sigma = 1)$ is found in each region? Be sure to draw a graph.

(a) $Z < -1.35$ (b) $Z > 1.48$ (c) $-0.4 < Z < 1.5$ (d) $|Z| > 2$

4.2 Area under the curve, Part II. What percent of a standard normal distribution $N(\mu = 0, \sigma = 1)$ is found in each region? Be sure to draw a graph.

(a) $Z > -1.13$ (b) $Z < 0.18$ (c) $Z > 8$ (d) $|Z| < 0.5$

4.3 GRE scores, Part I. Sophia who took the Graduate Record Examination (GRE) scored 160 on the Verbal Reasoning section and 157 on the Quantitative Reasoning section. The mean score for Verbal Reasoning section for all test takers was 151 with a standard deviation of 7, and the mean score for the Quantitative Reasoning was 153 with a standard deviation of 7.67. Suppose that both distributions are nearly normal.

(a) Write down the short-hand for these two normal distributions.

(b) What is Sophia's Z-score on the Verbal Reasoning section? On the Quantitative Reasoning section? Draw a standard normal distribution curve and mark these two Z-scores.

(c) What do these Z-scores tell you?

(d) Relative to others, which section did she do better on?

(e) Find her percentile scores for the two exams.

(f) What percent of the test takers did better than her on the Verbal Reasoning section? On the Quantitative Reasoning section?

(g) Explain why simply comparing raw scores from the two sections could lead to an incorrect conclusion as to which section a student did better on.

(h) If the distributions of the scores on these exams are not nearly normal, would your answers to parts (b) - (f) change? Explain your reasoning.

4.4 Triathlon times, Part I. In triathlons, it is common for racers to be placed into age and gender groups. Friends Leo and Mary both completed the Hermosa Beach Triathlon, where Leo competed in the *Men, Ages 30 - 34* group while Mary competed in the *Women, Ages 25 - 29* group. Leo completed the race in 1:22:28 (4948 seconds), while Mary completed the race in 1:31:53 (5513 seconds). Obviously Leo finished faster, but they are curious about how they did within their respective groups. Can you help them? Here is some information on the performance of their groups:

- The finishing times of the *Men, Ages 30 - 34* group has a mean of 4313 seconds with a standard deviation of 583 seconds.

- The finishing times of the *Women, Ages 25 - 29* group has a mean of 5261 seconds with a standard deviation of 807 seconds.

- The distributions of finishing times for both groups are approximately Normal.

Remember: a better performance corresponds to a faster finish.

(a) Write down the short-hand for these two normal distributions.

(b) What are the Z-scores for Leo's and Mary's finishing times? What do these Z-scores tell you?

(c) Did Leo or Mary rank better in their respective groups? Explain your reasoning.

(d) What percent of the triathletes did Leo finish faster than in his group?

(e) What percent of the triathletes did Mary finish faster than in her group?

(f) If the distributions of finishing times are not nearly normal, would your answers to parts (b) - (e) change? Explain your reasoning.

4.5 GRE scores, Part II. In Exercise 4.3 we saw two distributions for GRE scores: $N(\mu = 151, \sigma = 7)$ for the verbal part of the exam and $N(\mu = 153, \sigma = 7.67)$ for the quantitative part. Use this information to compute each of the following:

(a) The score of a student who scored in the 80th percentile on the Quantitative Reasoning section.

(b) The score of a student who scored worse than 70% of the test takers in the Verbal Reasoning section.

4.6 Triathlon times, Part II. In Exercise 4.4 we saw two distributions for triathlon times: $N(\mu = 4313, \sigma = 583)$ for *Men, Ages 30 - 34* and $N(\mu = 5261, \sigma = 807)$ for the *Women, Ages 25 - 29* group. Times are listed in seconds. Use this information to compute each of the following:

(a) The cutoff time for the fastest 5% of athletes in the men's group, i.e. those who took the shortest 5% of time to finish.

(b) The cutoff time for the slowest 10% of athletes in the women's group.

4.7 LA weather, Part I. The average daily high temperature in June in LA is 77°F with a standard deviation of 5°F. Suppose that the temperatures in June closely follow a normal distribution.

(a) What is the probability of observing an 83°F temperature or higher in LA during a randomly chosen day in June?

(b) How cool are the coldest 10% of the days (days with lowest average high temperature) during June in LA?

4.8 CAPM. The Capital Asset Pricing Model (CAPM) is a financial model that assumes returns on a portfolio are normally distributed. Suppose a portfolio has an average annual return of 14.7% (i.e. an average gain of 14.7%) with a standard deviation of 33%. A return of 0% means the value of the portfolio doesn't change, a negative return means that the portfolio loses money, and a positive return means that the portfolio gains money.

(a) What percent of years does this portfolio lose money, i.e. have a return less than 0%?

(b) What is the cutoff for the highest 15% of annual returns with this portfolio?

4.9 LA weather, Part II. Exercise 4.7 states that average daily high temperature in June in LA is 77°F with a standard deviation of 5°F, and it can be assumed that they to follow a normal distribution. We use the following equation to convert °F (Fahrenheit) to °C (Celsius):

$$C = (F - 32) \times \frac{5}{9}.$$

(a) Write the probability model for the distribution of temperature in °C in June in LA.

(b) What is the probability of observing a 28°C (which roughly corresponds to 83°F) temperature or higher in June in LA? Calculate using the °C model from part (a).

(c) Did you get the same answer or different answers in part (b) of this question and part (a) of Exercise 4.7? Are you surprised? Explain.

(d) Estimate the IQR of the temperatures (in °C) in June in LA.

4.10 Find the SD. Cholesterol levels for women aged 20 to 34 follow an approximately normal distribution with mean 185 milligrams per deciliter (mg/dl). Women with cholesterol levels above 220 mg/dl are considered to have high cholesterol and about 18.5% of women fall into this category. What is the standard deviation of the distribution of cholesterol levels for women aged 20 to 34?

4.2   Geometric distribution

How long should we expect to flip a coin until it turns up `heads`? Or how many times should we expect to roll a die until we get a 1? These questions can be answered using the geometric distribution. We first formalize each trial – such as a single coin flip or die toss – using the Bernoulli distribution, and then we combine these with our tools from probability (Chapter 3) to construct the geometric distribution.

4.2.1   Bernoulli distribution

Many health insurance plans in the United States have a deductible, where the insured individual is responsible for costs up to the deductible, and then the costs above the deductible are shared between the individual and insurance company for the remainder of the year.

Suppose a health insurance company found that 70% of the people they insure stay below their deductible in any given year. Each of these people can be thought of as a **trial**. We label a person a **success** if her healthcare costs do not exceed the deductible. We label a person a **failure** if she does exceed her deductible in the year. Because 70% of the individuals will not hit their deductible, we denote the **probability of a success** as $p = 0.7$. The probability of a failure is sometimes denoted with $q = 1 - p$, which would be 0.3 for the insurance example.

When an individual trial only has two possible outcomes, often labeled as `success` or `failure`, it is called a **Bernoulli random variable**. We chose to label a person who does not hit her deductible as a "success" and all others as "failures". However, we could just as easily have reversed these labels. The mathematical framework we will build does not depend on which outcome is labeled a success and which a failure, as long as we are consistent.

Bernoulli random variables are often denoted as 1 for a success and 0 for a failure. In addition to being convenient in entering data, it is also mathematically handy. Suppose we observe ten trials:

$$1\ 1\ 1\ 0\ 1\ 0\ 0\ 1\ 1\ 0$$

Then the **sample proportion**, $\hat{p}$, is the sample mean of these observations:

$$\hat{p} = \frac{\#\text{ of successes}}{\#\text{ of trials}} = \frac{1+1+1+0+1+0+0+1+1+0}{10} = 0.6$$

This mathematical inquiry of Bernoulli random variables can be extended even further. Because 0 and 1 are numerical outcomes, we can define the mean and standard deviation of a Bernoulli random variable. (See Exercises 4.15 and 4.16.)

BERNOULLI RANDOM VARIABLE

If X is a random variable that takes value 1 with probability of success p and 0 with probability $1 - p$, then X is a Bernoulli random variable with mean and standard deviation

$$\mu = p \qquad\qquad \sigma = \sqrt{p(1-p)}$$

In general, it is useful to think about a Bernoulli random variable as a random process with only two outcomes: a success or failure. Then we build our mathematical framework using the numerical labels 1 and 0 for successes and failures, respectively.

4.2.2 Geometric distribution

The **geometric distribution** is used to describe how many trials it takes to observe a success. Let's first look at an example.

EXAMPLE 4.22

Suppose we are working at the insurance company and need to find a case where the person did not exceed her (or his) deductible as a case study. If the probability a person will not exceed her deductible is 0.7 and we are drawing people at random, what are the chances that the first person will not have exceeded her deductible, i.e. be a success? The second person? The third? What about we pull $n-1$ cases before we find the first success, i.e. the first success is the nth person? (If the first success is the fifth person, then we say $n = 5$.)

The probability of stopping after the first person is just the chance the first person will not hit her (or his) deductible: 0.7. The probability the second person is the first to hit her deductible is

$$P(\text{second person is the first to hit deductible})$$
$$= P(\text{the first won't, the second will}) = (0.3)(0.7) = 0.21$$

Likewise, the probability it will be the third case is $(0.3)(0.3)(0.7) = 0.063$.

If the first success is on the nth person, then there are $n-1$ failures and finally 1 success, which corresponds to the probability $(0.3)^{n-1}(0.7)$. This is the same as $(1 - 0.7)^{n-1}(0.7)$.

Example 4.22 illustrates what the **geometric distribution**, which describes the waiting time until a success for **independent and identically distributed (iid)** Bernoulli random variables. In this case, the *independence* aspect just means the individuals in the example don't affect each other, and *identical* means they each have the same probability of success.

The geometric distribution from Example 4.22 is shown in Figure 4.8. In general, the probabilities for a geometric distribution decrease **exponentially** fast.

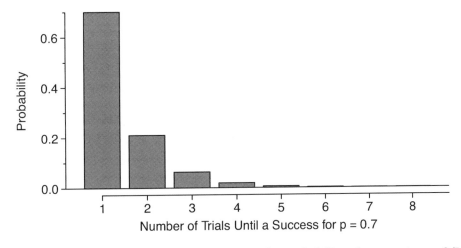

Figure 4.8: The geometric distribution when the probability of success is $p = 0.7$.

While this text will not derive the formulas for the mean (expected) number of trials needed to find the first success or the standard deviation or variance of this distribution, we present general formulas for each.

GEOMETRIC DISTRIBUTION

If the probability of a success in one trial is p and the probability of a failure is $1 - p$, then the probability of finding the first success in the nth trial is given by

$$(1 - p)^{n-1} p$$

The mean (i.e. expected value), variance, and standard deviation of this wait time are given by

$$\mu = \frac{1}{p} \qquad\qquad \sigma^2 = \frac{1 - p}{p^2} \qquad\qquad \sigma = \sqrt{\frac{1 - p}{p^2}}$$

It is no accident that we use the symbol μ for both the mean and expected value. The mean and the expected value are one and the same.

It takes, on average, $1/p$ trials to get a success under the geometric distribution. This mathematical result is consistent with what we would expect intuitively. If the probability of a success is high (e.g. 0.8), then we don't usually wait very long for a success: $1/0.8 = 1.25$ trials on average. If the probability of a success is low (e.g. 0.1), then we would expect to view many trials before we see a success: $1/0.1 = 10$ trials.

GUIDED PRACTICE 4.23

(G)

The probability that a particular case would not exceed their deductible is said to be 0.7. If we were to examine cases until we found one that where the person did not hit her deductible, how many cases should we expect to check?[17]

EXAMPLE 4.24

What is the chance that we would find the first success within the first 3 cases?

(E)

This is the chance it is the first ($n = 1$), second ($n = 2$), or third ($n = 3$) case is the first success, which are three disjoint outcomes. Because the individuals in the sample are randomly sampled from a large population, they are independent. We compute the probability of each case and add the separate results:

$$P(n = 1, 2, \text{ or } 3)$$
$$= P(n = 1) + P(n = 2) + P(n = 3)$$
$$= (0.3)^{1-1}(0.7) + (0.3)^{2-1}(0.7) + (0.3)^{3-1}(0.7)$$
$$= 0.973$$

There is a probability of 0.973 that we would find a successful case within 3 cases.

GUIDED PRACTICE 4.25

(G)

Determine a more clever way to solve Example 4.24. Show that you get the same result.[18]

---

[17]We would expect to see about $1/0.7 \approx 1.43$ individuals to find the first success.

[18]First find the probability of the complement: $P(\text{no success in first 3 trials}) = 0.3^3 = 0.027$. Next, compute one minus this probability: $1 - P(\text{no success in 3 trials}) = 1 - 0.027 = 0.973$.

EXAMPLE 4.26

Suppose a car insurer has determined that 88% of its drivers will not exceed their deductible in a given year. If someone at the company were to randomly draw driver files until they found one that had not exceeded their deductible, what is the expected number of drivers the insurance employee must check? What is the standard deviation of the number of driver files that must be drawn?

In this example, a success is again when someone will not exceed the insurance deductible, which has probability $p = 0.88$. The expected number of people to be checked is $1/p = 1/0.88 = 1.14$ and the standard deviation is $\sqrt{(1-p)/p^2} = 0.39$.

GUIDED PRACTICE 4.27

Using the results from Example 4.26, $\mu = 1.14$ and $\sigma = 0.39$, would it be appropriate to use the normal model to find what proportion of experiments would end in 3 or fewer trials?[19]

The independence assumption is crucial to the geometric distribution's accurate description of a scenario. Mathematically, we can see that to construct the probability of the success on the nth trial, we had to use the Multiplication Rule for Independent Processes. It is no simple task to generalize the geometric model for dependent trials.

---

[19]No. The geometric distribution is always right skewed and can never be well-approximated by the normal model.

Exercises

4.11 Is it Bernoulli? Determine if each trial can be considered an independent Bernoulli trial for the following situations.

(a) Cards dealt in a hand of poker.

(b) Outcome of each roll of a die.

4.12 With and without replacement. In the following situations assume that half of the specified population is male and the other half is female.

(a) Suppose you're sampling from a room with 10 people. What is the probability of sampling two females in a row when sampling with replacement? What is the probability when sampling without replacement?

(b) Now suppose you're sampling from a stadium with 10,000 people. What is the probability of sampling two females in a row when sampling with replacement? What is the probability when sampling without replacement?

(c) We often treat individuals who are sampled from a large population as independent. Using your findings from parts (a) and (b), explain whether or not this assumption is reasonable.

4.13 Eye color, Part I. A husband and wife both have brown eyes but carry genes that make it possible for their children to have brown eyes (probability 0.75), blue eyes (0.125), or green eyes (0.125).

(a) What is the probability the first blue-eyed child they have is their third child? Assume that the eye colors of the children are independent of each other.

(b) On average, how many children would such a pair of parents have before having a blue-eyed child? What is the standard deviation of the number of children they would expect to have until the first blue-eyed child?

4.14 Defective rate. A machine that produces a special type of transistor (a component of computers) has a 2% defective rate. The production is considered a random process where each transistor is independent of the others.

(a) What is the probability that the 10th transistor produced is the first with a defect?

(b) What is the probability that the machine produces no defective transistors in a batch of 100?

(c) On average, how many transistors would you expect to be produced before the first with a defect? What is the standard deviation?

(d) Another machine that also produces transistors has a 5% defective rate where each transistor is produced independent of the others. On average how many transistors would you expect to be produced with this machine before the first with a defect? What is the standard deviation?

(e) Based on your answers to parts (c) and (d), how does increasing the probability of an event affect the mean and standard deviation of the wait time until success?

4.15 Bernoulli, the mean. Use the probability rules from Section 3.4 to derive the mean of a Bernoulli random variable, i.e. a random variable X that takes value 1 with probability p and value 0 with probability $1 - p$. That is, compute the expected value of a generic Bernoulli random variable.

4.16 Bernoulli, the standard deviation. Use the probability rules from Section 3.4 to derive the standard deviation of a Bernoulli random variable, i.e. a random variable X that takes value 1 with probability p and value 0 with probability $1 - p$. That is, compute the square root of the variance of a generic Bernoulli random variable.

4.3 Binomial distribution

The **binomial distribution** is used to describe the number of successes in a fixed number of trials. This is different from the geometric distribution, which described the number of trials we must wait before we observe a success.

4.3.1 The binomial distribution

Let's again imagine ourselves back at the insurance agency where 70% of individuals do not exceed their deductible.

EXAMPLE 4.28

Suppose the insurance agency is considering a random sample of four individuals they insure. What is the chance exactly one of them will exceed the deductible and the other three will not? Let's call the four people Ariana (A), Brittany (B), Carlton (C), and Damian (D) for convenience.

Let's consider a scenario where one person exceeds the deductible:

$$P(A = \texttt{exceed}, \ B = \texttt{not}, \ C = \texttt{not}, \ D = \texttt{not})$$
$$= P(A = \texttt{exceed}) \ P(B = \texttt{not}) \ P(C = \texttt{not}) \ P(D = \texttt{not})$$
$$= (0.3)(0.7)(0.7)(0.7)$$
$$= (0.7)^3(0.3)^1$$
$$= 0.103$$

But there are three other scenarios: Brittany, Carlton, or Damian could have been the one to exceed the deductible. In each of these cases, the probability is again $(0.7)^3(0.3)^1$. These four scenarios exhaust all the possible ways that exactly one of these four people could have exceeded the deductible, so the total probability is $4 \times (0.7)^3(0.3)^1 = 0.412$.

GUIDED PRACTICE 4.29

Verify that the scenario where Brittany is the only one exceed the deductible has probability $(0.7)^3(0.3)^1$. [20]

The scenario outlined in Example 4.28 is an example of a binomial distribution scenario. The **binomial distribution** describes the probability of having exactly k successes in n independent Bernoulli trials with probability of a success p (in Example 4.28, $n = 4$, $k = 3$, $p = 0.7$). We would like to determine the probabilities associated with the binomial distribution more generally, i.e. we want a formula where we can use n, k, and p to obtain the probability. To do this, we reexamine each part of Example 4.28.

There were four individuals who could have been the one to exceed the deductible, and each of these four scenarios had the same probability. Thus, we could identify the final probability as

$$[\# \text{ of scenarios}] \times P(\text{single scenario})$$

The first component of this equation is the number of ways to arrange the $k = 3$ successes among the $n = 4$ trials. The second component is the probability of any of the four (equally probable) scenarios.

---

[20] $P(A = \texttt{not}, \ B = \texttt{exceed}, \ C = \texttt{not}, \ D = \texttt{not}) = (0.7)(0.3)(0.7)(0.7) = (0.7)^3(0.3)^1$.

Consider P(single scenario) under the general case of k successes and $n - k$ failures in the n trials. In any such scenario, we apply the Multiplication Rule for independent events:

$$p^k(1-p)^{n-k}$$

This is our general formula for P(single scenario).

Secondly, we introduce a general formula for the number of ways to choose k successes in n trials, i.e. arrange k successes and $n - k$ failures:

$$\binom{n}{k} = \frac{n!}{k!(n-k)!}$$

The quantity $\binom{n}{k}$ is read **n choose k**.[21] The exclamation point notation (e.g. $k!$) denotes a **factorial** expression.

$$0! = 1$$
$$1! = 1$$
$$2! = 2 \times 1 = 2$$
$$3! = 3 \times 2 \times 1 = 6$$
$$4! = 4 \times 3 \times 2 \times 1 = 24$$
$$\vdots$$
$$n! = n \times (n-1) \times \ldots \times 3 \times 2 \times 1$$

Using the formula, we can compute the number of ways to choose $k = 3$ successes in $n = 4$ trials:

$$\binom{4}{3} = \frac{4!}{3!(4-3)!} = \frac{4!}{3!1!} = \frac{4 \times 3 \times 2 \times 1}{(3 \times 2 \times 1)(1)} = 4$$

This result is exactly what we found by carefully thinking of each possible scenario in Example 4.28.

Substituting n choose k for the number of scenarios and $p^k(1-p)^{n-k}$ for the single scenario probability yields the general binomial formula.

BINOMIAL DISTRIBUTION

Suppose the probability of a single trial being a success is p. Then the probability of observing exactly k successes in n independent trials is given by

$$\binom{n}{k}p^k(1-p)^{n-k} = \frac{n!}{k!(n-k)!}p^k(1-p)^{n-k}$$

The mean, variance, and standard deviation of the number of observed successes are

$$\mu = np \qquad\qquad \sigma^2 = np(1-p) \qquad\qquad \sigma = \sqrt{np(1-p)}$$

IS IT BINOMIAL? FOUR CONDITIONS TO CHECK.

(1) The trials are independent.
(2) The number of trials, n, is fixed.
(3) Each trial outcome can be classified as a *success* or *failure*.
(4) The probability of a success, p, is the same for each trial.

---

[21]Other notation for n choose k includes ${}_nC_k$, C_n^k, and $C(n, k)$.

EXAMPLE 4.30

What is the probability that 3 of 8 randomly selected individuals will have exceeded the insurance deductible, i.e. that 5 of 8 will not exceed the deductible? Recall that 70% of individuals will not exceed the deductible.

We would like to apply the binomial model, so we check the conditions. The number of trials is fixed ($n = 8$) (condition 2) and each trial outcome can be classified as a success or failure (condition 3). Because the sample is random, the trials are independent (condition 1) and the probability of a success is the same for each trial (condition 4).

In the outcome of interest, there are $k = 5$ successes in $n = 8$ trials (recall that a success is an individual who does *not* exceed the deductible, and the probability of a success is $p = 0.7$. So the probability that 5 of 8 will not exceed the deductible and 3 will exceed the deductible is given by

$$\binom{8}{5}(0.7)^5(1 - 0.7)^{8-5} = \frac{8!}{5!(8-5)!}(0.7)^5(1 - 0.7)^{8-5}$$

$$= \frac{8!}{5!3!}(0.7)^5(0.3)^3$$

Dealing with the factorial part:

$$\frac{8!}{5!3!} = \frac{8 \times 7 \times 6 \times 5 \times 4 \times 3 \times 2 \times 1}{(5 \times 4 \times 3 \times 2 \times 1)(3 \times 2 \times 1)} = \frac{8 \times 7 \times 6}{3 \times 2 \times 1} = 56$$

Using $(0.7)^5(0.3)^3 \approx 0.00454$, the final probability is about $56 \times 0.00454 \approx 0.254$.

COMPUTING BINOMIAL PROBABILITIES

The first step in using the binomial model is to check that the model is appropriate. The second step is to identify n, p, and k. As the last stage use software or the formulas to determine the probability, then interpret the results.

If you must do calculations by hand, it's often useful to cancel out as many terms as possible in the top and bottom of the binomial coefficient.

GUIDED PRACTICE 4.31

If we randomly sampled 40 case files from the insurance agency discussed earlier, how many of the cases would you expect to not have exceeded the deductible in a given year? What is the standard deviation of the number that would not have exceeded the deductible?[22]

GUIDED PRACTICE 4.32

The probability that a random smoker will develop a severe lung condition in his or her lifetime is about 0.3. If you have 4 friends who smoke, are the conditions for the binomial model satisfied?[23]

---

[22]We are asked to determine the expected number (the mean) and the standard deviation, both of which can be directly computed from the formulas: $\mu = np = 40 \times 0.7 = 28$ and $\sigma = \sqrt{np(1-p)} = \sqrt{40 \times 0.7 \times 0.3} = 2.9$. Because very roughly 95% of observations fall within 2 standard deviations of the mean (see Section 2.1.4), we would probably observe at least 22 but fewer than 34 individuals in our sample who would not exceed the deductible.

[23]One possible answer: if the friends know each other, then the independence assumption is probably not satisfied. For example, acquaintances may have similar smoking habits, or those friends might make a pact to quit together.

GUIDED PRACTICE 4.33

Suppose these four friends do not know each other and we can treat them as if they were a random sample from the population. Is the binomial model appropriate? What is the probability that[24]

(a) None of them will develop a severe lung condition?

(b) One will develop a severe lung condition?

(c) That no more than one will develop a severe lung condition?

GUIDED PRACTICE 4.34

What is the probability that at least 2 of your 4 smoking friends will develop a severe lung condition in their lifetimes?[25]

GUIDED PRACTICE 4.35

Suppose you have 7 friends who are smokers and they can be treated as a random sample of smokers.[26]

(a) How many would you expect to develop a severe lung condition, i.e. what is the mean?

(b) What is the probability that at most 2 of your 7 friends will develop a severe lung condition.

Next we consider the first term in the binomial probability, n choose k under some special scenarios.

GUIDED PRACTICE 4.36

Why is it true that $\binom{n}{0} = 1$ and $\binom{n}{n} = 1$ for any number n?[27]

GUIDED PRACTICE 4.37

How many ways can you arrange one success and $n - 1$ failures in n trials? How many ways can you arrange $n - 1$ successes and one failure in n trials?[28]

---

[24]To check if the binomial model is appropriate, we must verify the conditions. (i) Since we are supposing we can treat the friends as a random sample, they are independent. (ii) We have a fixed number of trials ($n = 4$). (iii) Each outcome is a success or failure. (iv) The probability of a success is the same for each trials since the individuals are like a random sample ($p = 0.3$ if we say a "success" is someone getting a lung condition, a morbid choice). Compute parts (a) and (b) using the binomial formula: $P(0) = \binom{4}{0}(0.3)^0(0.7)^4 = 1 \times 1 \times 0.7^4 = 0.2401$, $P(1) = \binom{4}{1}(0.3)^1(0.7)^3 = 0.4116$. Note: $0! = 1$. Part (c) can be computed as the sum of parts (a) and (b): $P(0) + P(1) = 0.2401 + 0.4116 = 0.6517$. That is, there is about a 65% chance that no more than one of your four smoking friends will develop a severe lung condition.

[25]The complement (no more than one will develop a severe lung condition) as computed in Guided Practice 4.33 as 0.6517, so we compute one minus this value: 0.3483.

[26](a) $\mu = 0.3 \times 7 = 2.1$. (b) $P(0, 1, \text{or } 2 \text{ develop severe lung condition}) = P(k = 0) + P(k = 1) + P(k = 2) = 0.6471$.

[27]Frame these expressions into words. How many different ways are there to arrange 0 successes and n failures in n trials? (1 way.) How many different ways are there to arrange n successes and 0 failures in n trials? (1 way.)

[28]One success and $n - 1$ failures: there are exactly n unique places we can put the success, so there are n ways to arrange one success and $n - 1$ failures. A similar argument is used for the second question. Mathematically, we show these results by verifying the following two equations:

$$\binom{n}{1} = n, \qquad \binom{n}{n-1} = n$$

4.3.2 Normal approximation to the binomial distribution

The binomial formula is cumbersome when the sample size (n) is large, particularly when we consider a range of observations. In some cases we may use the normal distribution as an easier and faster way to estimate binomial probabilities.

EXAMPLE 4.38

Approximately 15% of the US population smokes cigarettes. A local government believed their community had a lower smoker rate and commissioned a survey of 400 randomly selected individuals. The survey found that only 42 of the 400 participants smoke cigarettes. If the true proportion of smokers in the community was really 15%, what is the probability of observing 42 or fewer smokers in a sample of 400 people?

We leave the usual verification that the four conditions for the binomial model are valid as an exercise.

The question posed is equivalent to asking, what is the probability of observing $k = 0$, 1, 2, ..., or 42 smokers in a sample of $n = 400$ when $p = 0.15$? We can compute these 43 different probabilities and add them together to find the answer:

$$P(k = 0 \text{ or } k = 1 \text{ or } \cdots \text{ or } k = 42)$$
$$= P(k = 0) + P(k = 1) + \cdots + P(k = 42)$$
$$= 0.0054$$

If the true proportion of smokers in the community is $p = 0.15$, then the probability of observing 42 or fewer smokers in a sample of $n = 400$ is 0.0054.

The computations in Example 4.38 are tedious and long. In general, we should avoid such work if an alternative method exists that is faster, easier, and still accurate. Recall that calculating probabilities of a range of values is much easier in the normal model. We might wonder, is it reasonable to use the normal model in place of the binomial distribution? Surprisingly, yes, if certain conditions are met.

GUIDED PRACTICE 4.39

Here we consider the binomial model when the probability of a success is $p = 0.10$. Figure 4.9 shows four hollow histograms for simulated samples from the binomial distribution using four different sample sizes: $n = 10, 30, 100, 300$. What happens to the shape of the distributions as the sample size increases? What distribution does the last hollow histogram resemble?[29]

NORMAL APPROXIMATION OF THE BINOMIAL DISTRIBUTION

The binomial distribution with probability of success p is nearly normal when the sample size n is sufficiently large that np and $n(1 - p)$ are both at least 10. The approximate normal distribution has parameters corresponding to the mean and standard deviation of the binomial distribution:

$$\mu = np \qquad\qquad \sigma = \sqrt{np(1 - p)}$$

The normal approximation may be used when computing the range of many possible successes. For instance, we may apply the normal distribution to the setting of Example 4.38.

---

[29]The distribution is transformed from a blocky and skewed distribution into one that rather resembles the normal distribution in last hollow histogram.

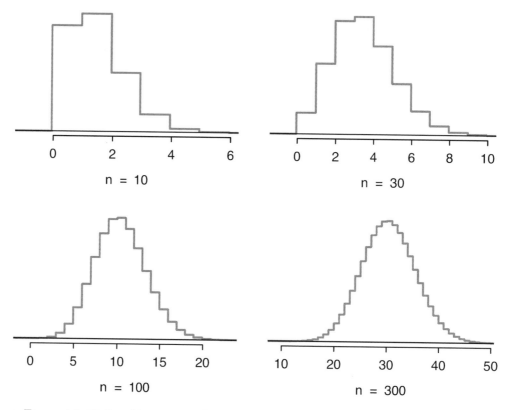

Figure 4.9: Hollow histograms of samples from the binomial model when $p = 0.10$. The sample sizes for the four plots are $n = 10$, 30, 100, and 300, respectively.

EXAMPLE 4.40

How can we use the normal approximation to estimate the probability of observing 42 or fewer smokers in a sample of 400, if the true proportion of smokers is $p = 0.15$?

Showing that the binomial model is reasonable was a suggested exercise in Example 4.38. We also verify that both np and $n(1 - p)$ are at least 10:

$$np = 400 \times 0.15 = 60 \qquad\qquad n(1 - p) = 400 \times 0.85 = 340$$

With these conditions checked, we may use the normal approximation in place of the binomial distribution using the mean and standard deviation from the binomial model:

$$\mu = np = 60 \qquad\qquad \sigma = \sqrt{np(1 - p)} = 7.14$$

We want to find the probability of observing 42 or fewer smokers using this model.

GUIDED PRACTICE 4.41

Use the normal model $N(\mu = 60, \sigma = 7.14)$ to estimate the probability of observing 42 or fewer smokers. Your answer should be approximately equal to the solution of Example 4.38: 0.0054. [30]

---

[30]Compute the Z-score first: $Z = \frac{42-60}{7.14} = -2.52$. The corresponding left tail area is 0.0059.

4.3.3 The normal approximation breaks down on small intervals

The normal approximation to the binomial distribution tends to perform poorly when estimating the probability of a small range of counts, even when the conditions are met.

Suppose we wanted to compute the probability of observing 49, 50, or 51 smokers in 400 when $p = 0.15$. With such a large sample, we might be tempted to apply the normal approximation and use the range 49 to 51. However, we would find that the binomial solution and the normal approximation notably differ:

Binomial: 0.0649   Normal: 0.0421

We can identify the cause of this discrepancy using Figure 4.10, which shows the areas representing the binomial probability (outlined) and normal approximation (shaded). Notice that the width of the area under the normal distribution is 0.5 units too slim on both sides of the interval.

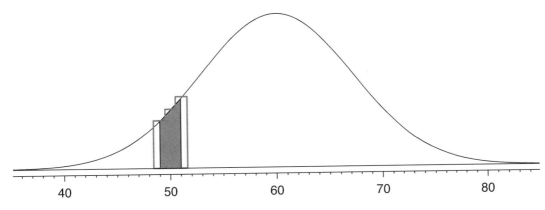

Figure 4.10: A normal curve with the area between 49 and 51 shaded. The outlined area represents the exact binomial probability.

IMPROVING THE NORMAL APPROXIMATION FOR THE BINOMIAL DISTRIBUTION

The normal approximation to the binomial distribution for intervals of values is usually improved if cutoff values are modified slightly. The cutoff values for the lower end of a shaded region should be reduced by 0.5, and the cutoff value for the upper end should be increased by 0.5.

The tip to add extra area when applying the normal approximation is most often useful when examining a range of observations. In the example above, the revised normal distribution estimate is 0.0633, much closer to the exact value of 0.0649. While it is possible to also apply this correction when computing a tail area, the benefit of the modification usually disappears since the total interval is typically quite wide.

Exercises

4.17 Underage drinking, Part I. Data collected by the Substance Abuse and Mental Health Services Administration (SAMSHA) suggests that 69.7% of 18-20 year olds consumed alcoholic beverages in any given year.[31]

(a) Suppose a random sample of ten 18-20 year olds is taken. Is the use of the binomial distribution appropriate for calculating the probability that exactly six consumed alcoholic beverages? Explain.

(b) Calculate the probability that exactly 6 out of 10 randomly sampled 18- 20 year olds consumed an alcoholic drink.

(c) What is the probability that exactly four out of ten 18-20 year olds have *not* consumed an alcoholic beverage?

(d) What is the probability that at most 2 out of 5 randomly sampled 18-20 year olds have consumed alcoholic beverages?

(e) What is the probability that at least 1 out of 5 randomly sampled 18-20 year olds have consumed alcoholic beverages?

4.18 Chicken pox, Part I. The National Vaccine Information Center estimates that 90% of Americans have had chickenpox by the time they reach adulthood.[32]

(a) Suppose we take a random sample of 100 American adults. Is the use of the binomial distribution appropriate for calculating the probability that exactly 97 out of 100 randomly sampled American adults had chickenpox during childhood? Explain.

(b) Calculate the probability that exactly 97 out of 100 randomly sampled American adults had chickenpox during childhood.

(c) What is the probability that exactly 3 out of a new sample of 100 American adults have *not* had chickenpox in their childhood?

(d) What is the probability that at least 1 out of 10 randomly sampled American adults have had chickenpox?

(e) What is the probability that at most 3 out of 10 randomly sampled American adults have *not* had chickenpox?

4.19 Underage drinking, Part II. We learned in Exercise 4.17 that about 70% of 18-20 year olds consumed alcoholic beverages in any given year. We now consider a random sample of fifty 18-20 year olds.

(a) How many people would you expect to have consumed alcoholic beverages? And with what standard deviation?

(b) Would you be surprised if there were 45 or more people who have consumed alcoholic beverages?

(c) What is the probability that 45 or more people in this sample have consumed alcoholic beverages? How does this probability relate to your answer to part (b)?

4.20 Chickenpox, Part II. We learned in Exercise 4.18 that about 90% of American adults had chickenpox before adulthood. We now consider a random sample of 120 American adults.

(a) How many people in this sample would you expect to have had chickenpox in their childhood? And with what standard deviation?

(b) Would you be surprised if there were 105 people who have had chickenpox in their childhood?

(c) What is the probability that 105 or fewer people in this sample have had chickenpox in their childhood? How does this probability relate to your answer to part (b)?

4.21 Game of dreidel. A dreidel is a four-sided spinning top with the Hebrew letters *nun, gimel, hei*, and *shin*, one on each side. Each side is equally likely to come up in a single spin of the dreidel. Suppose you spin a dreidel three times. Calculate the probability of getting

(a) at least one *nun*?

(b) exactly 2 *nuns*?

(c) exactly 1 *hei*?

(d) at most 2 *gimels*?

Photo by Staccabees, cropped (http://flic.kr/p/7gLZTf) CC BY 2.0 license

[31]SAMHSA, Office of Applied Studies, National Survey on Drug Use and Health, 2007 and 2008.
[32]National Vaccine Information Center, Chickenpox, The Disease & The Vaccine Fact Sheet.

4.22 Arachnophobia. A Gallup Poll found that 7% of teenagers (ages 13 to 17) suffer from arachnophobia and are extremely afraid of spiders. At a summer camp there are 10 teenagers sleeping in each tent. Assume that these 10 teenagers are independent of each other.[33]

(a) Calculate the probability that at least one of them suffers from arachnophobia.

(b) Calculate the probability that exactly 2 of them suffer from arachnophobia.

(c) Calculate the probability that at most 1 of them suffers from arachnophobia.

(d) If the camp counselor wants to make sure no more than 1 teenager in each tent is afraid of spiders, does it seem reasonable for him to randomly assign teenagers to tents?

4.23 Eye color, Part II. Exercise 4.13 introduces a husband and wife with brown eyes who have 0.75 probability of having children with brown eyes, 0.125 probability of having children with blue eyes, and 0.125 probability of having children with green eyes.

(a) What is the probability that their first child will have green eyes and the second will not?

(b) What is the probability that exactly one of their two children will have green eyes?

(c) If they have six children, what is the probability that exactly two will have green eyes?

(d) If they have six children, what is the probability that at least one will have green eyes?

(e) What is the probability that the first green eyed child will be the 4th child?

(f) Would it be considered unusual if only 2 out of their 6 children had brown eyes?

4.24 Sickle cell anemia. Sickle cell anemia is a genetic blood disorder where red blood cells lose their flexibility and assume an abnormal, rigid, "sickle" shape, which results in a risk of various complications. If both parents are carriers of the disease, then a child has a 25% chance of having the disease, 50% chance of being a carrier, and 25% chance of neither having the disease nor being a carrier. If two parents who are carriers of the disease have 3 children, what is the probability that

(a) two will have the disease?

(b) none will have the disease?

(c) at least one will neither have the disease nor be a carrier?

(d) the first child with the disease will the be 3rd child?

4.25 Exploring permutations. The formula for the number of ways to arrange n objects is $n! = n \times (n - 1) \times \cdots \times 2 \times 1$. This exercise walks you through the derivation of this formula for a couple of special cases.

A small company has five employees: Anna, Ben, Carl, Damian, and Eddy. There are five parking spots in a row at the company, none of which are assigned, and each day the employees pull into a random parking spot. That is, all possible orderings of the cars in the row of spots are equally likely.

(a) On a given day, what is the probability that the employees park in alphabetical order?

(b) If the alphabetical order has an equal chance of occurring relative to all other possible orderings, how many ways must there be to arrange the five cars?

(c) Now consider a sample of 8 employees instead. How many possible ways are there to order these 8 employees' cars?

4.26 Male children. While it is often assumed that the probabilities of having a boy or a girl are the same, the actual probability of having a boy is slightly higher at 0.51. Suppose a couple plans to have 3 kids.

(a) Use the binomial model to calculate the probability that two of them will be boys.

(b) Write out all possible orderings of 3 children, 2 of whom are boys. Use these scenarios to calculate the same probability from part (a) but using the addition rule for disjoint outcomes. Confirm that your answers from parts (a) and (b) match.

(c) If we wanted to calculate the probability that a couple who plans to have 8 kids will have 3 boys, briefly describe why the approach from part (b) would be more tedious than the approach from part (a).

---

[33]Gallup Poll, What Frightens America's Youth?, March 29, 2005.

4.4   Negative binomial distribution

The geometric distribution describes the probability of observing the first success on the nth trial. The **negative binomial distribution** is more general: it describes the probability of observing the kth success on the nth trial.

EXAMPLE 4.42

Each day a high school football coach tells his star kicker, Brian, that he can go home after he successfully kicks four 35 yard field goals. Suppose we say each kick has a probability p of being successful. If p is small – e.g. close to 0.1 – would we expect Brian to need many attempts before he successfully kicks his fourth field goal?

We are waiting for the fourth success ($k = 4$). If the probability of a success (p) is small, then the number of attempts (n) will probably be large. This means that Brian is more likely to need many attempts before he gets $k = 4$ successes. To put this another way, the probability of n being small is low.

To identify a negative binomial case, we check 4 conditions. The first three are common to the binomial distribution.

IS IT NEGATIVE BINOMIAL? FOUR CONDITIONS TO CHECK
(1) The trials are independent.
(2) Each trial outcome can be classified as a success or failure.
(3) The probability of a success (p) is the same for each trial.
(4) The last trial must be a success.

GUIDED PRACTICE 4.43

Suppose Brian is very diligent in his attempts and he makes each 35 yard field goal with probability $p = 0.8$. Take a guess at how many attempts he would need before making his fourth kick.[34]

EXAMPLE 4.44

In yesterday's practice, it took Brian only 6 tries to get his fourth field goal. Write out each of the possible sequence of kicks.

Because it took Brian six tries to get the fourth success, we know the last kick must have been a success. That leaves three successful kicks and two unsuccessful kicks (we label these as failures) that make up the first five attempts. There are ten possible sequences of these first five kicks, which are shown in Figure 4.11. If Brian achieved his fourth success ($k = 4$) on his sixth attempt ($n = 6$), then his order of successes and failures must be one of these ten possible sequences.

GUIDED PRACTICE 4.45

Each sequence in Figure 4.11 has exactly two failures and four successes with the last attempt always being a success. If the probability of a success is $p = 0.8$, find the probability of the first sequence.[35]

---

[34]One possible answer: since he is likely to make each field goal attempt, it will take him at least 4 attempts but probably not more than 6 or 7.
[35]The first sequence: $0.2 \times 0.2 \times 0.8 \times 0.8 \times 0.8 \times 0.8 = 0.0164$.

			Kick Attempt			
	1	2	3	4	5	6
1	F	F	S^1	S^2	S^3	S^4
2	F	S^1	F	S^2	S^3	S^4
3	F	S^1	S^2	F	S^3	S^4
4	F	S^1	S^2	S^3	F	S^4
5	S^1	F	F	S^2	S^3	S^4
6	S^1	F	S^2	F	S^3	S^4
7	S^1	F	S^2	S^3	F	S^4
8	S^1	S^2	F	F	S^3	S^4
9	S^1	S^2	F	S^3	F	S^4
10	S^1	S^2	S^3	F	F	S^4

Figure 4.11: The ten possible sequences when the fourth successful kick is on the sixth attempt.

If the probability Brian kicks a 35 yard field goal is $p = 0.8$, what is the probability it takes Brian exactly six tries to get his fourth successful kick? We can write this as

P(it takes Brian six tries to make four field goals)

$= P$(Brian makes three of his first five field goals, and he makes the sixth one)

$= P(1^{st}$ sequence OR 2nd sequence OR ... OR 10th sequence)

where the sequences are from Figure 4.11. We can break down this last probability into the sum of ten disjoint possibilities:

$P(1^{st}$ sequence OR 2nd sequence OR ... OR 10th sequence)

$= P(1^{st}$ sequence$) + P(2^{nd}$ sequence$) + \cdots + P(10^{th}$ sequence$)$

The probability of the first sequence was identified in Guided Practice 4.45 as 0.0164, and each of the other sequences have the same probability. Since each of the ten sequence has the same probability, the total probability is ten times that of any individual sequence.

The way to compute this negative binomial probability is similar to how the binomial problems were solved in Section 4.3. The probability is broken into two pieces:

P(it takes Brian six tries to make four field goals)

$= $ [Number of possible sequences] $\times P$(Single sequence)

Each part is examined separately, then we multiply to get the final result.

We first identify the probability of a single sequence. One particular case is to first observe all the failures ($n - k$ of them) followed by the k successes:

P(Single sequence)

$= P(n - k$ failures and then k successes)

$= (1 - p)^{n-k} p^k$

We must also identify the number of sequences for the general case. Above, ten sequences were identified where the fourth success came on the sixth attempt. These sequences were identified by fixing the last observation as a success and looking for all the ways to arrange the other observations. In other words, how many ways could we arrange $k-1$ successes in $n-1$ trials? This can be found using the n choose k coefficient but for $n-1$ and $k-1$ instead:

$$\binom{n-1}{k-1} = \frac{(n-1)!}{(k-1)!\,((n-1)-(k-1))!} = \frac{(n-1)!}{(k-1)!\,(n-k)!}$$

This is the number of different ways we can order $k-1$ successes and $n-k$ failures in $n-1$ trials. If the factorial notation (the exclamation point) is unfamiliar, see page 150.

NEGATIVE BINOMIAL DISTRIBUTION

The negative binomial distribution describes the probability of observing the kth success on the nth trial, where all trials are independent:

$$P(\text{the } k^{th} \text{ success on the } n^{th} \text{ trial}) = \binom{n-1}{k-1} p^k (1-p)^{n-k}$$

The value p represents the probability that an individual trial is a success.

EXAMPLE 4.46

Show using the formula for the negative binomial distribution that the probability Brian kicks his fourth successful field goal on the sixth attempt is 0.164.

The probability of a single success is $p = 0.8$, the number of successes is $k = 4$, and the number of necessary attempts under this scenario is $n = 6$.

$$\binom{n-1}{k-1} p^k (1-p)^{n-k} = \frac{5!}{3!2!}(0.8)^4(0.2)^2 = 10 \times 0.0164 = 0.164$$

GUIDED PRACTICE 4.47

The negative binomial distribution requires that each kick attempt by Brian is independent. Do you think it is reasonable to suggest that each of Brian's kick attempts are independent?[36]

GUIDED PRACTICE 4.48

Assume Brian's kick attempts are independent. What is the probability that Brian will kick his fourth field goal within 5 attempts?[37]

---

[36]Answers may vary. We cannot conclusively say they are or are not independent. However, many statistical reviews of athletic performance suggests such attempts are very nearly independent.

[37]If his fourth field goal ($k = 4$) is within five attempts, it either took him four or five tries ($n = 4$ or $n = 5$). We have $p = 0.8$ from earlier. Use the negative binomial distribution to compute the probability of $n = 4$ tries and $n = 5$ tries, then add those probabilities together:

$$P(n = 4 \text{ OR } n = 5) = P(n = 4) + P(n = 5)$$
$$= \binom{4-1}{4-1}0.8^4 + \binom{5-1}{4-1}(0.8)^4(1-0.8) = 1 \times 0.41 + 4 \times 0.082 = 0.41 + 0.33 = 0.74$$

BINOMIAL VERSUS NEGATIVE BINOMIAL

In the binomial case, we typically have a fixed number of trials and instead consider the number of successes. In the negative binomial case, we examine how many trials it takes to observe a fixed number of successes and require that the last observation be a success.

GUIDED PRACTICE 4.49

On 70% of days, a hospital admits at least one heart attack patient. On 30% of the days, no heart attack patients are admitted. Identify each case below as a binomial or negative binomial case, and compute the probability.[38]

(a) What is the probability the hospital will admit a heart attack patient on exactly three days this week?

(b) What is the probability the second day with a heart attack patient will be the fourth day of the week?

(c) What is the probability the fifth day of next month will be the first day with a heart attack patient?

---

[38]In each part, $p = 0.7$. (a) The number of days is fixed, so this is binomial. The parameters are $k = 3$ and $n = 7$: 0.097. (b) The last "success" (admitting a heart attack patient) is fixed to the last day, so we should apply the negative binomial distribution. The parameters are $k = 2$, $n = 4$: 0.132. (c) This problem is negative binomial with $k = 1$ and $n = 5$: 0.006. Note that the negative binomial case when $k = 1$ is the same as using the geometric distribution.

Exercises

4.27 Rolling a die. Calculate the following probabilities and indicate which probability distribution model is appropriate in each case. You roll a fair die 5 times. What is the probability of rolling

(a) the first 6 on the fifth roll?

(b) exactly three 6s?

(c) the third 6 on the fifth roll?

4.28 Playing darts. Calculate the following probabilities and indicate which probability distribution model is appropriate in each case. A very good darts player can hit the bull's eye (red circle in the center of the dart board) 65% of the time. What is the probability that he

(a) hits the bullseye for the 10th time on the 15th try?

(b) hits the bullseye 10 times in 15 tries?

(c) hits the first bullseye on the third try?

4.29 Sampling at school. For a sociology class project you are asked to conduct a survey on 20 students at your school. You decide to stand outside of your dorm's cafeteria and conduct the survey on a random sample of 20 students leaving the cafeteria after dinner one evening. Your dorm is comprised of 45% males and 55% females.

(a) Which probability model is most appropriate for calculating the probability that the 4th person you survey is the 2nd female? Explain.

(b) Compute the probability from part (a).

(c) The three possible scenarios that lead to 4th person you survey being the 2nd female are

$$\{M, M, F, F\}, \{M, F, M, F\}, \{F, M, M, F\}$$

One common feature among these scenarios is that the last trial is always female. In the first three trials there are 2 males and 1 female. Use the binomial coefficient to confirm that there are 3 ways of ordering 2 males and 1 female.

(d) Use the findings presented in part (c) to explain why the formula for the coefficient for the negative binomial is $\binom{n-1}{k-1}$ while the formula for the binomial coefficient is $\binom{n}{k}$.

4.30 Serving in volleyball. A not-so-skilled volleyball player has a 15% chance of making the serve, which involves hitting the ball so it passes over the net on a trajectory such that it will land in the opposing team's court. Suppose that her serves are independent of each other.

(a) What is the probability that on the 10th try she will make her 3rd successful serve?

(b) Suppose she has made two successful serves in nine attempts. What is the probability that her 10th serve will be successful?

(c) Even though parts (a) and (b) discuss the same scenario, the probabilities you calculated should be different. Can you explain the reason for this discrepancy?

4.5 Poisson distribution

EXAMPLE 4.50

There are about 8 million individuals in New York City. How many individuals might we expect to be hospitalized for acute myocardial infarction (AMI), i.e. a heart attack, each day? According to historical records, the average number is about 4.4 individuals. However, we would also like to know the approximate distribution of counts. What would a histogram of the number of AMI occurrences each day look like if we recorded the daily counts over an entire year?

A histogram of the number of occurrences of AMI on 365 days for NYC is shown in Figure 4.12.[39] The sample mean (4.38) is similar to the historical average of 4.4. The sample standard deviation is about 2, and the histogram indicates that about 70% of the data fall between 2.4 and 6.4. The distribution's shape is unimodal and skewed to the right.

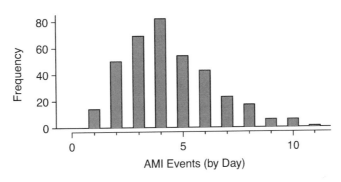

Figure 4.12: A histogram of the number of occurrences of AMI on 365 separate days in NYC.

The **Poisson distribution** is often useful for estimating the number of events in a large population over a unit of time. For instance, consider each of the following events:

- having a heart attack,
- getting married, and
- getting struck by lightning.

The Poisson distribution helps us describe the number of such events that will occur in a day for a fixed population if the individuals within the population are independent. The Poisson distribution could also be used over another unit of time, such as an hour or a week.

The histogram in Figure 4.12 approximates a Poisson distribution with rate equal to 4.4. The **rate** for a Poisson distribution is the average number of occurrences in a mostly-fixed population per unit of time. In Example 4.50, the time unit is a day, the population is all New York City residents, and the historical rate is 4.4. The parameter in the Poisson distribution is the rate – or how many events we expect to observe – and it is typically denoted by λ (the Greek letter *lambda*) or μ. Using the rate, we can describe the probability of observing exactly k events in a single unit of time.

---

[39]These data are simulated. In practice, we should check for an association between successive days.

POISSON DISTRIBUTION

Suppose we are watching for events and the number of observed events follows a Poisson distribution with rate λ. Then

$$P(\text{observe } k \text{ events}) = \frac{\lambda^k e^{-\lambda}}{k!}$$

where k may take a value 0, 1, 2, and so on, and $k!$ represents k-factorial, as described on page 150. The letter $e \approx 2.718$ is the base of the natural logarithm. The mean and standard deviation of this distribution are λ and $\sqrt{\lambda}$, respectively.

We will leave a rigorous set of conditions for the Poisson distribution to a later course. However, we offer a few simple guidelines that can be used for an initial evaluation of whether the Poisson model would be appropriate.

A random variable may follow a Poisson distribution if we are looking for the number of events, the population that generates such events is large, and the events occur independently of each other.

Even when events are not really independent – for instance, Saturdays and Sundays are especially popular for weddings – a Poisson model may sometimes still be reasonable if we allow it to have a different rate for different times. In the wedding example, the rate would be modeled as higher on weekends than on weekdays. The idea of modeling rates for a Poisson distribution against a second variable such as the day of week forms the foundation of some more advanced methods that fall in the realm of **generalized linear models**. In Chapters 8 and 9, we will discuss a foundation of linear models.

Exercises

4.31 Customers at a coffee shop. A coffee shop serves an average of 75 customers per hour during the morning rush.

(a) Which distribution have we studied that is most appropriate for calculating the probability of a given number of customers arriving within one hour during this time of day?

(b) What are the mean and the standard deviation of the number of customers this coffee shop serves in one hour during this time of day?

(c) Would it be considered unusually low if only 60 customers showed up to this coffee shop in one hour during this time of day?

(d) Calculate the probability that this coffee shop serves 70 customers in one hour during this time of day.

4.32 Stenographer's typos. A very skilled court stenographer makes one typographical error (typo) per hour on average.

(a) What probability distribution is most appropriate for calculating the probability of a given number of typos this stenographer makes in an hour?

(b) What are the mean and the standard deviation of the number of typos this stenographer makes?

(c) Would it be considered unusual if this stenographer made 4 typos in a given hour?

(d) Calculate the probability that this stenographer makes at most 2 typos in a given hour.

4.33 How many cars show up? For Monday through Thursday when there isn't a holiday, the average number of vehicles that visit a particular retailer between 2pm and 3pm each afternoon is 6.5, and the number of cars that show up on any given day follows a Poisson distribution.

(a) What is the probability that exactly 5 cars will show up next Monday?

(b) What is the probability that 0, 1, or 2 cars will show up next Monday between 2pm and 3pm?

(c) There is an average of 11.7 people who visit during those same hours from vehicles. Is it likely that the number of people visiting by car during this hour is also Poisson? Explain.

4.34 Lost baggage. Occasionally an airline will lose a bag. Suppose a small airline has found it can reasonably model the number of bags lost each weekday using a Poisson model with a mean of 2.2 bags.

(a) What is the probability that the airline will lose no bags next Monday?

(b) What is the probability that the airline will lose 0, 1, or 2 bags on next Monday?

(c) Suppose the airline expands over the course of the next 3 years, doubling the number of flights it makes, and the CEO asks you if it's reasonable for them to continue using the Poisson model with a mean of 2.2. What is an appropriate recommendation? Explain.

Chapter exercises

4.35 Roulette winnings. In the game of roulette, a wheel is spun and you place bets on where it will stop. One popular bet is that it will stop on a red slot; such a bet has an 18/38 chance of winning. If it stops on red, you double the money you bet. If not, you lose the money you bet. Suppose you play 3 times, each time with a $1 bet. Let Y represent the total amount won or lost. Write a probability model for Y.

4.36 Speeding on the I-5, Part I. The distribution of passenger vehicle speeds traveling on the Interstate 5 Freeway (I-5) in California is nearly normal with a mean of 72.6 miles/hour and a standard deviation of 4.78 miles/hour.[40]

(a) What percent of passenger vehicles travel slower than 80 miles/hour?

(b) What percent of passenger vehicles travel between 60 and 80 miles/hour?

(c) How fast do the fastest 5% of passenger vehicles travel?

(d) The speed limit on this stretch of the I-5 is 70 miles/hour. Approximate what percentage of the passenger vehicles travel above the speed limit on this stretch of the I-5.

4.37 University admissions. Suppose a university announced that it admitted 2,500 students for the following year's freshman class. However, the university has dorm room spots for only 1,786 freshman students. If there is a 70% chance that an admitted student will decide to accept the offer and attend this university, what is the approximate probability that the university will not have enough dormitory room spots for the freshman class?

4.38 Speeding on the I-5, Part II. Exercise 4.36 states that the distribution of speeds of cars traveling on the Interstate 5 Freeway (I-5) in California is nearly normal with a mean of 72.6 miles/hour and a standard deviation of 4.78 miles/hour. The speed limit on this stretch of the I-5 is 70 miles/hour.

(a) A highway patrol officer is hidden on the side of the freeway. What is the probability that 5 cars pass and none are speeding? Assume that the speeds of the cars are independent of each other.

(b) On average, how many cars would the highway patrol officer expect to watch until the first car that is speeding? What is the standard deviation of the number of cars he would expect to watch?

4.39 Auto insurance premiums. Suppose a newspaper article states that the distribution of auto insurance premiums for residents of California is approximately normal with a mean of $1,650. The article also states that 25% of California residents pay more than $1,800.

(a) What is the Z-score that corresponds to the top 25% (or the 75th percentile) of the standard normal distribution?

(b) What is the mean insurance cost? What is the cutoff for the 75th percentile?

(c) Identify the standard deviation of insurance premiums in California.

4.40 SAT scores. SAT scores (out of 1600) are distributed normally with a mean of 1100 and a standard deviation of 200. Suppose a school council awards a certificate of excellence to all students who score at least 1350 on the SAT, and suppose we pick one of the recognized students at random. What is the probability this student's score will be at least 1500? (The material covered in Section 3.2 on conditional probability would be useful for this question.)

4.41 Married women. The American Community Survey estimates that 47.1% of women ages 15 years and over are married.[41]

(a) We randomly select three women between these ages. What is the probability that the third woman selected is the only one who is married?

(b) What is the probability that all three randomly selected women are married?

(c) On average, how many women would you expect to sample before selecting a married woman? What is the standard deviation?

(d) If the proportion of married women was actually 30%, how many women would you expect to sample before selecting a married woman? What is the standard deviation?

(e) Based on your answers to parts (c) and (d), how does decreasing the probability of an event affect the mean and standard deviation of the wait time until success?

---

[40]S. Johnson and D. Murray. "Empirical Analysis of Truck and Automobile Speeds on Rural Interstates: Impact of Posted Speed Limits". In: *Transportation Research Board 89th Annual Meeting.* 2010.
[41]U.S. Census Bureau, 2010 American Community Survey, Marital Status.

4.42 Survey response rate. Pew Research reported that the typical response rate to their surveys is only 9%. If for a particular survey 15,000 households are contacted, what is the probability that at least 1,500 will agree to respond?[42]

4.43 Overweight baggage. Suppose weights of the checked baggage of airline passengers follow a nearly normal distribution with mean 45 pounds and standard deviation 3.2 pounds. Most airlines charge a fee for baggage that weigh in excess of 50 pounds. Determine what percent of airline passengers incur this fee.

4.44 Heights of 10 year olds, Part I. Heights of 10 year olds, regardless of gender, closely follow a normal distribution with mean 55 inches and standard deviation 6 inches.

(a) What is the probability that a randomly chosen 10 year old is shorter than 48 inches?

(b) What is the probability that a randomly chosen 10 year old is between 60 and 65 inches?

(c) If the tallest 10% of the class is considered "very tall", what is the height cutoff for "very tall"?

4.45 Buying books on Ebay. Suppose you're considering buying your expensive chemistry textbook on Ebay. Looking at past auctions suggests that the prices of this textbook follow an approximately normal distribution with mean $89 and standard deviation $15.

(a) What is the probability that a randomly selected auction for this book closes at more than $100?

(b) Ebay allows you to set your maximum bid price so that if someone outbids you on an auction you can automatically outbid them, up to the maximum bid price you set. If you are only bidding on one auction, what are the advantages and disadvantages of setting a bid price too high or too low? What if you are bidding on multiple auctions?

(c) If you watched 10 auctions, roughly what percentile might you use for a maximum bid cutoff to be somewhat sure that you will win one of these ten auctions? Is it possible to find a cutoff point that will ensure that you win an auction?

(d) If you are willing to track up to ten auctions closely, about what price might you use as your maximum bid price if you want to be somewhat sure that you will buy one of these ten books?

4.46 Heights of 10 year olds, Part II. Heights of 10 year olds, regardless of gender, closely follow a normal distribution with mean 55 inches and standard deviation 6 inches.

(a) The height requirement for *Batman the Ride* at Six Flags Magic Mountain is 54 inches. What percent of 10 year olds cannot go on this ride?

(b) Suppose there are four 10 year olds. What is the chance that at least two of them will be able to ride *Batman the Ride*?

(c) Suppose you work at the park to help them better understand their customers' demographics, and you are counting people as they enter the park. What is the chance that the first 10 year old you see who can ride *Batman the Ride* is the 3rd 10 year old who enters the park?

(d) What is the chance that the fifth 10 year old you see who can ride *Batman the Ride* is the 12th 10 year old who enters the park?

4.47 Heights of 10 year olds, Part III. Heights of 10 year olds, regardless of gender, closely follow a normal distribution with mean 55 inches and standard deviation 6 inches.

(a) What fraction of 10 year olds are taller than 76 inches?

(b) If there are 2,000 10 year olds entering Six Flags Magic Mountain in a single day, then compute the expected number of 10 year olds who are at least 76 inches tall. (You may assume the heights of the 10-year olds are independent.)

(c) Using the binomial distribution, compute the probability that 0 of the 2,000 10 year olds will be at least 76 inches tall.

(d) The number of 10 year olds who enter Six Flags Magic Mountain and are at least 76 inches tall in a given day follows a Poisson distribution with mean equal to the value found in part (b). Use the Poisson distribution to identify the probability no 10 year old will enter the park who is 76 inches or taller.

4.48 Multiple choice quiz. In a multiple choice quiz there are 5 questions and 4 choices for each question (a, b, c, d). Robin has not studied for the quiz at all, and decides to randomly guess the answers. What is the probability that

(a) the first question she gets right is the 3rd question?

(b) she gets exactly 3 or exactly 4 questions right?

(c) she gets the majority of the questions right?

---

[42]Pew Research Center, Assessing the Representativeness of Public Opinion Surveys, May 15, 2012.

Chapter 5

Foundations for inference

Statistical inference is primarily concerned with understanding and quantifying the uncertainty of parameter estimates. While the equations and details change depending on the setting, the foundations for inference are the same throughout all of statistics.

We start with a familiar topic: the idea of using a sample proportion to estimate a population proportion. Next, we create what's called a *confidence interval*, which is a range of plausible values where we may find the true population value. Finally, we introduce the *hypothesis testing framework*, which allows us to formally evaluate claims about the population, such as whether a survey provides strong evidence that a candidate has the support of a majority of the voting population.

For videos, slides, and other resources, please visit
www.openintro.org/os

5.1   Point estimates and sampling variability

Companies such as Pew Research frequently conduct polls as a way to understand the state of public opinion or knowledge on many topics, including politics, scientific understanding, brand recognition, and more. The ultimate goal in taking a poll is generally to use the responses to estimate the opinion or knowledge of the broader population.

5.1.1   Point estimates and error

Suppose a poll suggested the US President's approval rating is 45%. We would consider 45% to be a **point estimate** of the approval rating we might see if we collected responses from the entire population. This entire-population response proportion is generally referred to as the **parameter** of interest. When the parameter is a proportion, it is often denoted by p, and we often refer to the sample proportion as $\hat{p}$ (pronounced p-hat[1]). Unless we collect responses from every individual in the population, p remains unknown, and we use $\hat{p}$ as our estimate of p. The difference we observe from the poll versus the parameter is called the **error** in the estimate. Generally, the error consists of two aspects: sampling error and bias.

Sampling error, sometimes called *sampling uncertainty*, describes how much an estimate will tend to vary from one sample to the next. For instance, the estimate from one sample might be 1% too low while in another it may be 3% too high. Much of statistics, including much of this book, is focused on understanding and quantifying sampling error, and we will find it useful to consider a sample's size to help us quantify this error; the **sample size** is often represented by the letter n.

Bias describes a systematic tendency to over- or under-estimate the true population value. For example, if we were taking a student poll asking about support for a new college stadium, we'd probably get a biased estimate of the stadium's level of student support by wording the question as, *Do you support your school by supporting funding for the new stadium?* We try to minimize bias through thoughtful data collection procedures, which were discussed in Chapter 1 and are the topic of many other books.

5.1.2   Understanding the variability of a point estimate

Suppose the proportion of American adults who support the expansion of solar energy is $p = 0.88$, which is our parameter of interest.[2] If we were to take a poll of 1000 American adults on this topic, the estimate would not be perfect, but how close might we expect the sample proportion in the poll would be to 88%? We want to understand, *how does the sample proportion $\hat{p}$ behave when the true population proportion is 0.88.*[3] Let's find out! We can simulate responses we would get from a simple random sample of 1000 American adults, which is only possible because we know the actual support for expanding solar energy is 0.88. Here's how we might go about constructing such a simulation:

1. There were about 250 million American adults in 2018. On 250 million pieces of paper, write "support" on 88% of them and "not" on the other 12%.

2. Mix up the pieces of paper and pull out 1000 pieces to represent our sample of 1000 American adults.

3. Compute the fraction of the sample that say "support".

Any volunteers to conduct this simulation? Probably not. Running this simulation with 250 million pieces of paper would be time-consuming and very costly, but we can simulate it using computer

---

[1]Not to be confused with *phat*, the slang term used for something cool, like this book.

[2]We haven't actually conducted a census to measure this value perfectly. However, a very large sample has suggested the actual level of support is about 88%.

[3]88% written as a proportion would be 0.88. It is common to switch between proportion and percent. However, formulas presented in this book always refer to the proportion, not the percent.

code; we've written a short program in Figure 5.1 in case you are curious what the computer code looks like. In this simulation, the sample gave a point estimate of $\hat{p}_1 = 0.894$. We know the population proportion for the simulation was $p = 0.88$, so we know the estimate had an error of $0.894 - 0.88 = +0.014$.

```
# 1. Create a set of 250 million entries, where 88% of them are "support"
#    and 12% are "not".
pop_size <- 250000000
possible_entries <- c(rep("support", 0.88 * pop_size), rep("not", 0.12 * pop_size))

# 2. Sample 1000 entries without replacement.
sampled_entries <- sample(possible_entries, size = 1000)

# 3. Compute p-hat: count the number that are "support", then divide by
#    the sample size.
sum(sampled_entries == "support") / 1000
```

Figure 5.1: For those curious, this is code for a single $\hat{p}$ simulation using the statistical software called **R**. Each line that starts with **#** is a **code comment**, which is used to describe in regular language what the code is doing. We've provided software labs in **R** at openintro.org/stat/labs for anyone interested in learning more.

One simulation isn't enough to get a great sense of the distribution of estimates we might expect in the simulation, so we should run more simulations. In a second simulation, we get $\hat{p}_2 = 0.885$, which has an error of $+0.005$. In another, $\hat{p}_3 = 0.878$ for an error of -0.002. And in another, an estimate of $\hat{p}_4 = 0.859$ with an error of -0.021. With the help of a computer, we've run the simulation 10,000 times and created a histogram of the results from all 10,000 simulations in Figure 5.2. This distribution of sample proportions is called a **sampling distribution**. We can characterize this sampling distribution as follows:

Center. The center of the distribution is $\bar{x}_{\hat{p}} = 0.880$, which is the same as the parameter. Notice that the simulation mimicked a simple random sample of the population, which is a straightforward sampling strategy that helps avoid sampling bias.

Spread. The standard deviation of the distribution is $s_{\hat{p}} = 0.010$. When we're talking about a sampling distribution or the variability of a point estimate, we typically use the term **standard error** rather than *standard deviation*, and the notation $SE_{\hat{p}}$ is used for the standard error associated with the sample proportion.

Shape. The distribution is symmetric and bell-shaped, and it *resembles a normal distribution.*

These findings are encouraging! When the population proportion is $p = 0.88$ and the sample size is $n = 1000$, the sample proportion $\hat{p}$ tends to give a pretty good estimate of the population proportion. We also have the interesting observation that the histogram resembles a normal distribution.

SAMPLING DISTRIBUTIONS ARE NEVER OBSERVED, BUT WE KEEP THEM IN MIND

In real-world applications, we never actually observe the sampling distribution, yet it is useful to always think of a point estimate as coming from such a hypothetical distribution. Understanding the sampling distribution will help us characterize and make sense of the point estimates that we do observe.

EXAMPLE 5.1

If we used a much smaller sample size of $n = 50$, would you guess that the standard error for $\hat{p}$ would be larger or smaller than when we used $n = 1000$?

Intuitively, it seems like more data is better than less data, and generally that is correct! The typical error when $p = 0.88$ and $n = 50$ would be larger than the error we would expect when $n = 1000$.

Example 5.1 highlights an important property we will see again and again: a bigger sample tends to provide a more precise point estimate than a smaller sample.

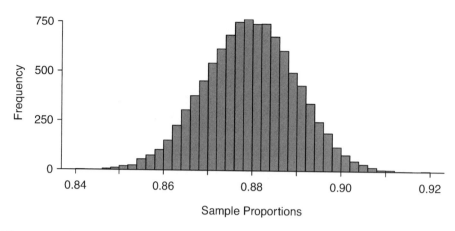

Figure 5.2: A histogram of 10,000 sample proportions, where each sample is taken from a population where the population proportion is 0.88 and the sample size is $n = 1000$.

5.1.3 Central Limit Theorem

The distribution in Figure 5.2 looks an awful lot like a normal distribution. That is no anomaly; it is the result of a general principle called the **Central Limit Theorem**.

CENTRAL LIMIT THEOREM AND THE SUCCESS-FAILURE CONDITION

When observations are independent and the sample size is sufficiently large, the sample proportion $\hat{p}$ will tend to follow a normal distribution with the following mean and standard error:

$$\mu_{\hat{p}} = p \qquad\qquad SE_{\hat{p}} = \sqrt{\frac{p(1-p)}{n}}$$

In order for the Central Limit Theorem to hold, the sample size is typically considered sufficiently large when $np \geq 10$ and $n(1-p) \geq 10$, which is called the **success-failure condition**.

The Central Limit Theorem is incredibly important, and it provides a foundation for much of statistics. As we begin applying the Central Limit Theorem, be mindful of the two technical conditions: the observations must be independent, and the sample size must be sufficiently large such that $np \geq 10$ and $n(1-p) \geq 10$.

EXAMPLE 5.2

Earlier we estimated the mean and standard error of $\hat{p}$ using simulated data when $p = 0.88$ and $n = 1000$. Confirm that the Central Limit Theorem applies and the sampling distribution is approximately normal.

Independence. There are $n = 1000$ observations for each sample proportion $\hat{p}$, and each of those observations are independent draws. *The most common way for observations to be considered independent is if they are from a simple random sample.*

Success-failure condition. We can confirm the sample size is sufficiently large by checking the success-failure condition and confirming the two calculated values are greater than 10:

$$np = 1000 \times 0.88 = 880 \geq 10 \qquad\qquad n(1-p) = 1000 \times (1 - 0.88) = 120 \geq 10$$

The independence and success-failure conditions are both satisfied, so the Central Limit Theorem applies, and it's reasonable to model $\hat{p}$ using a normal distribution.

HOW TO VERIFY SAMPLE OBSERVATIONS ARE INDEPENDENT

Subjects in an experiment are considered independent if they undergo random assignment to the treatment groups.

If the observations are from a simple random sample, then they are independent.

If a sample is from a seemingly random process, e.g. an occasional error on an assembly line, checking independence is more difficult. In this case, use your best judgement.

An additional condition that is sometimes added for samples from a population is that they are no larger than 10% of the population. When the sample exceeds 10% of the population size, the methods we discuss tend to overestimate the sampling error slightly versus what we would get using more advanced methods.[4] This is very rarely an issue, and when it is an issue, our methods tend to be conservative, so we consider this additional check as optional.

EXAMPLE 5.3

Compute the theoretical mean and standard error of $\hat{p}$ when $p = 0.88$ and $n = 1000$, according to the Central Limit Theorem.

The mean of the $\hat{p}$'s is simply the population proportion: $\mu_{\hat{p}} = 0.88$.

The calculation of the standard error of $\hat{p}$ uses the following formula:

$$SE_{\hat{p}} = \sqrt{\frac{p(1-p)}{n}} = \sqrt{\frac{0.88(1-0.88)}{1000}} = 0.010$$

EXAMPLE 5.4

Estimate how frequently the sample proportion $\hat{p}$ should be within 0.02 (2%) of the population value, $p = 0.88$. Based on Examples 5.2 and 5.3, we know that the distribution is approximately $N(\mu_{\hat{p}} = 0.88, SE_{\hat{p}} = 0.010)$.

After so much practice in Section 4.1, this normal distribution example will hopefully feel familiar! We would like to understand the fraction of $\hat{p}$'s between 0.86 and 0.90:

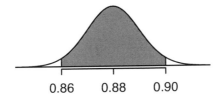

With $\mu_{\hat{p}} = 0.88$ and $SE_{\hat{p}} = 0.010$, we can compute the Z-score for both the left and right cutoffs:

$$Z_{0.86} = \frac{0.86 - 0.88}{0.010} = -2 \qquad\qquad Z_{0.90} = \frac{0.90 - 0.88}{0.010} = 2$$

We can use either statistical software, a graphing calculator, or a table to find the areas to the tails, and in any case we will find that they are each 0.0228. The total tail areas are $2 \times 0.0228 = 0.0456$, which leaves the shaded area of 0.9544. That is, about 95.44% of the sampling distribution in Figure 5.2 is within ± 0.02 of the population proportion, $p = 0.88$.

---

[4]For example, we could use what's called the **finite population correction factor**: if the sample is of size n and the population size is N, then we can multiple the typical standard error formula by $\sqrt{\frac{N-n}{N-1}}$ to obtain a smaller, more precise estimate of the actual standard error. When $n < 0.1 \times N$, this correction factor is relatively small.

GUIDED PRACTICE 5.5

In Example 5.1 we discussed how a smaller sample would tend to produce a less reliable estimate. Explain how this intuition is reflected in the formula for $SE_{\hat{p}} = \sqrt{\frac{p(1-p)}{n}}$.[5]

5.1.4   Applying the Central Limit Theorem to a real-world setting

We do not actually know the population proportion unless we conduct an expensive poll of all individuals in the population. Our earlier value of $p = 0.88$ was based on a Pew Research conducted a poll of 1000 American adults that found $\hat{p} = 0.887$ of them favored expanding solar energy. The researchers might have wondered: does the sample proportion from the poll approximately follow a normal distribution? We can check the conditions from the Central Limit Theorem:

Independence. The poll is a simple random sample of American adults, which means that the observations are independent.

Success-failure condition. To check this condition, we need the population proportion, p, to check if both np and $n(1 - p)$ are greater than 10. However, we do not actually know p, which is exactly why the pollsters would take a sample! In cases like these, we often use $\hat{p}$ as our next best way to check the success-failure condition:

$$n\hat{p} = 1000 \times 0.887 = 887 \qquad\qquad n(1 - \hat{p}) = 1000 \times (1 - 0.887) = 113$$

The sample proportion $\hat{p}$ acts as a reasonable substitute for p during this check, and each value in this case is well above the minimum of 10.

This **substitution approximation** of using $\hat{p}$ in place of p is also useful when computing the standard error of the sample proportion:

$$SE_{\hat{p}} = \sqrt{\frac{p(1 - p)}{n}} \approx \sqrt{\frac{\hat{p}(1 - \hat{p})}{n}} = \sqrt{\frac{0.887(1 - 0.887)}{1000}} = 0.010$$

This substitution technique is sometimes referred to as the "plug-in principle". In this case, $SE_{\hat{p}}$ didn't change enough to be detected using only 3 decimal places versus when we completed the calculation with 0.88 earlier. The computed standard error tends to be reasonably stable even when observing slightly different proportions in one sample or another.

---

[5]Since the sample size n is in the denominator (on the bottom) of the fraction, a bigger sample size means the entire expression when calculated will tend to be smaller. That is, a larger sample size would correspond to a smaller standard error.

5.1.5 More details regarding the Central Limit Theorem

We've applied the Central Limit Theorem in numerous examples so far this chapter:

When observations are independent and the sample size is sufficiently large, the distribution of $\hat{p}$ resembles a normal distribution with

$$\mu_{\hat{p}} = p \qquad\qquad SE_{\hat{p}} = \sqrt{\frac{p(1-p)}{n}}$$

The sample size is considered sufficiently large when $np \geq 10$ and $n(1-p) \geq 10$.

In this section, we'll explore the success-failure condition and seek to better understand the Central Limit Theorem.

An interesting question to answer is, *what happens when $np < 10$ or $n(1-p) < 10$?* As we did in Section 5.1.2, we can simulate drawing samples of different sizes where, say, the true proportion is $p = 0.25$. Here's a sample of size 10:

no, no, yes, yes, no, no, no, no, no, no

In this sample, we observe a sample proportion of yeses of $\hat{p} = \frac{2}{10} = 0.2$. We can simulate many such proportions to understand the sampling distribution of $\hat{p}$ when $n = 10$ and $p = 0.25$, which we've plotted in Figure 5.3 alongside a normal distribution with the same mean and variability. These distributions have a number of important differences.

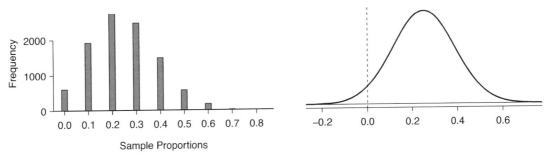

Figure 5.3: Left: simulations of $\hat{p}$ when the sample size is $n = 10$ and the population proportion is $p = 0.25$. Right: a normal distribution with the same mean (0.25) and standard deviation (0.137).

	Unimodal?	Smooth?	Symmetric?
Normal: $N(0.25, 0.14)$	**Yes**	**Yes**	**Yes**
$n = 10$, $p = 0.25$	**Yes**	*No*	*No*

Notice that the success-failure condition was not satisfied when $n = 10$ and $p = 0.25$:

$$np = 10 \times 0.25 = 2.5 \qquad\qquad n(1-p) = 10 \times 0.75 = 7.5$$

This single sampling distribution does not show that the success-failure condition is the perfect guideline, but we have found that the guideline did correctly identify that a normal distribution might not be appropriate.

We can complete several additional simulations, shown in Figures 5.4 and 5.5, and we can see some trends:

1. When either np or $n(1-p)$ is small, the distribution is more **discrete**, i.e. *not continuous*.

2. When np or $n(1-p)$ is smaller than 10, the skew in the distribution is more noteworthy.

3. The larger both np *and* $n(1-p)$, the more normal the distribution. This may be a little harder to see for the larger sample size in these plots as the variability also becomes much smaller.

4. When np and $n(1-p)$ are both very large, the distribution's discreteness is hardly evident, and the distribution looks much more like a normal distribution.

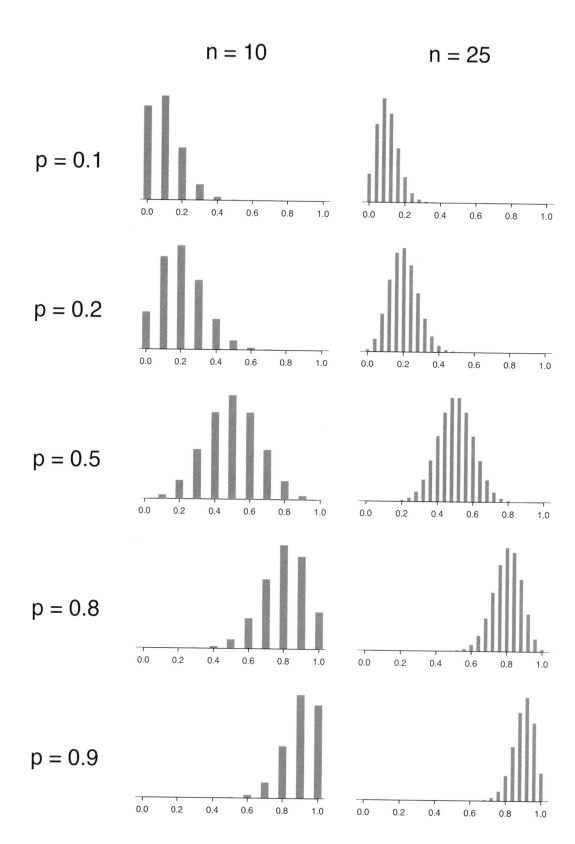

Figure 5.4: Sampling distributions for several scenarios of p and n.
Rows: $p = 0.10$, $p = 0.20$, $p = 0.50$, $p = 0.80$, and $p = 0.90$.
Columns: $n = 10$ and $n = 25$.

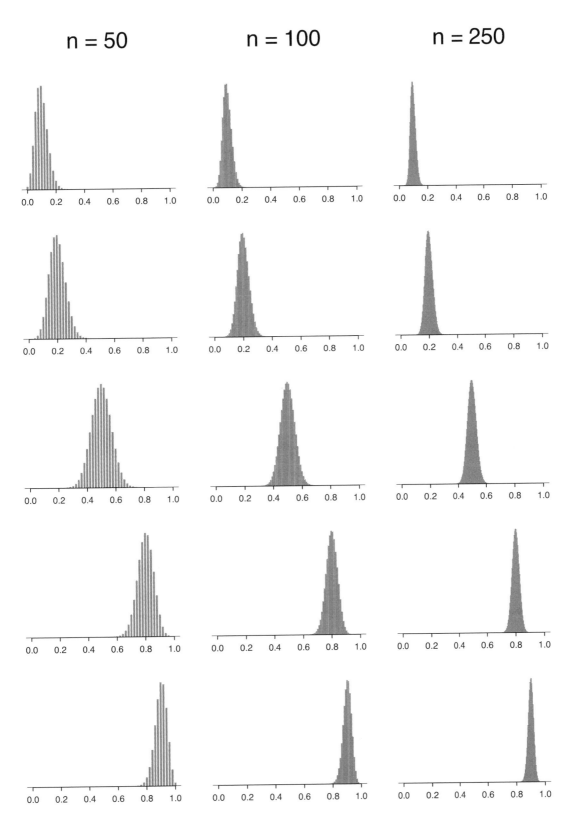

Figure 5.5: Sampling distributions for several scenarios of p and n.
Rows: $p = 0.10$, $p = 0.20$, $p = 0.50$, $p = 0.80$, and $p = 0.90$.
Columns: $n = 50$, $n = 100$, and $n = 250$.

So far we've only focused on the skew and discreteness of the distributions. We haven't considered how the mean and standard error of the distributions change. Take a moment to look back at the graphs, and pay attention to three things:

1. The centers of the distribution are always at the population proportion, p, that was used to generate the simulation. Because the sampling distribution of $\hat{p}$ is always centered at the population parameter p, it means the sample proportion $\hat{p}$ is **unbiased** when the data are independent and drawn from such a population.

2. For a particular population proportion p, the variability in the sampling distribution decreases as the sample size n becomes larger. This will likely align with your intuition: an estimate based on a larger sample size will tend to be more accurate.

3. For a particular sample size, the variability will be largest when $p = 0.5$. The differences may be a little subtle, so take a close look. This reflects the role of the proportion p in the standard error formula: $SE = \sqrt{\frac{p(1-p)}{n}}$. The standard error is largest when $p = 0.5$.

At no point will the distribution of $\hat{p}$ look *perfectly* normal, since $\hat{p}$ will always be take discrete values (x/n). It is always a matter of degree, and we will use the standard success-failure condition with minimums of 10 for np and $n(1-p)$ as our guideline within this book.

5.1.6 Extending the framework for other statistics

The strategy of using a sample statistic to estimate a parameter is quite common, and it's a strategy that we can apply to other statistics besides a proportion. For instance, if we want to estimate the average salary for graduates from a particular college, we could survey a random sample of recent graduates; in that example, we'd be using a sample mean $\bar{x}$ to estimate the population mean μ for all graduates. As another example, if we want to estimate the difference in product prices for two websites, we might take a random sample of products available on both sites, check the prices on each, and use then compute the average difference; this strategy certainly would give us some idea of the actual difference through a point estimate.

While this chapter emphases a single proportion context, we'll encounter many different contexts throughout this book where these methods will be applied. The principles and general ideas are the same, even if the details change a little. We've also sprinkled some other contexts into the exercises to help you start thinking about how the ideas generalize.

Exercises

5.1 Identify the parameter, Part I. For each of the following situations, state whether the parameter of interest is a mean or a proportion. It may be helpful to examine whether individual responses are numerical or categorical.

(a) In a survey, one hundred college students are asked how many hours per week they spend on the Internet.

(b) In a survey, one hundred college students are asked: "What percentage of the time you spend on the Internet is part of your course work?"

(c) In a survey, one hundred college students are asked whether or not they cited information from Wikipedia in their papers.

(d) In a survey, one hundred college students are asked what percentage of their total weekly spending is on alcoholic beverages.

(e) In a sample of one hundred recent college graduates, it is found that 85 percent expect to get a job within one year of their graduation date.

5.2 Identify the parameter, Part II. For each of the following situations, state whether the parameter of interest is a mean or a proportion.

(a) A poll shows that 64% of Americans personally worry a great deal about federal spending and the budget deficit.

(b) A survey reports that local TV news has shown a 17% increase in revenue within a two year period while newspaper revenues decreased by 6.4% during this time period.

(c) In a survey, high school and college students are asked whether or not they use geolocation services on their smart phones.

(d) In a survey, smart phone users are asked whether or not they use a web-based taxi service.

(e) In a survey, smart phone users are asked how many times they used a web-based taxi service over the last year.

5.3 Quality control. As part of a quality control process for computer chips, an engineer at a factory randomly samples 212 chips during a week of production to test the current rate of chips with severe defects. She finds that 27 of the chips are defective.

(a) What population is under consideration in the data set?

(b) What parameter is being estimated?

(c) What is the point estimate for the parameter?

(d) What is the name of the statistic can we use to measure the uncertainty of the point estimate?

(e) Compute the value from part (d) for this context.

(f) The historical rate of defects is 10%. Should the engineer be surprised by the observed rate of defects during the current week?

(g) Suppose the true population value was found to be 10%. If we use this proportion to recompute the value in part (e) using $p = 0.1$ instead of $\hat{p}$, does the resulting value change much?

5.4 Unexpected expense. In a random sample 765 adults in the United States, 322 say they could not cover a $400 unexpected expense without borrowing money or going into debt.

(a) What population is under consideration in the data set?

(b) What parameter is being estimated?

(c) What is the point estimate for the parameter?

(d) What is the name of the statistic can we use to measure the uncertainty of the point estimate?

(e) Compute the value from part (d) for this context.

(f) A cable news pundit thinks the value is actually 50%. Should she be surprised by the data?

(g) Suppose the true population value was found to be 40%. If we use this proportion to recompute the value in part (e) using $p = 0.4$ instead of $\hat{p}$, does the resulting value change much?

5.5 Repeated water samples. A nonprofit wants to understand the fraction of households that have elevated levels of lead in their drinking water. They expect at least 5% of homes will have elevated levels of lead, but not more than about 30%. They randomly sample 800 homes and work with the owners to retrieve water samples, and they compute the fraction of these homes with elevated lead levels. They repeat this 1,000 times and build a distribution of sample proportions.

(a) What is this distribution called?

(b) Would you expect the shape of this distribution to be symmetric, right skewed, or left skewed? Explain your reasoning.

(c) If the proportions are distributed around 8%, what is the variability of the distribution?

(d) What is the formal name of the value you computed in (c)?

(e) Suppose the researchers' budget is reduced, and they are only able to collect 250 observations per sample, but they can still collect 1,000 samples. They build a new distribution of sample proportions. How will the variability of this new distribution compare to the variability of the distribution when each sample contained 800 observations?

5.6 Repeated student samples. Of all freshman at a large college, 16% made the dean's list in the current year. As part of a class project, students randomly sample 40 students and check if those students made the list. They repeat this 1,000 times and build a distribution of sample proportions.

(a) What is this distribution called?

(b) Would you expect the shape of this distribution to be symmetric, right skewed, or left skewed? Explain your reasoning.

(c) Calculate the variability of this distribution.

(d) What is the formal name of the value you computed in (c)?

(e) Suppose the students decide to sample again, this time collecting 90 students per sample, and they again collect 1,000 samples. They build a new distribution of sample proportions. How will the variability of this new distribution compare to the variability of the distribution when each sample contained 40 observations?

5.2 Confidence intervals for a proportion

The sample proportion $\hat{p}$ provides a single plausible value for the population proportion p. However, the sample proportion isn't perfect and will have some *standard error* associated with it. When stating an estimate for the population proportion, it is better practice to provide a plausible *range of values* instead of supplying just the point estimate.

5.2.1 Capturing the population parameter

Using only a point estimate is like fishing in a murky lake with a spear. We can throw a spear where we saw a fish, but we will probably miss. On the other hand, if we toss a net in that area, we have a good chance of catching the fish. A **confidence interval** is like fishing with a net, and it represents a range of plausible values where we are likely to find the population parameter.

If we report a point estimate $\hat{p}$, we probably will not hit the exact population proportion. On the other hand, if we report a range of plausible values, representing a confidence interval, we have a good shot at capturing the parameter.

GUIDED PRACTICE 5.6

Ⓖ

If we want to be very certain we capture the population proportion in an interval, should we use a wider interval or a smaller interval?[6]

5.2.2 Constructing a 95% confidence interval

Our sample proportion $\hat{p}$ is the most plausible value of the population proportion, so it makes sense to build a confidence interval around this point estimate. The standard error provides a guide for how large we should make the confidence interval.

The standard error represents the standard deviation of the point estimate, and when the Central Limit Theorem conditions are satisfied, the point estimate closely follows a normal distribution. In a normal distribution, 95% of the data is within 1.96 standard deviations of the mean. Using this principle, we can construct a confidence interval that extends 1.96 standard errors from the sample proportion to be **95% confident** that the interval captures the population proportion:

$$\text{point estimate} \ \pm \ 1.96 \times SE$$

$$\hat{p} \ \pm \ 1.96 \times \sqrt{\frac{p(1-p)}{n}}$$

But what does "95% confident" mean? Suppose we took many samples and built a 95% confidence interval from each. Then about 95% of those intervals would contain the parameter, p. Figure 5.6 shows the process of creating 25 intervals from 25 samples from the simulation in Section 5.1.2, where 24 of the resulting confidence intervals contain the simulation's population proportion of $p = 0.88$, and one interval does not.

---

[6]If we want to be more certain we will capture the fish, we might use a wider net. Likewise, we use a wider confidence interval if we want to be more certain that we capture the parameter.

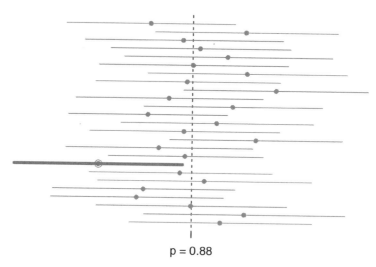

$p = 0.88$

Figure 5.6: Twenty-five point estimates and confidence intervals from the simulations in Section 5.1.2. These intervals are shown relative to the population proportion $p = 0.88$. Only 1 of these 25 intervals did not capture the population proportion, and this interval has been bolded.

EXAMPLE 5.7

In Figure 5.6, one interval does not contain $p = 0.88$. Does this imply that the population proportion used in the simulation could not have been $p = 0.88$?

Just as some observations naturally occur more than 1.96 standard deviations from the mean, some point estimates will be more than 1.96 standard errors from the parameter of interest. A confidence interval only provides a plausible range of values. While we might say other values are implausible based on the data, this does not mean they are impossible.

95% CONFIDENCE INTERVAL FOR A PARAMETER

When the distribution of a point estimate qualifies for the Central Limit Theorem and therefore closely follows a normal distribution, we can construct a 95% confidence interval as

$$\text{point estimate} \pm 1.96 \times SE$$

EXAMPLE 5.8

In Section 5.1 we learned about a Pew Research poll where 88.7% of a random sample of 1000 American adults supported expanding the role of solar power. Compute and interpret a 95% confidence interval for the population proportion.

We earlier confirmed that $\hat{p}$ follows a normal distribution and has a standard error of $SE_{\hat{p}} = 0.010$. To compute the 95% confidence interval, plug the point estimate $\hat{p} = 0.887$ and standard error into the 95% confidence interval formula:

$$\hat{p} \pm 1.96 \times SE_{\hat{p}} \quad \rightarrow \quad 0.887 \pm 1.96 \times 0.010 \quad \rightarrow \quad (0.8674, 0.9066)$$

We are 95% confident that the actual proportion of American adults who support expanding solar power is between 86.7% and 90.7%. (It's common to round to the nearest percentage point or nearest tenth of a percentage point when reporting a confidence interval.)

5.2.3 Changing the confidence level

Suppose we want to consider confidence intervals where the confidence level is higher than 95%, such as a confidence level of 99%. Think back to the analogy about trying to catch a fish: if we want to be more sure that we will catch the fish, we should use a wider net. To create a 99% confidence level, we must also widen our 95% interval. On the other hand, if we want an interval with lower confidence, such as 90%, we could use a slightly narrower interval than our original 95% interval.

The 95% confidence interval structure provides guidance in how to make intervals with different confidence levels. The general 95% confidence interval for a point estimate that follows a normal distribution is

$$\text{point estimate} \ \pm \ 1.96 \times SE$$

There are three components to this interval: the point estimate, "1.96", and the standard error. The choice of $1.96 \times SE$ was based on capturing 95% of the data since the estimate is within 1.96 standard errors of the parameter about 95% of the time. The choice of 1.96 corresponds to a 95% confidence level.

GUIDED PRACTICE 5.9

If X is a normally distributed random variable, what is the probability of the value X being within 2.58 standard deviations of the mean?[7]

Guided Practice 5.9 highlights that 99% of the time a normal random variable will be within 2.58 standard deviations of the mean. To create a 99% confidence interval, change 1.96 in the 95% confidence interval formula to be 2.58. That is, the formula for a 99% confidence interval is

$$\text{point estimate} \ \pm \ 2.58 \times SE$$

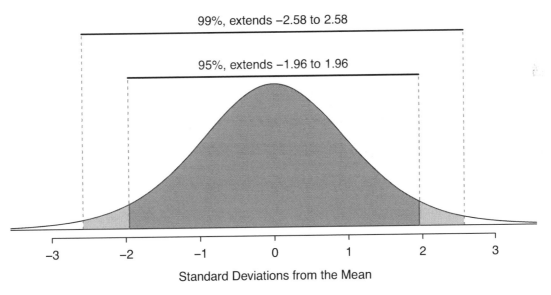

Figure 5.7: The area between $-z^\star$ and $z^\star$ increases as $z^\star$ becomes larger. If the confidence level is 99%, we choose $z^\star$ such that 99% of a normal normal distribution is between $-z^\star$ and $z^\star$, which corresponds to 0.5% in the lower tail and 0.5% in the upper tail: $z^\star = 2.58$.

---

[7]This is equivalent to asking how often the Z-score will be larger than -2.58 but less than 2.58. For a picture, see Figure 5.7. To determine this probability, we can use statistical software, a calculator, or a table to look up -2.58 and 2.58 for a normal distribution: 0.0049 and 0.9951. Thus, there is a $0.9951 - 0.0049 \approx 0.99$ probability that an unobserved normal random variable X will be within 2.58 standard deviations of μ.

This approach – using the Z-scores in the normal model to compute confidence levels – is appropriate when a point estimate such as $\hat{p}$ is associated with a normal distribution. For some other point estimates, a normal model is not a good fit; in these cases, we'll use alternative distributions that better represent the sampling distribution.

CONFIDENCE INTERVAL USING ANY CONFIDENCE LEVEL

If a point estimate closely follows a normal model with standard error SE, then a confidence interval for the population parameter is

$$\text{point estimate} \pm z^{\star} \times SE$$

where $z^{\star}$ corresponds to the confidence level selected.

Figure 5.7 provides a picture of how to identify $z^{\star}$ based on a confidence level. We select $z^{\star}$ so that the area between -$z^{\star}$ and $z^{\star}$ in the standard normal distribution, $N(0,1)$, corresponds to the confidence level.

MARGIN OF ERROR

In a confidence interval, $z^{\star} \times SE$ is called the **margin of error**.

EXAMPLE 5.10

Use the data in Example 5.8 to create a 90% confidence interval for the proportion of American adults that support expanding the use of solar power. We have already verified conditions for normality.

We first find $z^{\star}$ such that 90% of the distribution falls between -$z^{\star}$ and $z^{\star}$ in the standard normal distribution, $N(\mu = 0, \sigma = 1)$. We can do this using a graphing calculator, statistical software, or a probability table by looking for an upper tail of 5% (the other 5% is in the lower tail): $z^{\star} = 1.65$. The 90% confidence interval can then be computed as

$$\hat{p} \pm 1.65 \times SE_{\hat{p}} \quad \rightarrow \quad 0.887 \pm 1.65 \times 0.0100 \quad \rightarrow \quad (0.8705, 0.9035)$$

That is, we are 90% confident that 87.1% to 90.4% of American adults supported the expansion of solar power in 2018.

CONFIDENCE INTERVAL FOR A SINGLE PROPORTION

Once you've determined a one-proportion confidence interval would be helpful for an application, there are four steps to constructing the interval:

Prepare. Identify $\hat{p}$ and n, and determine what confidence level you wish to use.

Check. Verify the conditions to ensure $\hat{p}$ is nearly normal. For one-proportion confidence intervals, use $\hat{p}$ in place of p to check the success-failure condition.

Calculate. If the conditions hold, compute SE using $\hat{p}$, find $z^{\star}$, and construct the interval.

Conclude. Interpret the confidence interval in the context of the problem.

5.2.4 More case studies

In New York City on October 23rd, 2014, a doctor who had recently been treating Ebola patients in Guinea went to the hospital with a slight fever and was subsequently diagnosed with Ebola. Soon thereafter, an NBC 4 New York/The Wall Street Journal/Marist Poll found that 82% of New Yorkers favored a "mandatory 21-day quarantine for anyone who has come in contact with an Ebola patient". This poll included responses of 1,042 New York adults between Oct 26th and 28th, 2014.

EXAMPLE 5.11

What is the point estimate in this case, and is it reasonable to use a normal distribution to model that point estimate?

The point estimate, based on a sample of size $n = 1042$, is $\hat{p} = 0.82$. To check whether $\hat{p}$ can be reasonably modeled using a normal distribution, we check independence (the poll is based on a simple random sample) and the success-failure condition ($1042 \times \hat{p} \approx 854$ and $1042 \times (1 - \hat{p}) \approx 188$, both easily greater than 10). With the conditions met, we are assured that the sampling distribution of $\hat{p}$ can be reasonably modeled using a normal distribution.

EXAMPLE 5.12

Estimate the standard error of $\hat{p} = 0.82$ from the Ebola survey.

We'll use the substitution approximation of $p \approx \hat{p} = 0.82$ to compute the standard error:

$$SE_{\hat{p}} = \sqrt{\frac{p(1-p)}{n}} \approx \sqrt{\frac{0.82(1-0.82)}{1042}} = 0.012$$

EXAMPLE 5.13

Construct a 95% confidence interval for p, the proportion of New York adults who supported a quarantine for anyone who has come into contact with an Ebola patient.

Using the standard error $SE = 0.012$ from Example 5.12, the point estimate 0.82, and $z^{\star} = 1.96$ for a 95% confidence level, the confidence interval is

$$\text{point estimate} \pm z^{\star} \times SE \quad \rightarrow \quad 0.82 \pm 1.96 \times 0.012 \quad \rightarrow \quad (0.796, 0.844)$$

We are 95% confident that the proportion of New York adults in October 2014 who supported a quarantine for anyone who had come into contact with an Ebola patient was between 0.796 and 0.844.

GUIDED PRACTICE 5.14

Answer the following two questions about the confidence interval from Example 5.13:[8]

(a) What does 95% confident mean in this context?

(b) Do you think the confidence interval is still valid for the opinions of New Yorkers today?

---

[8](a) If we took many such samples and computed a 95% confidence interval for each, then about 95% of those intervals would contain the actual proportion of New York adults who supported a quarantine for anyone who has come into contact with an Ebola patient.
(b) Not necessarily. The poll was taken at a time where there was a huge public safety concern. Now that people have had some time to step back, they may have changed their opinions. We would need to run a new poll if we wanted to get an estimate of the current proportion of New York adults who would support such a quarantine period.

GUIDED PRACTICE 5.15

In the Pew Research poll about solar energy, they also inquired about other forms of energy, and 84.8% of the 1000 respondents supported expanding the use of wind turbines.[9]

(a) Is it reasonable to model the proportion of US adults who support expanding wind turbines using a normal distribution?

(b) Create a 99% confidence interval for the level of American support for expanding the use of wind turbines for power generation.

We can also construct confidence intervals for other parameters, such as a population mean. In these cases, a confidence interval would be computed in a similar way to that of a single proportion: a point estimate plus/minus some margin of error. We'll dive into these details in later chapters.

5.2.5 Interpreting confidence intervals

In each of the examples, we described the confidence intervals by putting them into the context of the data and also using somewhat formal language:

Solar. We are 90% confident that 87.1% to 90.4% of American adults support the expansion of solar power in 2018.

Ebola. We are 95% confident that the proportion of New York adults in October 2014 who supported a quarantine for anyone who had come into contact with an Ebola patient was between 0.796 and 0.844.

Wind Turbine. We are 99% confident the proportion of Americans adults that support expanding the use of wind turbines is between 81.9% and 87.7% in 2018.

First, notice that the statements are always about the population parameter, which considers *all* American adults for the energy polls or *all* New York adults for the quarantine poll.

We also avoided another common mistake: *incorrect* language might try to describe the confidence interval as capturing the population parameter with a certain probability. Making a probability interpretation is a common error: while it might be useful to think of it as a probability, the confidence level only quantifies how plausible it is that the parameter is in the given interval.

Another important consideration of confidence intervals is that they are *only about the population parameter*. A confidence interval says nothing about individual observations or point estimates. Confidence intervals only provide a plausible range for population parameters.

Lastly, keep in mind the methods we discussed only apply to sampling error, not to bias. If a data set is collected in a way that will tend to systematically under-estimate (or over-estimate) the population parameter, the techniques we have discussed will not address that problem. Instead, we rely on careful data collection procedures to help protect against bias in the examples we have considered, which is a common practice employed by data scientists to combat bias.

GUIDED PRACTICE 5.16

Consider the 90% confidence interval for the solar energy survey: 87.1% to 90.4%. If we ran the survey again, can we say that we're 90% confident that the new survey's proportion will be between 87.1% and 90.4%?[10]

---

[9](a) The survey was a random sample and counts are both ≥ 10 ($1000 \times 0.848 = 848$ and $1000 \times 0.152 = 152$), so independence and the success-failure condition are satisfied, and $\hat{p} = 0.848$ can be modeled using a normal distribution. (b) Guided Practice 5.15 confirmed that $\hat{p}$ closely follows a normal distribution, so we can use the C.I. formula:

$$\text{point estimate} \pm z^{\star} \times SE$$

In this case, the point estimate is $\hat{p} = 0.848$. For a 99% confidence interval, $z^{\star} = 2.58$. Computing the standard error: $SE_{\hat{p}} = \sqrt{\frac{0.848(1-0.848)}{1000}} = 0.0114$. Finally, we compute the interval as $0.848 \pm 2.58 \times 0.0114 \rightarrow (0.8186, 0.8774)$. It is also important to *always* provide an interpretation for the interval: we are 99% confident the proportion of American adults that support expanding the use of wind turbines in 2018 is between 81.9% and 87.7%.

[10] No, a confidence interval only provides a range of plausible values for a parameter, not future point estimates.

Exercises

5.7 Chronic illness, Part I. In 2013, the Pew Research Foundation reported that "45% of U.S. adults report that they live with one or more chronic conditions".[11] However, this value was based on a sample, so it may not be a perfect estimate for the population parameter of interest on its own. The study reported a standard error of about 1.2%, and a normal model may reasonably be used in this setting. Create a 95% confidence interval for the proportion of U.S. adults who live with one or more chronic conditions. Also interpret the confidence interval in the context of the study.

5.8 Twitter users and news, Part I. A poll conducted in 2013 found that 52% of U.S. adult Twitter users get at least some news on Twitter.[12]. The standard error for this estimate was 2.4%, and a normal distribution may be used to model the sample proportion. Construct a 99% confidence interval for the fraction of U.S. adult Twitter users who get some news on Twitter, and interpret the confidence interval in context.

5.9 Chronic illness, Part II. In 2013, the Pew Research Foundation reported that "45% of U.S. adults report that they live with one or more chronic conditions", and the standard error for this estimate is 1.2%. Identify each of the following statements as true or false. Provide an explanation to justify each of your answers.

(a) We can say with certainty that the confidence interval from Exercise 5.7 contains the true percentage of U.S. adults who suffer from a chronic illness.

(b) If we repeated this study 1,000 times and constructed a 95% confidence interval for each study, then approximately 950 of those confidence intervals would contain the true fraction of U.S. adults who suffer from chronic illnesses.

(c) The poll provides statistically significant evidence (at the $\alpha = 0.05$ level) that the percentage of U.S. adults who suffer from chronic illnesses is below 50%.

(d) Since the standard error is 1.2%, only 1.2% of people in the study communicated uncertainty about their answer.

5.10 Twitter users and news, Part II. A poll conducted in 2013 found that 52% of U.S. adult Twitter users get at least some news on Twitter, and the standard error for this estimate was 2.4%. Identify each of the following statements as true or false. Provide an explanation to justify each of your answers.

(a) The data provide statistically significant evidence that more than half of U.S. adult Twitter users get some news through Twitter. Use a significance level of $\alpha = 0.01$. (This part uses concepts from Section 5.3 and will be corrected in a future edition.)

(b) Since the standard error is 2.4%, we can conclude that 97.6% of all U.S. adult Twitter users were included in the study.

(c) If we want to reduce the standard error of the estimate, we should collect less data.

(d) If we construct a 90% confidence interval for the percentage of U.S. adults Twitter users who get some news through Twitter, this confidence interval will be wider than a corresponding 99% confidence interval.

---

[11]Pew Research Center, Washington, D.C. The Diagnosis Difference, November 26, 2013.

[12]Pew Research Center, Washington, D.C. Twitter News Consumers: Young, Mobile and Educated, November 4, 2013.

5.11   Waiting at an ER, Part I. A hospital administrator hoping to improve wait times decides to estimate the average emergency room waiting time at her hospital. She collects a simple random sample of 64 patients and determines the time (in minutes) between when they checked in to the ER until they were first seen by a doctor. A 95% confidence interval based on this sample is (128 minutes, 147 minutes), which is based on the normal model for the mean. Determine whether the following statements are true or false, and explain your reasoning.

(a) We are 95% confident that the average waiting time of these 64 emergency room patients is between 128 and 147 minutes.

(b) We are 95% confident that the average waiting time of all patients at this hospital's emergency room is between 128 and 147 minutes.

(c) 95% of random samples have a sample mean between 128 and 147 minutes.

(d) A 99% confidence interval would be narrower than the 95% confidence interval since we need to be more sure of our estimate.

(e) The margin of error is 9.5 and the sample mean is 137.5.

(f) In order to decrease the margin of error of a 95% confidence interval to half of what it is now, we would need to double the sample size. (Hint: the margin of error for a mean scales in the same way with sample size as the margin of error for a proportion.)

5.12   Mental health. The General Social Survey asked the question: "For how many days during the past 30 days was your mental health, which includes stress, depression, and problems with emotions, not good?" Based on responses from 1,151 US residents, the survey reported a 95% confidence interval of 3.40 to 4.24 days in 2010.

(a) Interpret this interval in context of the data.

(b) What does "95% confident" mean? Explain in the context of the application.

(c) Suppose the researchers think a 99% confidence level would be more appropriate for this interval. Will this new interval be smaller or wider than the 95% confidence interval?

(d) If a new survey were to be done with 500 Americans, do you think the standard error of the estimate be larger, smaller, or about the same.

5.13   Website registration. A website is trying to increase registration for first-time visitors, exposing 1% of these visitors to a new site design. Of 752 randomly sampled visitors over a month who saw the new design, 64 registered.

(a) Check any conditions required for constructing a confidence interval.

(b) Compute the standard error.

(c) Construct and interpret a 90% confidence interval for the fraction of first-time visitors of the site who would register under the new design (assuming stable behaviors by new visitors over time).

5.14   Coupons driving visits. A store randomly samples 603 shoppers over the course of a year and finds that 142 of them made their visit because of a coupon they'd received in the mail. Construct a 95% confidence interval for the fraction of all shoppers during the year whose visit was because of a coupon they'd received in the mail.

5.3 Hypothesis testing for a proportion

The following question comes from a book written by Hans Rosling, Anna Rosling Rönnlund, and Ola Rosling called *Factfulness*:

How many of the world's 1 year old children today have been vaccinated against some disease:

    a. 20%

    b. 50%

    c. 80%

Write down what your answer (or guess), and when you're ready, find the answer in the footnote.[13]

In this section, we'll be exploring how people with a 4-year college degree perform on this and other world health questions as we learn about hypothesis tests, which are a framework used to rigorously evaluate competing ideas and claims.

5.3.1 Hypothesis testing framework

We're interested in understanding how much people know about world health and development. If we take a multiple choice world health question, then we might like to understand if

H_0: People never learn these particular topics and their responses are simply equivalent to random guesses.

H_A: People have knowledge that helps them do better than random guessing, or perhaps, they have false knowledge that leads them to actually do worse than random guessing.

These competing ideas are called **hypotheses**. We call H_0 the null hypothesis and H_A the alternative hypothesis. When there is a subscript 0 like in H_0, data scientists pronounce it as "nought" (e.g. H_0 is pronounced "H-nought").

NULL AND ALTERNATIVE HYPOTHESES

The **null hypothesis** (H_0) often represents a skeptical perspective or a claim to be tested. The **alternative hypothesis** (H_A) represents an alternative claim under consideration and is often represented by a range of possible parameter values.

Our job as data scientists is to play the role of a skeptic: before we buy into the alternative hypothesis, we need to see strong supporting evidence.

The null hypothesis often represents a skeptical position or a perspective of "no difference". In our first example, we'll consider whether the typical person does any different than random guessing on Roslings' question about infant vaccinations.

The alternative hypothesis generally represents a new or stronger perspective. In the case of the question about infant vaccinations, it would certainly be interesting to learn whether people do better than random guessing, since that would mean that the typical person knows something about world health statistics. It would also be very interesting if we learned that people do *worse* than random guessing, which would suggest people believe incorrect information about world health.

The hypothesis testing framework is a very general tool, and we often use it without a second thought. If a person makes a somewhat unbelievable claim, we are initially skeptical. However, if there is sufficient evidence that supports the claim, we set aside our skepticism and reject the null hypothesis in favor of the alternative. The hallmarks of hypothesis testing are also found in the US court system.

---

[13]The correct answer is (c): 80% of the world's 1 year olds have been vaccinated against some disease.

GUIDED PRACTICE 5.17

A US court considers two possible claims about a defendant: she is either innocent or guilty. If we set these claims up in a hypothesis framework, which would be the null hypothesis and which the alternative?[14]

Jurors examine the evidence to see whether it convincingly shows a defendant is guilty. Even if the jurors leave unconvinced of guilt beyond a reasonable doubt, this does not mean they believe the defendant is innocent. This is also the case with hypothesis testing: *even if we fail to reject the null hypothesis, we typically do not accept the null hypothesis as true.* Failing to find strong evidence for the alternative hypothesis is not equivalent to accepting the null hypothesis.

When considering Roslings' question about infant vaccination, the null hypothesis represents the notion that the people we will be considering – college-educated adults – are as accurate as random guessing. That is, the proportion p of respondents who pick the correct answer, that 80% of 1 year olds have been vaccinated against some disease, is about 33.3% (or 1-in-3 if wanting to be perfectly precise). The alternative hypothesis is that this proportion is something other than 33.3%. While it's helpful to write these hypotheses in words, it can be useful to write them using mathematical notation:

H_0: $p = 0.333$

H_A: $p \neq 0.333$

In this hypothesis setup, we want to make a conclusion about the population parameter p. The value we are comparing the parameter to is called the **null value**, which in this case is 0.333. It's common to label the null value with the same symbol as the parameter but with a subscript '0'. That is, in this case, the null value is $p_0 = 0.333$ (pronounced "p-nought equals 0.333").

EXAMPLE 5.18

It may seem impossible that the proportion of people who get the correct answer is *exactly* 33.3%. If we don't believe the null hypothesis, should we simply reject it?

No. While we may not buy into the notion that the proportion is exactly 33.3%, the hypothesis testing framework requires that there be strong evidence before we reject the null hypothesis and conclude something more interesting.

After all, even if we don't believe the proportion is *exactly* 33.3%, that doesn't really tell us anything useful! We would still be stuck with the original question: do people do better or worse than random guessing on Roslings' question? Without data that strongly points in one direction or the other, it is both uninteresting and pointless to reject H_0.

GUIDED PRACTICE 5.19

Another example of a real-world hypothesis testing situation is evaluating whether a new drug is better or worse than an existing drug at treating a particular disease. What should we use for the null and alternative hypotheses in this case?[15]

---

[14]The jury considers whether the evidence is so convincing (strong) that there is no reasonable doubt regarding the person's guilt; in such a case, the jury rejects innocence (the null hypothesis) and concludes the defendant is guilty (alternative hypothesis).

[15]The null hypothesis (H_0) in this case is the declaration of *no difference*: the drugs are equally effective. The alternative hypothesis (H_A) is that the new drug performs differently than the original, i.e. it could perform better or worse.

5.3.2 Testing hypotheses using confidence intervals

We will use the `rosling_responses` data set to evaluate the hypothesis test evaluating whether college-educated adults who get the question about infant vaccination correct is different from 33.3%. This data set summarizes the answers of 50 college-educated adults. Of these 50 adults, 24% of respondents got the question correct that 80% of 1 year olds have been vaccinated against some disease.

Up until now, our discussion has been philosophical. However, now that we have data, we might ask ourselves: does the data provide strong evidence that the proportion of all college-educated adults who would answer this question correctly is different than 33.3%?

We learned in Section 5.1 that there is fluctuation from one sample to another, and it is unlikely that our sample proportion, $\hat{p}$, will exactly equal p, but we want to make a conclusion about p. We have a nagging concern: is this deviation of 24% from 33.3% simply due to chance, or does the data provide strong evidence that the population proportion is different from 33.3%?

In Section 5.2, we learned how to quantify the uncertainty in our estimate using confidence intervals. The same method for measuring variability can be useful for the hypothesis test.

EXAMPLE 5.20

Check whether it is reasonable to construct a confidence interval for p using the sample data, and if so, construct a 95% confidence interval.

The conditions are met for $\hat{p}$ to be approximately normal: the data come from a simple random sample (satisfies independence), and $n\hat{p} = 12$ and $n(1 - \hat{p}) = 38$ are both at least 10 (success-failure condition).

To construct the confidence interval, we will need to identify the point estimate ($\hat{p} = 0.24$), the critical value for the 95% confidence level ($z^\star = 1.96$), and the standard error of $\hat{p}$ ($SE_{\hat{p}} = \sqrt{\hat{p}(1-\hat{p})/n} = 0.060$). With those pieces, the confidence interval for p can be constructed:

$$\hat{p} \pm z^\star \times SE_{\hat{p}}$$
$$0.24 \pm 1.96 \times 0.060$$
$$(0.122, 0.358)$$

We are 95% confident that the proportion of all college-educated adults to correctly answer this particular question about infant vaccination is between 12.2% and 35.8%.

Because the null value in the hypothesis test is $p_0 = 0.333$, which falls within the range of plausible values from the confidence interval, we cannot say the null value is implausible.[16] That is, the data do not provide sufficient evidence to reject the notion that the performance of college-educated adults was different than random guessing, and we do not reject the null hypothesis, H_0.

EXAMPLE 5.21

Explain why we cannot conclude that college-educated adults simply guessed on the infant vaccination question.

While we failed to reject H_0, that does not necessarily mean the null hypothesis is true. Perhaps there was an actual difference, but we were not able to detect it with the relatively small sample of 50.

DOUBLE NEGATIVES CAN SOMETIMES BE USED IN STATISTICS

In many statistical explanations, we use double negatives. For instance, we might say that the null hypothesis is *not implausible* or we *failed to reject* the null hypothesis. Double negatives are used to communicate that while we are not rejecting a position, we are also not saying it is correct.

---

[16]Arguably this method is slightly imprecise. As we'll see in a few pages, the standard error is often computed slightly differently in the context of a hypothesis test for a proportion.

GUIDED PRACTICE 5.22

Let's move onto a second question posed by the Roslings:

> *There are 2 billion children in the world today aged 0-15 years old, how many children will there be in year 2100 according to the United Nations?*

> a. *4 billion.*
>
> b. *3 billion.*
>
> c. *2 billion.*

Set up appropriate hypotheses to evaluate whether college-educated adults are better than random guessing on this question. Also, see if you can guess the correct answer before checking the answer in the footnote![17]

GUIDED PRACTICE 5.23

This time we took a larger sample of 228 college-educated adults, 34 (14.9%) selected the correct answer to the question in Guided Practice 5.22: 2 billion. Can we model the sample proportion using a normal distribution and construct a confidence interval?[18]

EXAMPLE 5.24

Compute a 95% confidence interval for the fraction of college-educated adults who answered the children-in-2100 question correctly, and evaluate the hypotheses in Guided Practice 5.22.

To compute the standard error, we'll again use $\hat{p}$ in place of p for the calculation:

$$SE_{\hat{p}} = \sqrt{\frac{\hat{p}(1 - \hat{p})}{n}} = \sqrt{\frac{0.149(1 - 0.149)}{228}} = 0.024$$

In Guided Practice 5.23, we found that $\hat{p}$ can be modeled using a normal distribution, which ensures a 95% confidence interval may be accurately constructed as

$$\hat{p} \pm z^{\star} \times SE \quad \rightarrow \quad 0.149 \pm 1.96 \times 0.024 \quad \rightarrow \quad (0.103, 0.195)$$

Because the null value, $p_0 = 0.333$, is not in the confidence interval, a population proportion of 0.333 is implausible and we reject the null hypothesis. That is, the data provide statistically significant evidence that the actual proportion of college adults who get the children-in-2100 question correct is different from random guessing. Because the entire 95% confidence interval is below 0.333, we can conclude college-educated adults do *worse* than random guessing on this question.

One subtle consideration is that we used a 95% confidence interval. What if we had used a 99% confidence level? Or even a 99.9% confidence level? It's possible to come to a different conclusion if using a different confidence level. Therefore, when we make a conclusion based on confidence interval, we should also be sure it is clear what confidence level we used.

The worse-than-random performance on this last question is not a fluke: there are many such world health questions where people do worse than random guessing. In general, the answers suggest that people tend to be more pessimistic about progress than reality suggests. This topic is discussed in much greater detail in the Roslings' book, *Factfulness*.

---

[17]The appropriate hypotheses are:

H_0: the proportion who get the answer correct is the same as random guessing: 1-in-3, or $p = 0.333$.

H_A: the proportion who get the answer correct is different than random guessing, $p \neq 0.333$.

The correct answer to the question is 2 billion. While the world population is projected to increase, the average age is also expected to rise. That is, the majority of the population growth will happen in older age groups, meaning people are projected to live longer in the future across much of the world.

[18]We check both conditions, which are satisfied, so it is reasonable to use a normal distribution for $\hat{p}$:

Independence. Since the data are from a simple random sample, the observations are independent.

Success-failure. We'll use $\hat{p}$ in place of p to check: $n\hat{p} = 34$ and $n(1 - \hat{p}) = 194$. Both are greater than 10, so the success-failure condition is satisfied.

5.3.3 Decision errors

Hypothesis tests are not flawless: we can make an incorrect decision in a statistical hypothesis test based on the data. For example, in the court system innocent people are sometimes wrongly convicted and the guilty sometimes walk free. One key distinction with statistical hypothesis tests is that we have the tools necessary to probabilistically quantify how often we make errors in our conclusions.

Recall that there are two competing hypotheses: the null and the alternative. In a hypothesis test, we make a statement about which one might be true, but we might choose incorrectly. There are four possible scenarios, which are summarized in Figure 5.8.

		Test conclusion	
		do not reject H_0	reject H_0 in favor of H_A
Truth	H_0 true	okay	**Type 1 Error**
	H_A true	**Type 2 Error**	okay

Figure 5.8: Four different scenarios for hypothesis tests.

A **Type 1 Error** is rejecting the null hypothesis when H_0 is actually true. A **Type 2 Error** is failing to reject the null hypothesis when the alternative is actually true.

GUIDED PRACTICE 5.25

In a US court, the defendant is either innocent (H_0) or guilty (H_A). What does a Type 1 Error represent in this context? What does a Type 2 Error represent? Figure 5.8 may be useful.[19]

EXAMPLE 5.26

How could we reduce the Type 1 Error rate in US courts? What influence would this have on the Type 2 Error rate?

To lower the Type 1 Error rate, we might raise our standard for conviction from "beyond a reasonable doubt" to "beyond a conceivable doubt" so fewer people would be wrongly convicted. However, this would also make it more difficult to convict the people who are actually guilty, so we would make more Type 2 Errors.

GUIDED PRACTICE 5.27

How could we reduce the Type 2 Error rate in US courts? What influence would this have on the Type 1 Error rate?[20]

Exercises 5.25-5.27 provide an important lesson: if we reduce how often we make one type of error, we generally make more of the other type.

Hypothesis testing is built around rejecting or failing to reject the null hypothesis. That is, we do not reject H_0 unless we have strong evidence. But what precisely does *strong evidence* mean? As a general rule of thumb, for those cases where the null hypothesis is actually true, we do not want to incorrectly reject H_0 more than 5% of the time. This corresponds to a **significance level** of 0.05. That is, if the null hypothesis is true, the significance level indicates how often the data lead us to incorrectly reject H_0. We often write the significance level using α (the Greek letter *alpha*): $\alpha = 0.05$. We discuss the appropriateness of different significance levels in Section 5.3.5.

---

[19]If the court makes a Type 1 Error, this means the defendant is innocent (H_0 true) but wrongly convicted. Note that a Type 1 Error is only possible if we've rejected the null hypothesis.

A Type 2 Error means the court failed to reject H_0 (i.e. failed to convict the person) when she was in fact guilty (H_A true). Note that a Type 2 Error is only possible if we have failed to reject the null hypothesis.

[20]To lower the Type 2 Error rate, we want to convict more guilty people. We could lower the standards for conviction from "beyond a reasonable doubt" to "beyond a little doubt". Lowering the bar for guilt will also result in more wrongful convictions, raising the Type 1 Error rate.

If we use a 95% confidence interval to evaluate a hypothesis test and the null hypothesis happens to be true, we will make an error whenever the point estimate is at least 1.96 standard errors away from the population parameter. This happens about 5% of the time (2.5% in each tail). Similarly, using a 99% confidence interval to evaluate a hypothesis is equivalent to a significance level of $\alpha = 0.01$.

A confidence interval is very helpful in determining whether or not to reject the null hypothesis. However, the confidence interval approach isn't always sustainable. In several sections, we will encounter situations where a confidence interval cannot be constructed. For example, if we wanted to evaluate the hypothesis that several proportions are equal, it isn't clear how to construct and compare many confidence intervals altogether.

Next we will introduce a statistic called the *p-value* to help us expand our statistical toolkit, which will enable us to both better understand the strength of evidence and work in more complex data scenarios in later sections.

5.3.4   Formal testing using p-values

The p-value is a way of quantifying the strength of the evidence against the null hypothesis and in favor of the alternative hypothesis. Statistical hypothesis testing typically uses the p-value method rather than making a decision based on confidence intervals.

P-VALUE

The **p-value** is the probability of observing data at least as favorable to the alternative hypothesis as our current data set, if the null hypothesis were true. We typically use a summary statistic of the data, in this section the sample proportion, to help compute the p-value and evaluate the hypotheses.

EXAMPLE 5.28

Pew Research asked a random sample of 1000 American adults whether they supported the increased usage of coal to produce energy. Set up hypotheses to evaluate whether a majority of American adults support or oppose the increased usage of coal.

The uninteresting result is that there is no majority either way: half of Americans support and the other half oppose expanding the use of coal to produce energy. The alternative hypothesis would be that there is a majority support or oppose (though we do not known which one!) expanding the use of coal. If p represents the proportion supporting, then we can write the hypotheses as

H_0: $p = 0.5$

H_A: $p \neq 0.5$

In this case, the null value is $p_0 = 0.5$.

When evaluating hypotheses for proportions using the p-value method, we will slightly modify how we check the success-failure condition and compute the standard error for the single proportion case. These changes aren't dramatic, but pay close attention to how we use the null value, p_0.

EXAMPLE 5.29

Pew Research's sample show that 37% of American adults support increased usage of coal. We now wonder, does 37% represent a real difference from the null hypothesis of 50%? What would the sampling distribution of $\hat{p}$ look like if the null hypothesis were true?

If the null hypothesis were true, the population proportion would be the null value, 0.5. We previously learned that the sampling distribution of $\hat{p}$ will be normal when two conditions are met:

Independence. The poll was based on a simple random sample, so independence is satisfied.

Success-failure. Based on the poll's sample size of $n = 1000$, the success-failure condition is met, since

$$np \overset{H_0}{=} 1000 \times 0.5 = 500 \qquad\qquad n(1-p) \overset{H_0}{=} 1000 \times (1 - 0.5) = 500$$

are both at least 10. Note that the success-failure condition was checked using the null value, $p_0 = 0.5$; this is the first procedural difference from confidence intervals.

If the null hypothesis were true, the sampling distribution indicates that a sample proportion based on $n = 1000$ observations would be normally distributed. Next, we can compute the standard error, where we will again use the null value $p_0 = 0.5$ in the calculation:

$$SE_{\hat{p}} = \sqrt{\frac{p(1-p)}{n}} \overset{H_0}{=} \sqrt{\frac{0.5 \times (1-0.5)}{1000}} = 0.016$$

This marks the other procedural difference from confidence intervals: since the sampling distribution is determined under the null proportion, the null value p_0 was used for the proportion in the calculation rather than $\hat{p}$.

Ultimately, if the null hypothesis were true, then the sample proportion should follow a normal distribution with mean 0.5 and a standard error of 0.016. This distribution is shown in Figure 5.9.

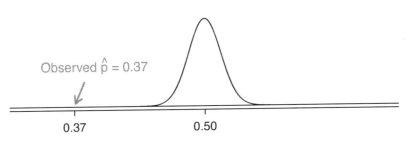

Figure 5.9: If the null hypothesis were true, this normal distribution describes the distribution of $\hat{p}$.

CHECKING SUCCESS-FAILURE AND COMPUTING $SE_{\hat{p}}$ FOR A HYPOTHESIS TEST

When using the p-value method to evaluate a hypothesis test, we check the conditions for $\hat{p}$ and construct the standard error using the null value, p_0, instead of using the sample proportion.

In a hypothesis test with a p-value, we are supposing the null hypothesis is true, which is a different mindset than when we compute a confidence interval. This is why we use p_0 instead of $\hat{p}$ when we check conditions and compute the standard error in this context.

When we identify the sampling distribution under the null hypothesis, it has a special name: the **null distribution**. The p-value represents the probability of the observed $\hat{p}$, or a $\hat{p}$ that is more extreme, if the null hypothesis were true. To find the p-value, we generally find the null distribution, and then we find a tail area in that distribution corresponding to our point estimate.

EXAMPLE 5.30

If the null hypothesis were true, determine the chance of finding $\hat{p}$ at least as far into the tails as 0.37 under the null distribution, which is a normal distribution with mean $\mu = 0.5$ and $SE = 0.016$.

This is a normal probability problem where $x = 0.37$. First, we draw a simple graph to represent the situation, similar to what is shown in Figure 5.9. Since $\hat{p}$ is so far out in the tail, we know the tail area is going to be very small. To find it, we start by computing the Z-score using the mean of 0.5 and the standard error of 0.016:

$$Z = \frac{0.37 - 0.5}{0.016} = -8.125$$

We can use software to find the tail area: 2.2×10^{-16} (0.00000000000000022). If using the normal probability table in Appendix C.1, we'd find that $Z = -8.125$ is off the table, so we would use the smallest area listed: 0.0002.

The potential $\hat{p}$'s in the upper tail beyond 0.63, which are shown in Figure 5.10, also represent observations at least as extreme as the observed value of 0.37. To account for these values that are also more extreme under the hypothesis setup, we double the lower tail to get an estimate of the p-value: 4.4×10^{-16} (or if using the table method, 0.0004).

The p-value represents the probability of observing such an extreme sample proportion by chance, if the null hypothesis were true.

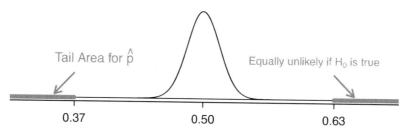

Figure 5.10: If H_0 were true, then the values above 0.63 are just as unlikely as values below 0.37.

EXAMPLE 5.31

How should we evaluate the hypotheses using the p-value of 4.4×10^{-16}? Use the standard significance level of $\alpha = 0.05$.

If the null hypothesis were true, there's only an incredibly small chance of observing such an extreme deviation of $\hat{p}$ from 0.5. This means one of the following must be true:

1. The null hypothesis is true, and we just happened to get observe something so extreme that only happens about once in every 23 quadrillion times (1 quadrillion = 1 million × 1 billion).

2. The alternative hypothesis is true, which would be consistent with observing a sample proportion far from 0.5.

The first scenario is laughably improbable, while the second scenario seems much more plausible.

Formally, when we evaluate a hypothesis test, we compare the p-value to the significance level, which in this case is $\alpha = 0.05$. Since the p-value is less than α, we reject the null hypothesis. That is, the data provide strong evidence against H_0. The data indicate the direction of the difference: a majority of Americans do not support expanding the use of coal-powered energy.

> **COMPARE THE P-VALUE TO α TO EVALUATE H_0**
>
> When the p-value is less than the significance level, α, reject H_0. We would report a conclusion that the data provide strong evidence supporting the alternative hypothesis.
>
> When the p-value is greater than α, do not reject H_0, and report that we do not have sufficient evidence to reject the null hypothesis.
>
> In either case, it is important to describe the conclusion in the context of the data.

GUIDED PRACTICE 5.32

Do a majority of Americans support or oppose nuclear arms reduction? Set up hypotheses to evaluate this question.[21]

EXAMPLE 5.33

A simple random sample of 1028 US adults in March 2013 show that 56% support nuclear arms reduction. Does this provide convincing evidence that a majority of Americans supported nuclear arms reduction at the 5% significance level?

First, check conditions:

Independence. The poll was of a simple random sample of US adults, meaning the observations are independent.

Success-failure. In a one-proportion hypothesis test, this condition is checked using the null proportion, which is $p_0 = 0.5$ in this context: $np_0 = n(1 - p_0) = 1028 \times 0.5 = 514 \geq 10$.

With these conditions verified, we can model $\hat{p}$ using a normal model.

Next the standard error can be computed. The null value p_0 is used again here, because this is a hypothesis test for a single proportion.

$$SE_{\hat{p}} = \sqrt{\frac{p_0(1 - p_0)}{n}} = \sqrt{\frac{0.5(1 - 0.5)}{1028}} = 0.0156$$

Based on the normal model, the test statistic can be computed as the Z-score of the point estimate:

$$Z = \frac{\text{point estimate} - \text{null value}}{SE} = \frac{0.56 - 0.50}{0.0156} = 3.85$$

It's generally helpful to draw null distribution and the tail areas of interest for computing the p-value:

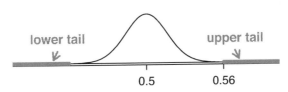

The upper tail area is about 0.0001, and we double this tail area to get the p-value: 0.0002. Because the p-value is smaller than 0.05, we reject H_0. The poll provides convincing evidence that a majority of Americans supported nuclear arms reduction efforts in March 2013.

---

[21]We would like to understand if a majority supports or opposes, or ultimately, if there is no difference. If p is the proportion of Americans who support nuclear arms reduction, then H_0: $p = 0.50$ and H_A: $p \neq 0.50$.

HYPOTHESIS TESTING FOR A SINGLE PROPORTION

Once you've determined a one-proportion hypothesis test is the correct procedure, there are four steps to completing the test:

Prepare. Identify the parameter of interest, list hypotheses, identify the significance level, and identify $\hat{p}$ and n.

Check. Verify conditions to ensure $\hat{p}$ is nearly normal under H_0. For one-proportion hypothesis tests, use the null value to check the success-failure condition.

Calculate. If the conditions hold, compute the standard error, again using p_0, compute the Z-score, and identify the p-value.

Conclude. Evaluate the hypothesis test by comparing the p-value to α, and provide a conclusion in the context of the problem.

5.3.5 Choosing a significance level

Choosing a significance level for a test is important in many contexts, and the traditional level is $\alpha = 0.05$. However, it can be helpful to adjust the significance level based on the application. We may select a level that is smaller or larger than 0.05 depending on the consequences of any conclusions reached from the test.

If making a Type 1 Error is dangerous or especially costly, we should choose a small significance level (e.g. 0.01). Under this scenario we want to be very cautious about rejecting the null hypothesis, so we demand very strong evidence favoring H_A before we would reject H_0.

If a Type 2 Error is relatively more dangerous or much more costly than a Type 1 Error, then we might choose a higher significance level (e.g. 0.10). Here we want to be cautious about failing to reject H_0 when the alternative hypothesis is actually true.

Additionally, if the cost of collecting data is small relative to the cost of a Type 2 Error, then it may also be a good strategy to collect more data. Under this strategy, the Type 2 Error can be reduced while not affecting the Type 1 Error rate. Of course, collecting extra data is often costly, so there is typically a cost-benefit analysis to be considered.

EXAMPLE 5.34

A car manufacturer is considering switching to a new, higher quality piece of equipment that constructs vehicle door hinges. They figure that they will save money in the long run if this new machine produces hinges that have flaws less than 0.2% of the time. However, if the hinges are flawed more than 0.2% of the time, they wouldn't get a good enough return-on-investment from the new piece of equipment, and they would lose money. Is there good reason to modify the significance level in such a hypothesis test?

The null hypothesis would be that the rate of flawed hinges is 0.2%, while the alternative is that it the rate is different than 0.2%. This decision is just one of many that have a marginal impact on the car and company. A significance level of 0.05 seems reasonable since neither a Type 1 or Type 2 Error should be dangerous or (relatively) much more expensive.

EXAMPLE 5.35

The same car manufacturer is considering a slightly more expensive supplier for parts related to safety, not door hinges. If the durability of these safety components is shown to be better than the current supplier, they will switch manufacturers. Is there good reason to modify the significance level in such an evaluation?

The null hypothesis would be that the suppliers' parts are equally reliable. Because safety is involved, the car company should be eager to switch to the slightly more expensive manufacturer (reject H_0), even if the evidence of increased safety is only moderately strong. A slightly larger significance level, such as $\alpha = 0.10$, might be appropriate.

GUIDED PRACTICE 5.36

A part inside of a machine is very expensive to replace. However, the machine usually functions properly even if this part is broken, so the part is replaced only if we are extremely certain it is broken based on a series of measurements. Identify appropriate hypotheses for this test (in plain language) and suggest an appropriate significance level.[22]

WHY IS 0.05 THE DEFAULT?

The $\alpha = 0.05$ threshold is most common. But why? Maybe the standard level should be smaller, or perhaps larger. If you're a little puzzled, you're reading with an extra critical eye – good job! We've made a 5-minute task to help clarify *why 0.05*:

www.openintro.org/why05

5.3.6 Statistical significance versus practical significance

When the sample size becomes larger, point estimates become more precise and any real differences in the mean and null value become easier to detect and recognize. Even a very small difference would likely be detected if we took a large enough sample. Sometimes researchers will take such large samples that even the slightest difference is detected, even differences where there is no practical value. In such cases, we still say the difference is **statistically significant**, but it is not **practically significant**. For example, an online experiment might identify that placing additional ads on a movie review website statistically significantly increases viewership of a TV show by 0.001%, but this increase might not have any practical value.

One role of a data scientist in conducting a study often includes planning the size of the study. The data scientist might first consult experts or scientific literature to learn what would be the smallest meaningful difference from the null value. She also would obtain other information, such as a very rough estimate of the true proportion p, so that she could roughly estimate the standard error. From here, she can suggest a sample size that is sufficiently large that, if there is a real difference that is meaningful, we could detect it. While larger sample sizes may still be used, these calculations are especially helpful when considering costs or potential risks, such as possible health impacts to volunteers in a medical study.

---

[22]Here the null hypothesis is that the part is not broken, and the alternative is that it is broken. If we don't have sufficient evidence to reject H_0, we would not replace the part. It sounds like failing to fix the part if it is broken (H_0 false, H_A true) is not very problematic, and replacing the part is expensive. Thus, we should require very strong evidence against H_0 before we replace the part. Choose a small significance level, such as $\alpha = 0.01$.

5.3.7 One-sided hypothesis tests (special topic)

So far we've only considered what are called **two-sided hypothesis tests**, where we care about detecting whether p is either above or below some null value p_0. There is a second type of hypothesis test called a **one-sided hypothesis test**. For a one-sided hypothesis test, the hypotheses take one of the following forms:

1. There's only value in detecting if the population parameter is *less than* some value p_0. In this case, the alternative hypothesis is written as $p < p_0$ for some null value p_0.

2. There's only value in detecting if the population parameter is *more than* some value p_0: In this case, the alternative hypothesis is written as $p > p_0$.

While we adjust the form of the alternative hypothesis, we continue to write the null hypothesis using an equals-sign in the one-sided hypothesis test case.

In the entire hypothesis testing procedure, there is only one difference in evaluating a one-sided hypothesis test vs a two-sided hypothesis test: how to compute the p-value. In a one-sided hypothesis test, we compute the p-value as the tail area in the *direction of the alternative hypothesis only*, meaning it is represented by a single tail area. Herein lies the reason why one-sided tests are sometimes interesting: if we don't have to double the tail area to get the p-value, then the p-value is smaller and the level of evidence required to identify an interesting finding in the direction of the alternative hypothesis goes down. However, one-sided tests aren't all sunshine and rainbows: the heavy price paid is that any interesting findings in the opposite direction must be disregarded.

EXAMPLE 5.37

In Section 1.1, we encountered an example where doctors were interested in determining whether stents would help people who had a high risk of stroke. The researchers believed the stents would help. Unfortunately, the data showed the opposite: patients who received stents actually did worse. Why was using a two-sided test so important in this context?

Before the study, researchers had reason to believe that stents would help patients since existing research suggested stents helped in patients with heart attacks. It would surely have been tempting to use a one-sided test in this situation, and had they done this, they would have limited their ability to identify potential harm to patients.

Example 5.37 highlights that using a one-sided hypothesis creates a risk of overlooking data supporting the opposite conclusion. We could have made a similar error when reviewing the Roslings' question data this section; if we had a pre-conceived notion that college-educated people wouldn't do worse than random guessing and so used a one-sided test, we would have missed the really interesting finding that many people have incorrect knowledge about global public health.

When might a one-sided test be appropriate to use? *Very rarely.* Should you ever find yourself considering using a one-sided test, carefully answer the following question:

> *What would I, or others, conclude if the data happens to go clearly in the opposite direction than my alternative hypothesis?*

If you or others would find any value in making a conclusion about the data that goes in the opposite direction of a one-sided test, then a two-sided hypothesis test should actually be used. These considerations can be subtle, so exercise caution. We will only apply two-sided tests in the rest of this book.

EXAMPLE 5.38

Why can't we simply run a one-sided test that goes in the direction of the data?

We've been building a careful framework that controls for the Type 1 Error, which is the significance level α in a hypothesis test. We'll use the $\alpha = 0.05$ below to keep things simple.

Imagine we could pick the one-sided test after we saw the data. What will go wrong?

- If $\hat{p}$ is *smaller* than the null value, then a one-sided test where $p < p_0$ would mean that any observation in the *lower* 5% tail of the null distribution would lead to us rejecting H_0.

- If $\hat{p}$ is *larger* than the null value, then a one-sided test where $p > p_0$ would mean that any observation in the *upper* 5% tail of the null distribution would lead to us rejecting H_0.

Then if H_0 were true, there's a 10% chance of being in one of the two tails, so our testing error is actually $\alpha = 0.10$, not 0.05. That is, not being careful about when to use one-sided tests effectively undermines the methods we're working so hard to develop and utilize.

Exercises

5.15 Identify hypotheses, Part I. Write the null and alternative hypotheses in words and then symbols for each of the following situations.

(a) A tutoring company would like to understand if most students tend to improve their grades (or not) after they use their services. They sample 200 of the students who used their service in the past year and ask them if their grades have improved or declined from the previous year.

(b) Employers at a firm are worried about the effect of March Madness, a basketball championship held each spring in the US, on employee productivity. They estimate that on a regular business day employees spend on average 15 minutes of company time checking personal email, making personal phone calls, etc. They also collect data on how much company time employees spend on such non-business activities during March Madness. They want to determine if these data provide convincing evidence that employee productivity changed during March Madness.

5.16 Identify hypotheses, Part II. Write the null and alternative hypotheses in words and using symbols for each of the following situations.

(a) Since 2008, chain restaurants in California have been required to display calorie counts of each menu item. Prior to menus displaying calorie counts, the average calorie intake of diners at a restaurant was 1100 calories. After calorie counts started to be displayed on menus, a nutritionist collected data on the number of calories consumed at this restaurant from a random sample of diners. Do these data provide convincing evidence of a difference in the average calorie intake of a diners at this restaurant?

(b) The state of Wisconsin would like to understand the fraction of its adult residents that consumed alcohol in the last year, specifically if the rate is different from the national rate of 70%. To help them answer this question, they conduct a random sample of 852 residents and ask them about their alcohol consumption.

5.17 Online communication. A study suggests that 60% of college student spend 10 or more hours per week communicating with others online. You believe that this is incorrect and decide to collect your own sample for a hypothesis test. You randomly sample 160 students from your dorm and find that 70% spent 10 or more hours a week communicating with others online. A friend of yours, who offers to help you with the hypothesis test, comes up with the following set of hypotheses. Indicate any errors you see.

$$H_0 : \hat{p} < 0.6$$
$$H_A : \hat{p} > 0.7$$

5.18 Married at 25. A study suggests that the 25% of 25 year olds have gotten married. You believe that this is incorrect and decide to collect your own sample for a hypothesis test. From a random sample of 25 year olds in census data with size 776, you find that 24% of them are married. A friend of yours offers to help you with setting up the hypothesis test and comes up with the following hypotheses. Indicate any errors you see.

$$H_0 : \hat{p} = 0.24$$
$$H_A : \hat{p} \neq 0.24$$

5.19 Cyberbullying rates. Teens were surveyed about cyberbullying, and 54% to 64% reported experiencing cyberbullying (95% confidence interval).[23] Answer the following questions based on this interval.

(a) A newspaper claims that a majority of teens have experienced cyberbullying. Is this claim supported by the confidence interval? Explain your reasoning.

(b) A researcher conjectured that 70% of teens have experienced cyberbullying. Is this claim supported by the confidence interval? Explain your reasoning.

(c) Without actually calculating the interval, determine if the claim of the researcher from part (b) would be supported based on a 90% confidence interval?

---

[23]Pew Research Center, A Majority of Teens Have Experienced Some Form of Cyberbullying. September 27, 2018.

5.20 Waiting at an ER, Part II. Exercise 5.11 provides a 95% confidence interval for the mean waiting time at an emergency room (ER) of (128 minutes, 147 minutes). Answer the following questions based on this interval.

(a) A local newspaper claims that the average waiting time at this ER exceeds 3 hours. Is this claim supported by the confidence interval? Explain your reasoning.

(b) The Dean of Medicine at this hospital claims the average wait time is 2.2 hours. Is this claim supported by the confidence interval? Explain your reasoning.

(c) Without actually calculating the interval, determine if the claim of the Dean from part (b) would be supported based on a 99% confidence interval?

5.21 Minimum wage, Part I. Do a majority of US adults believe raising the minimum wage will help the economy, or is there a majority who do not believe this? A Rasmussen Reports survey of 1,000 US adults found that 42% believe it will help the economy.[24] Conduct an appropriate hypothesis test to help answer the research question.

5.22 Getting enough sleep. 400 students were randomly sampled from a large university, and 289 said they did not get enough sleep. Conduct a hypothesis test to check whether this represents a statistically significant difference from 50%, and use a significance level of 0.01.

5.23 Working backwards, Part I. You are given the following hypotheses:

$$H_0 : p = 0.3$$
$$H_A : p \neq 0.3$$

We know the sample size is 90. For what sample proportion would the p-value be equal to 0.05? Assume that all conditions necessary for inference are satisfied.

5.24 Working backwards, Part II. You are given the following hypotheses:

$$H_0 : p = 0.9$$
$$H_A : p \neq 0.9$$

We know that the sample size is 1,429. For what sample proportion would the p-value be equal to 0.01? Assume that all conditions necessary for inference are satisfied.

5.25 Testing for Fibromyalgia. A patient named Diana was diagnosed with Fibromyalgia, a long-term syndrome of body pain, and was prescribed anti-depressants. Being the skeptic that she is, Diana didn't initially believe that anti-depressants would help her symptoms. However after a couple months of being on the medication she decides that the anti-depressants are working, because she feels like her symptoms are in fact getting better.

(a) Write the hypotheses in words for Diana's skeptical position when she started taking the anti-depressants.

(b) What is a Type 1 Error in this context?

(c) What is a Type 2 Error in this context?

5.26 Which is higher? In each part below, there is a value of interest and two scenarios (I and II). For each part, report if the value of interest is larger under scenario I, scenario II, or whether the value is equal under the scenarios.

(a) The standard error of $\hat{p}$ when (I) $n = 125$ or (II) $n = 500$.

(b) The margin of error of a confidence interval when the confidence level is (I) 90% or (II) 80%.

(c) The p-value for a Z-statistic of 2.5 calculated based on a (I) sample with $n = 500$ or based on a (II) sample with $n = 1000$.

(d) The probability of making a Type 2 Error when the alternative hypothesis is true and the significance level is (I) 0.05 or (II) 0.10.

---

[24]Rasmussen Reports survey, Most Favor Minimum Wage of $10.50 Or Higher, April 16, 2019.

Chapter exercises

5.27 Relaxing after work. The General Social Survey asked the question: "After an average work day, about how many hours do you have to relax or pursue activities that you enjoy?" to a random sample of 1,155 Americans.[25] A 95% confidence interval for the mean number of hours spent relaxing or pursuing activities they enjoy was (1.38, 1.92).

(a) Interpret this interval in context of the data.

(b) Suppose another set of researchers reported a confidence interval with a larger margin of error based on the same sample of 1,155 Americans. How does their confidence level compare to the confidence level of the interval stated above?

(c) Suppose next year a new survey asking the same question is conducted, and this time the sample size is 2,500. Assuming that the population characteristics, with respect to how much time people spend relaxing after work, have not changed much within a year. How will the margin of error of the 95% confidence interval constructed based on data from the new survey compare to the margin of error of the interval stated above?

5.28 Minimum wage, Part II. In Exercise 5.21, we learned that a Rasmussen Reports survey of 1,000 US adults found that 42% believe raising the minimum wage will help the economy. Construct a 99% confidence interval for the true proportion of US adults who believe this.

5.29 Testing for food safety. A food safety inspector is called upon to investigate a restaurant with a few customer reports of poor sanitation practices. The food safety inspector uses a hypothesis testing framework to evaluate whether regulations are not being met. If he decides the restaurant is in gross violation, its license to serve food will be revoked.

(a) Write the hypotheses in words.

(b) What is a Type 1 Error in this context?

(c) What is a Type 2 Error in this context?

(d) Which error is more problematic for the restaurant owner? Why?

(e) Which error is more problematic for the diners? Why?

(f) As a diner, would you prefer that the food safety inspector requires strong evidence or very strong evidence of health concerns before revoking a restaurant's license? Explain your reasoning.

5.30 True or false. Determine if the following statements are true or false, and explain your reasoning. If false, state how it could be corrected.

(a) If a given value (for example, the null hypothesized value of a parameter) is within a 95% confidence interval, it will also be within a 99% confidence interval.

(b) Decreasing the significance level (α) will increase the probability of making a Type 1 Error.

(c) Suppose the null hypothesis is $p = 0.5$ and we fail to reject H_0. Under this scenario, the true population proportion is 0.5.

(d) With large sample sizes, even small differences between the null value and the observed point estimate, a difference often called the effect size, will be identified as statistically significant.

5.31 Unemployment and relationship problems. A USA Today/Gallup poll asked a group of unemployed and underemployed Americans if they have had major problems in their relationships with their spouse or another close family member as a result of not having a job (if unemployed) or not having a full-time job (if underemployed). 27% of the 1,145 unemployed respondents and 25% of the 675 underemployed respondents said they had major problems in relationships as a result of their employment status.

(a) What are the hypotheses for evaluating if the proportions of unemployed and underemployed people who had relationship problems were different?

(b) The p-value for this hypothesis test is approximately 0.35. Explain what this means in context of the hypothesis test and the data.

---

[25]National Opinion Research Center, General Social Survey, 2018.

5.32 Nearsighted. It is believed that nearsightedness affects about 8% of all children. In a random sample of 194 children, 21 are nearsighted. Conduct a hypothesis test for the following question: do these data provide evidence that the 8% value is inaccurate?

5.33 Nutrition labels. The nutrition label on a bag of potato chips says that a one ounce (28 gram) serving of potato chips has 130 calories and contains ten grams of fat, with three grams of saturated fat. A random sample of 35 bags yielded a confidence interval for the number of calories per bag of 128.2 to 139.8 calories. Is there evidence that the nutrition label does not provide an accurate measure of calories in the bags of potato chips?

5.34 CLT for proportions. Define the term "sampling distribution" of the sample proportion, and describe how the shape, center, and spread of the sampling distribution change as the sample size increases when $p = 0.1$.

5.35 Practical vs. statistical significance. Determine whether the following statement is true or false, and explain your reasoning: "With large sample sizes, even small differences between the null value and the observed point estimate can be statistically significant."

5.36 Same observation, different sample size. Suppose you conduct a hypothesis test based on a sample where the sample size is $n = 50$, and arrive at a p-value of 0.08. You then refer back to your notes and discover that you made a careless mistake, the sample size should have been $n = 500$. Will your p-value increase, decrease, or stay the same? Explain.

5.37 Gender pay gap in medicine. A study examined the average pay for men and women entering the workforce as doctors for 21 different positions.[26]

(a) If each gender was equally paid, then we would expect about half of those positions to have men paid more than women and women would be paid more than men in the other half of positions. Write appropriate hypotheses to test this scenario.

(b) Men were, on average, paid more in 19 of those 21 positions. Complete a hypothesis test using your hypotheses from part (a).

---

[26]Lo Sasso AT et al. "The $16,819 Pay Gap For Newly Trained Physicians: The Unexplained Trend Of Men Earning More Than Women". In: *Health Affairs* 30.2 (2011).

Chapter 6

Inference for categorical data

In this chapter, we apply the methods and ideas from Chapter 5 in several contexts for categorical data. We'll start by revisiting what we learned for a single proportion, where the normal distribution can be used to model the uncertainty in the sample proportion. Next, we apply these same ideas to analyze the difference of two proportions using the normal model. Later in the chapter, we apply inference techniques to contingency tables; while we will use a different distribution in this context, the core ideas of hypothesis testing remain the same.

For videos, slides, and other resources, please visit
www.openintro.org/os

6.1   Inference for a single proportion

We encountered inference methods for a single proportion in Chapter 5, exploring point estimates, confidence intervals, and hypothesis tests. In this section, we'll do a review of these topics and also how to choose an appropriate sample size when collecting data for single proportion contexts.

6.1.1   Identifying when the sample proportion is nearly normal

A sample proportion $\hat{p}$ can be modeled using a normal distribution when the sample observations are independent and the sample size is sufficiently large.

SAMPLING DISTRIBUTION OF $\hat{p}$

The sampling distribution for $\hat{p}$ based on a sample of size n from a population with a true proportion p is nearly normal when:

1. The sample's observations are independent, e.g. are from a simple random sample.

2. We expected to see at least 10 successes and 10 failures in the sample, i.e. $np \geq 10$ and $n(1-p) \geq 10$. This is called the **success-failure condition**.

When these conditions are met, then the sampling distribution of $\hat{p}$ is nearly normal with mean p and standard error $SE = \sqrt{\frac{p(1-p)}{n}}$.

Typically we don't know the true proportion p, so we substitute some value to check conditions and estimate the standard error. For confidence intervals, the sample proportion $\hat{p}$ is used to check the success-failure condition and compute the standard error. For hypothesis tests, typically the null value – that is, the proportion claimed in the null hypothesis – is used in place of p.

6.1.2   Confidence intervals for a proportion

A confidence interval provides a range of plausible values for the parameter p, and when $\hat{p}$ can be modeled using a normal distribution, the confidence interval for p takes the form

$$\hat{p} \pm z^{\star} \times SE$$

EXAMPLE 6.1

A simple random sample of 826 payday loan borrowers was surveyed to better understand their interests around regulation and costs. 70% of the responses supported new regulations on payday lenders. Is it reasonable to model $\hat{p} = 0.70$ using a normal distribution?

The data are a random sample, so the observations are independent and representative of the population of interest.

We also must check the success-failure condition, which we do using $\hat{p}$ in place of p when computing a confidence interval:

Support: $np \approx 826 \times 0.70 = 578$ Not: $n(1-p) \approx 826 \times (1 - 0.70) = 248$

Since both values are at least 10, we can use the normal distribution to model $\hat{p}$.

GUIDED PRACTICE 6.2

Estimate the standard error of $\hat{p} = 0.70$. Because p is unknown and the standard error is for a confidence interval, use $\hat{p}$ in place of p in the formula.[1]

EXAMPLE 6.3

Construct a 95% confidence interval for p, the proportion of payday borrowers who support increased regulation for payday lenders.

Using the point estimate 0.70, $z^\star = 1.96$ for a 95% confidence interval, and the standard error $SE = 0.016$ from Guided Practice 6.2, the confidence interval is

$$\text{point estimate} \pm z^\star \times SE \quad \rightarrow \quad 0.70 \pm 1.96 \times 0.016 \quad \rightarrow \quad (0.669, 0.731)$$

We are 95% confident that the true proportion of payday borrowers who supported regulation at the time of the poll was between 0.669 and 0.731.

CONFIDENCE INTERVAL FOR A SINGLE PROPORTION

Once you've determined a one-proportion confidence interval would be helpful for an application, there are four steps to constructing the interval:

Prepare. Identify $\hat{p}$ and n, and determine what confidence level you wish to use.

Check. Verify the conditions to ensure $\hat{p}$ is nearly normal. For one-proportion confidence intervals, use $\hat{p}$ in place of p to check the success-failure condition.

Calculate. If the conditions hold, compute SE using $\hat{p}$, find $z^\star$, and construct the interval.

Conclude. Interpret the confidence interval in the context of the problem.

For additional one-proportion confidence interval examples, see Section 5.2.

6.1.3 Hypothesis testing for a proportion

One possible regulation for payday lenders is that they would be required to do a credit check and evaluate debt payments against the borrower's finances. We would like to know: would borrowers support this form of regulation?

GUIDED PRACTICE 6.4

Set up hypotheses to evaluate whether borrowers have a majority support or majority opposition for this type of regulation.[2]

To apply the normal distribution framework in the context of a hypothesis test for a proportion, the independence and success-failure conditions must be satisfied. In a hypothesis test, the success-failure condition is checked using the null proportion: we verify np_0 and $n(1 - p_0)$ are at least 10, where p_0 is the null value.

---

[1] $SE = \sqrt{\frac{p(1-p)}{n}} \approx \sqrt{\frac{0.70(1-0.70)}{826}} = 0.016$.
[2] H_0: $p = 0.50$. H_A: $p \neq 0.50$.

GUIDED PRACTICE 6.5

Do payday loan borrowers support a regulation that would require lenders to pull their credit report and evaluate their debt payments? From a random sample of 826 borrowers, 51% said they would support such a regulation. Is it reasonable to model $\hat{p} = 0.51$ using a normal distribution for a hypothesis test here?[3]

EXAMPLE 6.6

Using the hypotheses and data from Guided Practice 6.4 and 6.5, evaluate whether the poll provides convincing evidence that a majority of payday loan borrowers support a new regulation that would require lenders to pull credit reports and evaluate debt payments.

With hypotheses already set up and conditions checked, we can move onto calculations. The standard error in the context of a one-proportion hypothesis test is computed using the null value, p_0:

$$SE = \sqrt{\frac{p_0(1 - p_0)}{n}} = \sqrt{\frac{0.5(1 - 0.5)}{826}} = 0.017$$

A picture of the normal model is shown below with the p-value represented by the shaded region.

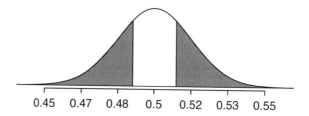

Based on the normal model, the test statistic can be computed as the Z-score of the point estimate:

$$Z = \frac{\text{point estimate} - \text{null value}}{SE} = \frac{0.51 - 0.50}{0.017} = 0.59$$

The single tail area is 0.2776, and the p-value, represented by both tail areas together, is 0.5552. Because the p-value is larger than 0.05, we do not reject H_0. The poll does not provide convincing evidence that a majority of payday loan borrowers support or oppose regulations around credit checks and evaluation of debt payments.

HYPOTHESIS TESTING FOR A SINGLE PROPORTION

Once you've determined a one-proportion hypothesis test is the correct procedure, there are four steps to completing the test:

Prepare. Identify the parameter of interest, list hypotheses, identify the significance level, and identify $\hat{p}$ and n.

Check. Verify conditions to ensure $\hat{p}$ is nearly normal under H_0. For one-proportion hypothesis tests, use the null value to check the success-failure condition.

Calculate. If the conditions hold, compute the standard error, again using p_0, compute the Z-score, and identify the p-value.

Conclude. Evaluate the hypothesis test by comparing the p-value to α, and provide a conclusion in the context of the problem.

For additional one-proportion hypothesis test examples, see Section 5.3.

---

[3]Independence holds since the poll is based on a random sample. The success-failure condition also holds, which is checked using the null value ($p_0 = 0.5$) from H_0: $np_0 = 826 \times 0.5 = 413$, $n(1 - p_0) = 826 \times 0.5 = 413$.

6.1.4 When one or more conditions aren't met

We've spent a lot of time discussing conditions for when $\hat{p}$ can be reasonably modeled by a normal distribution. What happens when the success-failure condition fails? What about when the independence condition fails? In either case, the general ideas of confidence intervals and hypothesis tests remain the same, but the strategy or technique used to generate the interval or p-value change.

When the success-failure condition isn't met for a hypothesis test, we can simulate the null distribution of $\hat{p}$ using the null value, p_0. The simulation concept is similar to the ideas used in the malaria case study presented in Section 2.3, and an online section outlines this strategy:

www.openintro.org/r?go=stat_sim_prop_ht

For a confidence interval when the success-failure condition isn't met, we can use what's called the **Clopper-Pearson interval**. The details are beyond the scope of this book. However, there are many internet resources covering this topic.

The independence condition is a more nuanced requirement. When it isn't met, it is important to understand how and why it isn't met. For example, if we took a cluster sample (see Section 1.3), suitable statistical methods are available but would be beyond the scope of even most second or third courses in statistics. On the other hand, we'd be stretched to find any method that we could confidently apply to correct the inherent biases of data from a convenience sample.

While this book is scoped to well-constrained statistical problems, do remember that this is just the first book in what is a large library of statistical methods that are suitable for a very wide range of data and contexts.

6.1.5 Choosing a sample size when estimating a proportion

When collecting data, we choose a sample size suitable for the purpose of the study. Often times this means choosing a sample size large enough that the **margin of error** – which is the part we add and subtract from the point estimate in a confidence interval – is sufficiently small that the sample is useful. For example, our task might be to find a sample size n so that the sample proportion is within ± 0.04 of the actual proportion in a 95% confidence interval.

EXAMPLE 6.7

A university newspaper is conducting a survey to determine what fraction of students support a $200 per year increase in fees to pay for a new football stadium. How big of a sample is required to ensure the margin of error is smaller than 0.04 using a 95% confidence level?

The margin of error for a sample proportion is

$$z^\star \sqrt{\frac{p(1-p)}{n}}$$

Our goal is to find the smallest sample size n so that this margin of error is smaller than 0.04. For a 95% confidence level, the value $z^\star$ corresponds to 1.96:

$$1.96 \times \sqrt{\frac{p(1-p)}{n}} < 0.04$$

There are two unknowns in the equation: p and n. If we have an estimate of p, perhaps from a prior survey, we could enter in that value and solve for n. If we have no such estimate, we must use some other value for p. It turns out that the margin of error is largest when p is 0.5, so we typically use this *worst case value* if no estimate of the proportion is available:

$$1.96 \times \sqrt{\frac{0.5(1-0.5)}{n}} < 0.04$$

$$1.96^2 \times \frac{0.5(1-0.5)}{n} < 0.04^2$$

$$1.96^2 \times \frac{0.5(1-0.5)}{0.04^2} < n$$

$$600.25 < n$$

We would need over 600.25 participants, which means we need 601 participants or more, to ensure the sample proportion is within 0.04 of the true proportion with 95% confidence.

When an estimate of the proportion is available, we use it in place of the worst case proportion value, 0.5.

GUIDED PRACTICE 6.8

A manager is about to oversee the mass production of a new tire model in her factory, and she would like to estimate what proportion of these tires will be rejected through quality control. The quality control team has monitored the last three tire models produced by the factory, failing 1.7% of tires in the first model, 6.2% of the second model, and 1.3% of the third model. The manager would like to examine enough tires to estimate the failure rate of the new tire model to within about 1% with a 90% confidence level. There are three different failure rates to choose from. Perform the sample size computation for each separately, and identify three sample sizes to consider.[4]

EXAMPLE 6.9

The sample sizes vary widely in Guided Practice 6.8. Which of the three would you suggest using? What would influence your choice?

We could examine which of the old models is most like the new model, then choose the corresponding sample size. Or if two of the previous estimates are based on small samples while the other is based on a larger sample, we might consider the value corresponding to the larger sample. There are also other reasonable approaches.

Also observe that the success-failure condition would need to be checked in the final sample. For instance, if we sampled $n = 1584$ tires and found a failure rate of 0.5%, the normal approximation would not be reasonable, and we would require more advanced statistical methods for creating the confidence interval.

GUIDED PRACTICE 6.10

Suppose we want to continually track the support of payday borrowers for regulation on lenders, where we would conduct a new poll every month. Running such frequent polls is expensive, so we decide a wider margin of error of 5% for each individual survey would be acceptable. Based on the original sample of borrowers where 70% supported some form of regulation, how big should our monthly sample be for a margin of error of 0.05 with 95% confidence?[5]

---

[4]For a 90% confidence interval, $z^\star = 1.65$, and since an estimate of the proportion 0.017 is available, we'll use it in the margin of error formula:

$$1.65 \times \sqrt{\frac{0.017(1 - 0.017)}{n}} < 0.01 \quad \rightarrow \quad \frac{0.017(1 - 0.017)}{n} < \left(\frac{0.01}{1.65}\right)^2 \quad \rightarrow \quad 454.96 < n$$

For sample size calculations, we always round up, so the first tire model suggests 455 tires would be sufficient.

A similar computation can be accomplished using 0.062 and 0.013 for p, and you should verify that using these proportions results in minimum sample sizes of 1584 and 350 tires, respectively.

[5]We complete the same computations as before, except now we use 0.70 instead of 0.5 for p:

$$1.96 \times \sqrt{\frac{p(1 - p)}{n}} \approx 1.96 \times \sqrt{\frac{0.70(1 - 0.70)}{n}} \leq 0.05 \quad \rightarrow \quad n \geq 322.7$$

A sample size of 323 or more would be reasonable. (Reminder: always round up for sample size calculations!) Given that we plan to track this poll over time, we also may want to periodically repeat these calculations to ensure that we're being thoughtful in our sample size recommendations in case the baseline rate fluctuates.

Exercises

6.1 Vegetarian college students. Suppose that 8% of college students are vegetarians. Determine if the following statements are true or false, and explain your reasoning.

(a) The distribution of the sample proportions of vegetarians in random samples of size 60 is approximately normal since $n \geq 30$.

(b) The distribution of the sample proportions of vegetarian college students in random samples of size 50 is right skewed.

(c) A random sample of 125 college students where 12% are vegetarians would be considered unusual.

(d) A random sample of 250 college students where 12% are vegetarians would be considered unusual.

(e) The standard error would be reduced by one-half if we increased the sample size from 125 to 250.

6.2 Young Americans, Part I. About 77% of young adults think they can achieve the American dream. Determine if the following statements are true or false, and explain your reasoning.[6]

(a) The distribution of sample proportions of young Americans who think they can achieve the American dream in samples of size 20 is left skewed.

(b) The distribution of sample proportions of young Americans who think they can achieve the American dream in random samples of size 40 is approximately normal since $n \geq 30$.

(c) A random sample of 60 young Americans where 85% think they can achieve the American dream would be considered unusual.

(d) A random sample of 120 young Americans where 85% think they can achieve the American dream would be considered unusual.

6.3 Orange tabbies. Suppose that 90% of orange tabby cats are male. Determine if the following statements are true or false, and explain your reasoning.

(a) The distribution of sample proportions of random samples of size 30 is left skewed.

(b) Using a sample size that is 4 times as large will reduce the standard error of the sample proportion by one-half.

(c) The distribution of sample proportions of random samples of size 140 is approximately normal.

(d) The distribution of sample proportions of random samples of size 280 is approximately normal.

6.4 Young Americans, Part II. About 25% of young Americans have delayed starting a family due to the continued economic slump. Determine if the following statements are true or false, and explain your reasoning.[7]

(a) The distribution of sample proportions of young Americans who have delayed starting a family due to the continued economic slump in random samples of size 12 is right skewed.

(b) In order for the distribution of sample proportions of young Americans who have delayed starting a family due to the continued economic slump to be approximately normal, we need random samples where the sample size is at least 40.

(c) A random sample of 50 young Americans where 20% have delayed starting a family due to the continued economic slump would be considered unusual.

(d) A random sample of 150 young Americans where 20% have delayed starting a family due to the continued economic slump would be considered unusual.

(e) Tripling the sample size will reduce the standard error of the sample proportion by one-third.

---

[6]A. Vaughn. "Poll finds young adults optimistic, but not about money". In: *Los Angeles Times* (2011).
[7]Demos.org. "The State of Young America: The Poll". In: (2011).

6.5 Gender equality. The General Social Survey asked a random sample of 1,390 Americans the following question: "On the whole, do you think it should or should not be the government's responsibility to promote equality between men and women?" 82% of the respondents said it "should be". At a 95% confidence level, this sample has 2% margin of error. Based on this information, determine if the following statements are true or false, and explain your reasoning.[8]

(a) We are 95% confident that between 80% and 84% of Americans in this sample think it's the government's responsibility to promote equality between men and women.

(b) We are 95% confident that between 80% and 84% of all Americans think it's the government's responsibility to promote equality between men and women.

(c) If we considered many random samples of 1,390 Americans, and we calculated 95% confidence intervals for each, 95% of these intervals would include the true population proportion of Americans who think it's the government's responsibility to promote equality between men and women.

(d) In order to decrease the margin of error to 1%, we would need to quadruple (multiply by 4) the sample size.

(e) Based on this confidence interval, there is sufficient evidence to conclude that a majority of Americans think it's the government's responsibility to promote equality between men and women.

6.6 Elderly drivers. The Marist Poll published a report stating that 66% of adults nationally think licensed drivers should be required to retake their road test once they reach 65 years of age. It was also reported that interviews were conducted on 1,018 American adults, and that the margin of error was 3% using a 95% confidence level.[9]

(a) Verify the margin of error reported by The Marist Poll.

(b) Based on a 95% confidence interval, does the poll provide convincing evidence that *more than* 70% of the population think that licensed drivers should be required to retake their road test once they turn 65?

6.7 Fireworks on July 4th. A local news outlet reported that 56% of 600 randomly sampled Kansas residents planned to set off fireworks on July 4th. Determine the margin of error for the 56% point estimate using a 95% confidence level.[10]

6.8 Life rating in Greece. Greece has faced a severe economic crisis since the end of 2009. A Gallup poll surveyed 1,000 randomly sampled Greeks in 2011 and found that 25% of them said they would rate their lives poorly enough to be considered "suffering".[11]

(a) Describe the population parameter of interest. What is the value of the point estimate of this parameter?

(b) Check if the conditions required for constructing a confidence interval based on these data are met.

(c) Construct a 95% confidence interval for the proportion of Greeks who are "suffering".

(d) Without doing any calculations, describe what would happen to the confidence interval if we decided to use a higher confidence level.

(e) Without doing any calculations, describe what would happen to the confidence interval if we used a larger sample.

6.9 Study abroad. A survey on 1,509 high school seniors who took the SAT and who completed an optional web survey shows that 55% of high school seniors are fairly certain that they will participate in a study abroad program in college.[12]

(a) Is this sample a representative sample from the population of all high school seniors in the US? Explain your reasoning.

(b) Let's suppose the conditions for inference are met. Even if your answer to part (a) indicated that this approach would not be reliable, this analysis may still be interesting to carry out (though not report). Construct a 90% confidence interval for the proportion of high school seniors (of those who took the SAT) who are fairly certain they will participate in a study abroad program in college, and interpret this interval in context.

(c) What does "90% confidence" mean?

(d) Based on this interval, would it be appropriate to claim that the majority of high school seniors are fairly certain that they will participate in a study abroad program in college?

[8]National Opinion Research Center, General Social Survey, 2018.

[9]Marist Poll, Road Rules: Re-Testing Drivers at Age 65?, March 4, 2011.

[10]Survey USA, News Poll #19333, data collected on June 27, 2012.

[11]Gallup World, More Than One in 10 "Suffering" Worldwide, data collected throughout 2011.

[12]studentPOLL, College-Bound Students' Interests in Study Abroad and Other International Learning Activities, January 2008.

6.10   Legalization of marijuana, Part I. The General Social Survey asked 1,578 US residents: "Do you think the use of marijuana should be made legal, or not?" 61% of the respondents said it should be made legal.[13]

(a) Is 61% a sample statistic or a population parameter? Explain.

(b) Construct a 95% confidence interval for the proportion of US residents who think marijuana should be made legal, and interpret it in the context of the data.

(c) A critic points out that this 95% confidence interval is only accurate if the statistic follows a normal distribution, or if the normal model is a good approximation. Is this true for these data? Explain.

(d) A news piece on this survey's findings states, "Majority of Americans think marijuana should be legalized." Based on your confidence interval, is this news piece's statement justified?

6.11   National Health Plan, Part I. A *Kaiser Family Foundation* poll for US adults in 2019 found that 79% of Democrats, 55% of Independents, and 24% of Republicans supported a generic "National Health Plan". There were 347 Democrats, 298 Republicans, and 617 Independents surveyed.[14]

(a) A political pundit on TV claims that a majority of Independents support a National Health Plan. Do these data provide strong evidence to support this type of statement?

(b) Would you expect a confidence interval for the proportion of Independents who oppose the public option plan to include 0.5? Explain.

6.12   Is college worth it? Part I. Among a simple random sample of 331 American adults who do not have a four-year college degree and are not currently enrolled in school, 48% said they decided not to go to college because they could not afford school.[15]

(a) A newspaper article states that only a minority of the Americans who decide not to go to college do so because they cannot afford it and uses the point estimate from this survey as evidence. Conduct a hypothesis test to determine if these data provide strong evidence supporting this statement.

(b) Would you expect a confidence interval for the proportion of American adults who decide not to go to college because they cannot afford it to include 0.5? Explain.

6.13   Taste test. Some people claim that they can tell the difference between a diet soda and a regular soda in the first sip. A researcher wanting to test this claim randomly sampled 80 such people. He then filled 80 plain white cups with soda, half diet and half regular through random assignment, and asked each person to take one sip from their cup and identify the soda as diet or regular. 53 participants correctly identified the soda.

(a) Do these data provide strong evidence that these people are any better or worse than random guessing at telling the difference between diet and regular soda?

(b) Interpret the p-value in this context.

6.14   Is college worth it? Part II. Exercise 6.12 presents the results of a poll where 48% of 331 Americans who decide to not go to college do so because they cannot afford it.

(a) Calculate a 90% confidence interval for the proportion of Americans who decide to not go to college because they cannot afford it, and interpret the interval in context.

(b) Suppose we wanted the margin of error for the 90% confidence level to be about 1.5%. How large of a survey would you recommend?

6.15   National Health Plan, Part II. Exercise 6.11 presents the results of a poll evaluating support for a generic "National Health Plan" in the US in 2019, reporting that 55% of Independents are supportive. If we wanted to estimate this number to within 1% with 90% confidence, what would be an appropriate sample size?

6.16   Legalize Marijuana, Part II. As discussed in Exercise 6.10, the General Social Survey reported a sample where about 61% of US residents thought marijuana should be made legal. If we wanted to limit the margin of error of a 95% confidence interval to 2%, about how many Americans would we need to survey?

---

[13]National Opinion Research Center, General Social Survey, 2018.

[14]Kaiser Family Foundation, The Public On Next Steps For The ACA And Proposals To Expand Coverage, data collected between Jan 9-14, 2019.

[15]Pew Research Center Publications, Is College Worth It?, data collected between March 15-29, 2011.

6.2 Difference of two proportions

We would like to extend the methods from Section 6.1 to apply confidence intervals and hypothesis tests to differences in population proportions: $p_1 - p_2$. In our investigations, we'll identify a reasonable point estimate of $p_1 - p_2$ based on the sample, and you may have already guessed its form: $\hat{p}_1 - \hat{p}_2$. Next, we'll apply the same processes we used in the single-proportion context: we verify that the point estimate can be modeled using a normal distribution, we compute the estimate's standard error, and we apply our inferential framework.

6.2.1 Sampling distribution of the difference of two proportions

Like with $\hat{p}$, the difference of two sample proportions $\hat{p}_1 - \hat{p}_2$ can be modeled using a normal distribution when certain conditions are met. First, we require a broader independence condition, and secondly, the success-failure condition must be met by both groups.

CONDITIONS FOR THE SAMPLING DISTRIBUTION OF $\hat{p}_1 - \hat{p}_2$ TO BE NORMAL

The difference $\hat{p}_1 - \hat{p}_2$ can be modeled using a normal distribution when

- *Independence, extended.* The data are independent within and between the two groups. Generally this is satisfied if the data come from two independent random samples or if the data come from a randomized experiment.
- *Success-failure condition.* The success-failure condition holds for both groups, where we check successes and failures in each group separately.

When these conditions are satisfied, the standard error of $\hat{p}_1 - \hat{p}_2$ is

$$SE = \sqrt{\frac{p_1(1-p_1)}{n_1} + \frac{p_2(1-p_2)}{n_2}}$$

where p_1 and p_2 represent the population proportions, and n_1 and n_2 represent the sample sizes.

6.2.2 Confidence intervals for $p_1 - p_2$

We can apply the generic confidence interval formula for a difference of two proportions, where we use $\hat{p}_1 - \hat{p}_2$ as the point estimate and substitute the SE formula:

$$\text{point estimate} \pm z^\star \times SE \qquad \rightarrow \qquad \hat{p}_1 - \hat{p}_2 \pm z^\star \times \sqrt{\frac{p_1(1-p_1)}{n_1} + \frac{p_2(1-p_2)}{n_2}}$$

We can also follow the same Prepare, Check, Calculate, Conclude steps for computing a confidence interval or completing a hypothesis test. The details change a little, but the general approach remain the same. Think about these steps when you apply statistical methods.

EXAMPLE 6.11

We consider an experiment for patients who underwent cardiopulmonary resuscitation (CPR) for a heart attack and were subsequently admitted to a hospital. These patients were randomly divided into a treatment group where they received a blood thinner or the control group where they did not receive a blood thinner. The outcome variable of interest was whether the patients survived for at least 24 hours. The results are shown in Figure 6.1. Check whether we can model the difference in sample proportions using the normal distribution.

We first check for independence: since this is a randomized experiment, this condition is satisfied.

Next, we check the success-failure condition for each group. We have at least 10 successes and 10 failures in each experiment arm (11, 14, 39, 26), so this condition is also satisfied.

With both conditions satisfied, the difference in sample proportions can be reasonably modeled using a normal distribution for these data.

	Survived	Died	Total
Control	11	39	50
Treatment	14	26	40
Total	25	65	90

Figure 6.1: Results for the CPR study. Patients in the treatment group were given a blood thinner, and patients in the control group were not.

EXAMPLE 6.12

Create and interpret a 90% confidence interval of the difference for the survival rates in the CPR study.

We'll use p_t for the survival rate in the treatment group and p_c for the control group:

$$\hat{p}_t - \hat{p}_c = \frac{14}{40} - \frac{11}{50} = 0.35 - 0.22 = 0.13$$

We use the standard error formula provided on page 217. As with the one-sample proportion case, we use the sample estimates of each proportion in the formula in the confidence interval context:

$$SE \approx \sqrt{\frac{0.35(1-0.35)}{40} + \frac{0.22(1-0.22)}{50}} = 0.095$$

For a 90% confidence interval, we use $z^\star = 1.65$:

$$\text{point estimate} \pm z^\star \times SE \quad \rightarrow \quad 0.13 \pm 1.65 \times 0.095 \quad \rightarrow \quad (-0.027, 0.287)$$

We are 90% confident that blood thinners have a difference of -2.7% to +28.7% percentage point impact on survival rate for patients who are like those in the study. Because 0% is contained in the interval, we do not have enough information to say whether blood thinners help or harm heart attack patients who have been admitted after they have undergone CPR.

GUIDED PRACTICE 6.13

A 5-year experiment was conducted to evaluate the effectiveness of fish oils on reducing cardiovascular events, where each subject was randomized into one of two treatment groups. We'll consider heart attack outcomes in these patients:

	heart attack	no event	Total
fish oil	145	12788	12933
placebo	200	12738	12938

Create a 95% confidence interval for the effect of fish oils on heart attacks for patients who are well-represented by those in the study. Also interpret the interval in the context of the study.[16]

6.2.3 Hypothesis tests for the difference of two proportions

A mammogram is an X-ray procedure used to check for breast cancer. Whether mammograms should be used is part of a controversial discussion, and it's the topic of our next example where we learn about 2-proportion hypothesis tests when H_0 is $p_1 - p_2 = 0$ (or equivalently, $p_1 = p_2$).

A 30-year study was conducted with nearly 90,000 female participants. During a 5-year screening period, each woman was randomized to one of two groups: in the first group, women received regular mammograms to screen for breast cancer, and in the second group, women received regular non-mammogram breast cancer exams. No intervention was made during the following 25 years of the study, and we'll consider death resulting from breast cancer over the full 30-year period. Results from the study are summarized in Figure 6.2.

If mammograms are much more effective than non-mammogram breast cancer exams, then we would expect to see additional deaths from breast cancer in the control group. On the other hand, if mammograms are not as effective as regular breast cancer exams, we would expect to see an increase in breast cancer deaths in the mammogram group.

	Death from breast cancer?	
	Yes	No
Mammogram	500	44,425
Control	505	44,405

Figure 6.2: Summary results for breast cancer study.

GUIDED PRACTICE 6.14

Is this study an experiment or an observational study?[17]

---

[16] Because the patients were randomized, the subjects are independent, both within and between the two groups. The success-failure condition is also met for both groups as all counts are at least 10. This satisfies the conditions necessary to model the difference in proportions using a normal distribution.
Compute the sample proportions ($\hat{p}_{\text{fish oil}} = 0.0112$, $\hat{p}_{\text{placebo}} = 0.0155$), point estimate of the difference ($0.0112 - 0.0155 = -0.0043$), and standard error ($SE = \sqrt{\frac{0.0112 \times 0.9888}{12933} + \frac{0.0155 \times 0.9845}{12938}} = 0.00145$). Next, plug the values into the general formula for a confidence interval, where we'll use a 95% confidence level with $z^\star = 1.96$:

$$-0.0043 \pm 1.96 \times 0.00145 \quad \rightarrow \quad (-0.0071, -0.0015)$$

We are 95% confident that fish oils decreases heart attacks by 0.15 to 0.71 percentage points (off of a baseline of about 1.55%) over a 5-year period for subjects who are similar to those in the study. Because the interval is entirely below 0, the data provide strong evidence that fish oil supplements reduce heart attacks in patients like those in the study.
[17] This is an experiment. Patients were randomized to receive mammograms or a standard breast cancer exam. We will be able to make causal conclusions based on this study.

GUIDED PRACTICE 6.15

Set up hypotheses to test whether there was a difference in breast cancer deaths in the mammogram and control groups.[18]

In Example 6.16, we will check the conditions for using a normal distribution to analyze the results of the study. The details are very similar to that of confidence intervals. However, when the null hypothesis is that $p_1 - p_2 = 0$, we use a special proportion called the **pooled proportion** to check the success-failure condition:

$$\hat{p}_{pooled} = \frac{\text{\# of patients who died from breast cancer in the entire study}}{\text{\# of patients in the entire study}}$$
$$= \frac{500 + 505}{500 + 44{,}425 + 505 + 44{,}405}$$
$$= 0.0112$$

This proportion is an estimate of the breast cancer death rate across the entire study, and it's our best estimate of the proportions p_{mgm} and p_{ctrl} *if the null hypothesis is true that* $p_{mgm} = p_{ctrl}$. We will also use this pooled proportion when computing the standard error.

EXAMPLE 6.16

Is it reasonable to model the difference in proportions using a normal distribution in this study?

Because the patients are randomized, they can be treated as independent, both within and between groups. We also must check the success-failure condition for each group. Under the null hypothesis, the proportions p_{mgm} and p_{ctrl} are equal, so we check the success-failure condition with our best estimate of these values under H_0, the pooled proportion from the two samples, $\hat{p}_{pooled} = 0.0112$:

$$\hat{p}_{pooled} \times n_{mgm} = 0.0112 \times 44{,}925 = 503 \qquad (1 - \hat{p}_{pooled}) \times n_{mgm} = 0.9888 \times 44{,}925 = 44{,}422$$
$$\hat{p}_{pooled} \times n_{ctrl} = 0.0112 \times 44{,}910 = 503 \qquad (1 - \hat{p}_{pooled}) \times n_{ctrl} = 0.9888 \times 44{,}910 = 44{,}407$$

The success-failure condition is satisfied since all values are at least 10. With both conditions satisfied, we can safely model the difference in proportions using a normal distribution.

USE THE POOLED PROPORTION WHEN H_0 IS $p_1 - p_2 = 0$

When the null hypothesis is that the proportions are equal, use the pooled proportion ($\hat{p}_{pooled}$) to verify the success-failure condition and estimate the standard error:

$$\hat{p}_{pooled} = \frac{\text{number of "successes"}}{\text{number of cases}} = \frac{\hat{p}_1 n_1 + \hat{p}_2 n_2}{n_1 + n_2}$$

Here $\hat{p}_1 n_1$ represents the number of successes in sample 1 since

$$\hat{p}_1 = \frac{\text{number of successes in sample 1}}{n_1}$$

Similarly, $\hat{p}_2 n_2$ represents the number of successes in sample 2.

In Example 6.16, the pooled proportion was used to check the success-failure condition.[19] In the next example, we see the second place where the pooled proportion comes into play: the standard error calculation.

---

[18]H_0: the breast cancer death rate for patients screened using mammograms is the same as the breast cancer death rate for patients in the control, $p_{mgm} - p_{ctrl} = 0$.
H_A: the breast cancer death rate for patients screened using mammograms is different than the breast cancer death rate for patients in the control, $p_{mgm} - p_{ctrl} \neq 0$.

[19]For an example of a two-proportion hypothesis test that does not require the success-failure condition to be met, see Section 2.3.

EXAMPLE 6.17

Compute the point estimate of the difference in breast cancer death rates in the two groups, and use the pooled proportion $\hat{p}_{pooled} = 0.0112$ to calculate the standard error.

The point estimate of the difference in breast cancer death rates is

$$\hat{p}_{mgm} - \hat{p}_{ctrl} = \frac{500}{500 + 44,425} - \frac{505}{505 + 44,405}$$
$$= 0.01113 - 0.01125$$
$$= -0.00012$$

The breast cancer death rate in the mammogram group was 0.012% less than in the control group. Next, the standard error is calculated *using the pooled proportion, $\hat{p}_{pooled}$*:

$$SE = \sqrt{\frac{\hat{p}_{pooled}(1 - \hat{p}_{pooled})}{n_{mgm}} + \frac{\hat{p}_{pooled}(1 - \hat{p}_{pooled})}{n_{ctrl}}} = 0.00070$$

EXAMPLE 6.18

Using the point estimate $\hat{p}_{mgm} - \hat{p}_{ctrl} = -0.00012$ and standard error $SE = 0.00070$, calculate a p-value for the hypothesis test and write a conclusion.

Just like in past tests, we first compute a test statistic and draw a picture:

$$Z = \frac{\text{point estimate} - \text{null value}}{SE} = \frac{-0.00012 - 0}{0.00070} = -0.17$$

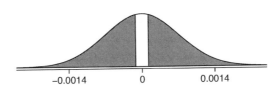

The lower tail area is 0.4325, which we double to get the p-value: 0.8650. Because this p-value is larger than 0.05, we do not reject the null hypothesis. That is, the difference in breast cancer death rates is reasonably explained by chance, and we do not observe benefits or harm from mammograms relative to a regular breast exam.

Can we conclude that mammograms have no benefits or harm? Here are a few considerations to keep in mind when reviewing the mammogram study as well as any other medical study:

- We do not accept the null hypothesis, which means we don't have sufficient evidence to conclude that mammograms reduce or increase breast cancer deaths.
- If mammograms are helpful or harmful, the data suggest the effect isn't very large.
- Are mammograms more or less expensive than a non-mammogram breast exam? If one option is much more expensive than the other and doesn't offer clear benefits, then we should lean towards the less expensive option.
- The study's authors also found that mammograms led to overdiagnosis of breast cancer, which means some breast cancers were found (or thought to be found) but that these cancers would not cause symptoms during patients' lifetimes. That is, something else would kill the patient before breast cancer symptoms appeared. This means some patients may have been treated for breast cancer unnecessarily, and this treatment is another cost to consider. It is also important to recognize that overdiagnosis can cause unnecessary physical or emotional harm to patients.

These considerations highlight the complexity around medical care and treatment recommendations. Experts and medical boards who study medical treatments use considerations like those above to provide their best recommendation based on the current evidence.

6.2.4   More on 2-proportion hypothesis tests (special topic)

When we conduct a 2-proportion hypothesis test, usually H_0 is $p_1 - p_2 = 0$. However, there are rare situations where we want to check for some difference in p_1 and p_2 that is some value other than 0. For example, maybe we care about checking a null hypothesis where $p_1 - p_2 = 0.1$. In contexts like these, we generally use $\hat{p}_1$ and $\hat{p}_2$ to check the success-failure condition and construct the standard error.

GUIDED PRACTICE 6.19

A quadcopter company is considering a new manufacturer for rotor blades. The new manufacturer would be more expensive, but they claim their higher-quality blades are more reliable, with 3% more blades passing inspection than their competitor. Set up appropriate hypotheses for the test.[20]

Figure 6.3: A Phantom quadcopter.

---

[20]H_0: The higher-quality blades will pass inspection 3% more frequently than the standard-quality blades. $p_{highQ} - p_{standard} = 0.03$. H_A: The higher-quality blades will pass inspection some amount different than 3% more often than the standard-quality blades. $p_{highQ} - p_{standard} \neq 0.03$.

EXAMPLE 6.20

The quality control engineer from Guided Practice 6.19 collects a sample of blades, examining 1000 blades from each company, and she finds that 899 blades pass inspection from the current supplier and 958 pass inspection from the prospective supplier. Using these data, evaluate the hypotheses from Guided Practice 6.19 with a significance level of 5%.

First, we check the conditions. The sample is not necessarily random, so to proceed we must assume the blades are all independent; for this sample we will suppose this assumption is reasonable, but the engineer would be more knowledgeable as to whether this assumption is appropriate. The success-failure condition also holds for each sample. Thus, the difference in sample proportions, $0.958 - 0.899 = 0.059$, can be said to come from a nearly normal distribution.

The standard error is computed using the two sample proportions since we do not use a pooled proportion for this context:

$$SE = \sqrt{\frac{0.958(1 - 0.958)}{1000} + \frac{0.899(1 - 0.899)}{1000}} = 0.0114$$

In this hypothesis test, because the null is that $p_1 - p_2 = 0.03$, the sample proportions were used for the standard error calculation rather than a pooled proportion.

Next, we compute the test statistic and use it to find the p-value, which is depicted in Figure 6.4.

$$Z = \frac{\text{point estimate} - \text{null value}}{SE} = \frac{0.059 - 0.03}{0.0114} = 2.54$$

Using a standard normal distribution for this test statistic, we identify the right tail area as 0.006, and we double it to get the p-value: 0.012. We reject the null hypothesis because 0.012 is less than 0.05. Since we observed a larger-than-3% increase in blades that pass inspection, we have statistically significant evidence that the higher-quality blades pass inspection *more than* 3% as often as the currently used blades, exceeding the company's claims.

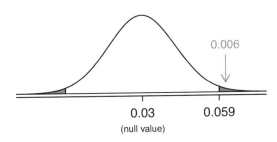

Figure 6.4: Distribution of the test statistic if the null hypothesis was true. The p-value is represented by the shaded areas.

6.2.5 Examining the standard error formula (special topic)

This subsection covers more theoretical topics that offer deeper insights into the origins of the standard error formula for the difference of two proportions. Ultimately, all of the standard error formulas we encounter in this chapter and in Chapter 7 can be derived from the probability principles of Section 3.4.

The formula for the standard error of the difference in two proportions can be deconstructed into the formulas for the standard errors of the individual sample proportions. Recall that the standard error of the individual sample proportions $\hat{p}_1$ and $\hat{p}_2$ are

$$SE_{\hat{p}_1} = \sqrt{\frac{p_1(1-p_1)}{n_1}} \qquad\qquad SE_{\hat{p}_2} = \sqrt{\frac{p_2(1-p_2)}{n_2}}$$

The standard error of the difference of two sample proportions can be deconstructed from the standard errors of the separate sample proportions:

$$SE_{\hat{p}_1-\hat{p}_2} = \sqrt{SE_{\hat{p}_1}^2 + SE_{\hat{p}_2}^2} = \sqrt{\frac{p_1(1-p_1)}{n_1} + \frac{p_2(1-p_2)}{n_2}}$$

This special relationship follows from probability theory.

GUIDED PRACTICE 6.21

Prerequisite: Section 3.4. We can rewrite the equation above in a different way:

$$SE_{\hat{p}_1-\hat{p}_2}^2 = SE_{\hat{p}_1}^2 + SE_{\hat{p}_2}^2$$

Explain where this formula comes from using the formula for the variability of the sum of two random variables.[21]

---

[21]The standard error squared represents the variance of the estimate. If X and Y are two random variables with variances σ_x^2 and σ_y^2, then the variance of $X - Y$ is $\sigma_x^2 + \sigma_y^2$. Likewise, the variance corresponding to $\hat{p}_1 - \hat{p}_2$ is $\sigma_{\hat{p}_1}^2 + \sigma_{\hat{p}_2}^2$. Because $\sigma_{\hat{p}_1}^2$ and $\sigma_{\hat{p}_2}^2$ are just another way of writing $SE_{\hat{p}_1}^2$ and $SE_{\hat{p}_2}^2$, the variance associated with $\hat{p}_1 - \hat{p}_2$ may be written as $SE_{\hat{p}_1}^2 + SE_{\hat{p}_2}^2$.

Exercises

6.17 Social experiment, Part I. A "social experiment" conducted by a TV program questioned what people do when they see a very obviously bruised woman getting picked on by her boyfriend. On two different occasions at the same restaurant, the same couple was depicted. In one scenario the woman was dressed "provocatively" and in the other scenario the woman was dressed "conservatively". The table below shows how many restaurant diners were present under each scenario, and whether or not they intervened.

		Scenario		Total
		Provocative	Conservative	
Intervene	Yes	5	15	20
	No	15	10	25
	Total	20	25	45

Explain why the sampling distribution of the difference between the proportions of interventions under provocative and conservative scenarios does not follow an approximately normal distribution.

6.18 Heart transplant success. The Stanford University Heart Transplant Study was conducted to determine whether an experimental heart transplant program increased lifespan. Each patient entering the program was officially designated a heart transplant candidate, meaning that he was gravely ill and might benefit from a new heart. Patients were randomly assigned into treatment and control groups. Patients in the treatment group received a transplant, and those in the control group did not. The table below displays how many patients survived and died in each group.[22]

	control	treatment
alive	4	24
dead	30	45

Suppose we are interested in estimating the difference in survival rate between the control and treatment groups using a confidence interval. Explain why we cannot construct such an interval using the normal approximation. What might go wrong if we constructed the confidence interval despite this problem?

6.19 Gender and color preference. A study asked 1,924 male and 3,666 female undergraduate college students their favorite color. A 95% confidence interval for the difference between the proportions of males and females whose favorite color is black ($p_{male} - p_{female}$) was calculated to be (0.02, 0.06). Based on this information, determine if the following statements are true or false, and explain your reasoning for each statement you identify as false.[23]

(a) We are 95% confident that the true proportion of males whose favorite color is black is 2% lower to 6% higher than the true proportion of females whose favorite color is black.

(b) We are 95% confident that the true proportion of males whose favorite color is black is 2% to 6% higher than the true proportion of females whose favorite color is black.

(c) 95% of random samples will produce 95% confidence intervals that include the true difference between the population proportions of males and females whose favorite color is black.

(d) We can conclude that there is a significant difference between the proportions of males and females whose favorite color is black and that the difference between the two sample proportions is too large to plausibly be due to chance.

(e) The 95% confidence interval for ($p_{female} - p_{male}$) cannot be calculated with only the information given in this exercise.

---

[22]B. Turnbull et al. "Survivorship of Heart Transplant Data". In: *Journal of the American Statistical Association* 69 (1974), pp. 74–80.

[23]L. Ellis and C. Ficek. "Color preferences according to gender and sexual orientation". In: *Personality and Individual Differences* 31.8 (2001), pp. 1375–1379.

6.20 Government shutdown. The United States federal government shutdown of 2018–2019 occurred from December 22, 2018 until January 25, 2019, a span of 35 days. A Survey USA poll of 614 randomly sampled Americans during this time period reported that 48% of those who make less than $40,000 per year and 55% of those who make $40,000 or more per year said the government shutdown has not at all affected them personally. A 95% confidence interval for $(p_{<40K} - p_{\geq 40K})$, where p is the proportion of those who said the government shutdown has not at all affected them personally, is (-0.16, 0.02). Based on this information, determine if the following statements are true or false, and explain your reasoning if you identify the statement as false.[24]

(a) At the 5% significance level, the data provide convincing evidence of a real difference in the proportion who are not affected personally between Americans who make less than $40,000 annually and Americans who make $40,000 annually.

(b) We are 95% confident that 16% more to 2% fewer Americans who make less than $40,000 per year are not at all personally affected by the government shutdown compared to those who make $40,000 or more per year.

(c) A 90% confidence interval for $(p_{<40K} - p_{\geq 40K})$ would be wider than the $(-0.16, 0.02)$ interval.

(d) A 95% confidence interval for $(p_{\geq 40K} - p_{<40K})$ is (-0.02, 0.16).

6.21 National Health Plan, Part III. Exercise 6.11 presents the results of a poll evaluating support for a generically branded "National Health Plan" in the United States. 79% of 347 Democrats and 55% of 617 Independents support a National Health Plan.

(a) Calculate a 95% confidence interval for the difference between the proportion of Democrats and Independents who support a National Health Plan $(p_D - p_I)$, and interpret it in this context. We have already checked conditions for you.

(b) True or false: If we had picked a random Democrat and a random Independent at the time of this poll, it is more likely that the Democrat would support the National Health Plan than the Independent.

6.22 Sleep deprivation, CA vs. OR, Part I. According to a report on sleep deprivation by the Centers for Disease Control and Prevention, the proportion of California residents who reported insufficient rest or sleep during each of the preceding 30 days is 8.0%, while this proportion is 8.8% for Oregon residents. These data are based on simple random samples of 11,545 California and 4,691 Oregon residents. Calculate a 95% confidence interval for the difference between the proportions of Californians and Oregonians who are sleep deprived and interpret it in context of the data.[25]

6.23 Offshore drilling, Part I. A survey asked 827 randomly sampled registered voters in California "Do you support? Or do you oppose? Drilling for oil and natural gas off the Coast of California? Or do you not know enough to say?" Below is the distribution of responses, separated based on whether or not the respondent graduated from college.[26]

(a) What percent of college graduates and what percent of the non-college graduates in this sample do not know enough to have an opinion on drilling for oil and natural gas off the Coast of California?

(b) Conduct a hypothesis test to determine if the data provide strong evidence that the proportion of college graduates who do not have an opinion on this issue is different than that of non-college graduates.

	College Grad	
	Yes	No
Support	154	132
Oppose	180	126
Do not know	104	131
Total	438	389

6.24 Sleep deprivation, CA vs. OR, Part II. Exercise 6.22 provides data on sleep deprivation rates of Californians and Oregonians. The proportion of California residents who reported insufficient rest or sleep during each of the preceding 30 days is 8.0%, while this proportion is 8.8% for Oregon residents. These data are based on simple random samples of 11,545 California and 4,691 Oregon residents.

(a) Conduct a hypothesis test to determine if these data provide strong evidence the rate of sleep deprivation is different for the two states. (Reminder: Check conditions)

(b) It is possible the conclusion of the test in part (a) is incorrect. If this is the case, what type of error was made?

[24]Survey USA, News Poll #24568, data collected on April 21, 2019.
[25]CDC, Perceived Insufficient Rest or Sleep Among Adults — United States, 2008.
[26]Survey USA, Election Poll #16804, data collected July 8-11, 2010.

6.25 Offshore drilling, Part II. Results of a poll evaluating support for drilling for oil and natural gas off the coast of California were introduced in Exercise 6.23.

	College Grad	
	Yes	No
Support	154	132
Oppose	180	126
Do not know	104	131
Total	438	389

(a) What percent of college graduates and what percent of the non-college graduates in this sample support drilling for oil and natural gas off the Coast of California?

(b) Conduct a hypothesis test to determine if the data provide strong evidence that the proportion of college graduates who support off-shore drilling in California is different than that of non-college graduates.

6.26 Full body scan, Part I. A news article reports that "Americans have differing views on two potentially inconvenient and invasive practices that airports could implement to uncover potential terrorist attacks." This news piece was based on a survey conducted among a random sample of 1,137 adults nationwide, where one of the questions on the survey was "Some airports are now using 'full-body' digital x-ray machines to electronically screen passengers in airport security lines. Do you think these new x-ray machines should or should not be used at airports?" Below is a summary of responses based on party affiliation.[27]

		Party Affiliation		
		Republican	Democrat	Independent
Answer	Should	264	299	351
	Should not	38	55	77
	Don't know/No answer	16	15	22
	Total	318	369	450

(a) Conduct an appropriate hypothesis test evaluating whether there is a difference in the proportion of Republicans and Democrats who think the full- body scans should be applied in airports. Assume that all relevant conditions are met.

(b) The conclusion of the test in part (a) may be incorrect, meaning a testing error was made. If an error was made, was it a Type 1 or a Type 2 Error? Explain.

6.27 Sleep deprived transportation workers. The National Sleep Foundation conducted a survey on the sleep habits of randomly sampled transportation workers and a control sample of non-transportation workers. The results of the survey are shown below.[28]

		Transportation Professionals			
	Control	Pilots	Truck Drivers	Train Operators	Bus/Taxi/Limo Drivers
Less than 6 hours of sleep	35	19	35	29	21
6 to 8 hours of sleep	193	132	117	119	131
More than 8 hours	64	51	51	32	58
Total	292	202	203	180	210

Conduct a hypothesis test to evaluate if these data provide evidence of a difference between the proportions of truck drivers and non-transportation workers (the control group) who get less than 6 hours of sleep per day, i.e. are considered sleep deprived.

---

[27]S. Condon. "Poll: 4 in 5 Support Full-Body Airport Scanners". In: *CBS News* (2010).
[28]National Sleep Foundation, 2012 Sleep in America Poll: Transportation Workers' Sleep, 2012.

6.28 Prenatal vitamins and Autism. Researchers studying the link between prenatal vitamin use and autism surveyed the mothers of a random sample of children aged 24 - 60 months with autism and conducted another separate random sample for children with typical development. The table below shows the number of mothers in each group who did and did not use prenatal vitamins during the three months before pregnancy (periconceptional period).[29]

| | | Autism | | |
		Autism	Typical development	Total
Periconceptional	No vitamin	111	70	181
prenatal vitamin	Vitamin	143	159	302
	Total	254	229	483

(a) State appropriate hypotheses to test for independence of use of prenatal vitamins during the three months before pregnancy and autism.

(b) Complete the hypothesis test and state an appropriate conclusion. (Reminder: Verify any necessary conditions for the test.)

(c) A New York Times article reporting on this study was titled "Prenatal Vitamins May Ward Off Autism". Do you find the title of this article to be appropriate? Explain your answer. Additionally, propose an alternative title.[30]

6.29 HIV in sub-Saharan Africa. In July 2008 the US National Institutes of Health announced that it was stopping a clinical study early because of unexpected results. The study population consisted of HIV-infected women in sub-Saharan Africa who had been given single dose Nevaripine (a treatment for HIV) while giving birth, to prevent transmission of HIV to the infant. The study was a randomized comparison of continued treatment of a woman (after successful childbirth) with Nevaripine vs Lopinavir, a second drug used to treat HIV. 240 women participated in the study; 120 were randomized to each of the two treatments. Twenty-four weeks after starting the study treatment, each woman was tested to determine if the HIV infection was becoming worse (an outcome called *virologic failure*). Twenty-six of the 120 women treated with Nevaripine experienced virologic failure, while 10 of the 120 women treated with the other drug experienced virologic failure.[31]

(a) Create a two-way table presenting the results of this study.

(b) State appropriate hypotheses to test for difference in virologic failure rates between treatment groups.

(c) Complete the hypothesis test and state an appropriate conclusion. (Reminder: Verify any necessary conditions for the test.)

6.30 An apple a day keeps the doctor away. A physical education teacher at a high school wanting to increase awareness on issues of nutrition and health asked her students at the beginning of the semester whether they believed the expression "an apple a day keeps the doctor away", and 40% of the students responded yes. Throughout the semester she started each class with a brief discussion of a study highlighting positive effects of eating more fruits and vegetables. She conducted the same apple-a-day survey at the end of the semester, and this time 60% of the students responded yes. Can she used a two-proportion method from this section for this analysis? Explain your reasoning.

---

[29]R.J. Schmidt et al. "Prenatal vitamins, one-carbon metabolism gene variants, and risk for autism". In: *Epidemiology* 22.4 (2011), p. 476.

[30]R.C. Rabin. "Patterns: Prenatal Vitamins May Ward Off Autism". In: *New York Times* (2011).

[31]S. Lockman et al. "Response to antiretroviral therapy after a single, peripartum dose of nevirapine". In: *Obstetrical & gynecological survey* 62.6 (2007), p. 361.

6.3 Testing for goodness of fit using chi-square

In this section, we develop a method for assessing a null model when the data are binned. This technique is commonly used in two circumstances:

- Given a sample of cases that can be classified into several groups, determine if the sample is representative of the general population.

- Evaluate whether data resemble a particular distribution, such as a normal distribution or a geometric distribution.

Each of these scenarios can be addressed using the same statistical test: a chi-square test.

In the first case, we consider data from a random sample of 275 jurors in a small county. Jurors identified their racial group, as shown in Figure 6.5, and we would like to determine if these jurors are racially representative of the population. If the jury is representative of the population, then the proportions in the sample should roughly reflect the population of eligible jurors, i.e. registered voters.

Race	White	Black	Hispanic	Other	Total
Representation in juries	205	26	25	19	275
Registered voters	0.72	0.07	0.12	0.09	1.00

Figure 6.5: Representation by race in a city's juries and population.

While the proportions in the juries do not precisely represent the population proportions, it is unclear whether these data provide convincing evidence that the sample is not representative. If the jurors really were randomly sampled from the registered voters, we might expect small differences due to chance. However, unusually large differences may provide convincing evidence that the juries were not representative.

A second application, assessing the fit of a distribution, is presented at the end of this section. Daily stock returns from the S&P500 for 25 years are used to assess whether stock activity each day is independent of the stock's behavior on previous days.

In these problems, we would like to examine all bins simultaneously, not simply compare one or two bins at a time, which will require us to develop a new test statistic.

6.3.1 Creating a test statistic for one-way tables

EXAMPLE 6.22

Of the people in the city, 275 served on a jury. If the individuals are randomly selected to serve on a jury, about how many of the 275 people would we expect to be white? How many would we expect to be black?

About 72% of the population is white, so we would expect about 72% of the jurors to be white: $0.72 \times 275 = 198$.

Similarly, we would expect about 7% of the jurors to be black, which would correspond to about $0.07 \times 275 = 19.25$ black jurors.

GUIDED PRACTICE 6.23

Twelve percent of the population is Hispanic and 9% represent other races. How many of the 275 jurors would we expect to be Hispanic or from another race? Answers can be found in Figure 6.6.

The sample proportion represented from each race among the 275 jurors was not a precise match for any ethnic group. While some sampling variation is expected, we would expect the

Race	White	Black	Hispanic	Other	Total
Observed data	205	26	25	19	275
Expected counts	198	19.25	33	24.75	275

Figure 6.6: Actual and expected make-up of the jurors.

sample proportions to be fairly similar to the population proportions if there is no bias on juries. We need to test whether the differences are strong enough to provide convincing evidence that the jurors are not a random sample. These ideas can be organized into hypotheses:

H_0: The jurors are a random sample, i.e. there is no racial bias in who serves on a jury, and the observed counts reflect natural sampling fluctuation.

H_A: The jurors are not randomly sampled, i.e. there is racial bias in juror selection.

To evaluate these hypotheses, we quantify how different the observed counts are from the expected counts. Strong evidence for the alternative hypothesis would come in the form of unusually large deviations in the groups from what would be expected based on sampling variation alone.

6.3.2   The chi-square test statistic

In previous hypothesis tests, we constructed a test statistic of the following form:

$$\frac{\text{point estimate} - \text{null value}}{\text{SE of point estimate}}$$

This construction was based on (1) identifying the difference between a point estimate and an expected value if the null hypothesis was true, and (2) standardizing that difference using the standard error of the point estimate. These two ideas will help in the construction of an appropriate test statistic for count data.

Our strategy will be to first compute the difference between the observed counts and the counts we would expect if the null hypothesis was true, then we will standardize the difference:

$$Z_1 = \frac{\text{observed white count} - \text{null white count}}{\text{SE of observed white count}}$$

The standard error for the point estimate of the count in binned data is the square root of the count under the null.[32] Therefore:

$$Z_1 = \frac{205 - 198}{\sqrt{198}} = 0.50$$

The fraction is very similar to previous test statistics: first compute a difference, then standardize it. These computations should also be completed for the black, Hispanic, and other groups:

Black

$$Z_2 = \frac{26 - 19.25}{\sqrt{19.25}} = 1.54$$

Hispanic

$$Z_3 = \frac{25 - 33}{\sqrt{33}} = -1.39$$

Other

$$Z_4 = \frac{19 - 24.75}{\sqrt{24.75}} = -1.16$$

We would like to use a single test statistic to determine if these four standardized differences are irregularly far from zero. That is, Z_1, Z_2, Z_3, and Z_4 must be combined somehow to help determine if they – as a group – tend to be unusually far from zero. A first thought might be to take the absolute value of these four standardized differences and add them up:

$$|Z_1| + |Z_2| + |Z_3| + |Z_4| = 4.58$$

---

[32]Using some of the rules learned in earlier chapters, we might think that the standard error would be $np(1-p)$, where n is the sample size and p is the proportion in the population. This would be correct if we were looking only at one count. However, we are computing many standardized differences and adding them together. It can be shown – though not here – that the square root of the count is a better way to standardize the count differences.

Indeed, this does give one number summarizing how far the actual counts are from what was expected. However, it is more common to add the squared values:

$$Z_1^2 + Z_2^2 + Z_3^2 + Z_4^2 = 5.89$$

Squaring each standardized difference before adding them together does two things:

- Any standardized difference that is squared will now be positive.
- Differences that already look unusual – e.g. a standardized difference of 2.5 – will become much larger after being squared.

The test statistic X^2, which is the sum of the Z^2 values, is generally used for these reasons. We can also write an equation for X^2 using the observed counts and null counts:

$$X^2 = \frac{(\text{observed count}_1 - \text{null count}_1)^2}{\text{null count}_1} + \cdots + \frac{(\text{observed count}_4 - \text{null count}_4)^2}{\text{null count}_4}$$

The final number X^2 summarizes how strongly the observed counts tend to deviate from the null counts. In Section 6.3.4, we will see that if the null hypothesis is true, then X^2 follows a new distribution called a *chi-square distribution*. Using this distribution, we will be able to obtain a p-value to evaluate the hypotheses.

6.3.3 The chi-square distribution and finding areas

The **chi-square distribution** is sometimes used to characterize data sets and statistics that are always positive and typically right skewed. Recall a normal distribution had two parameters – mean and standard deviation – that could be used to describe its exact characteristics. The chi-square distribution has just one parameter called **degrees of freedom (df)**, which influences the shape, center, and spread of the distribution.

GUIDED PRACTICE 6.24

Figure 6.7 shows three chi-square distributions.
(a) How does the center of the distribution change when the degrees of freedom is larger?
(b) What about the variability (spread)?
(c) How does the shape change?[33]

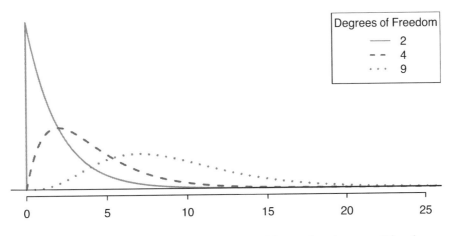

Figure 6.7: Three chi-square distributions with varying degrees of freedom.

[33](a) The center becomes larger. If took a careful look, we could see that the mean of each distribution is equal to the distribution's degrees of freedom. (b) The variability increases as the degrees of freedom increases. (c) The distribution is very strongly skewed for $df = 2$, and then the distributions become more symmetric for the larger degrees of freedom $df = 4$ and $df = 9$. We would see this trend continue if we examined distributions with even more larger degrees of freedom.

Figure 6.7 and Guided Practice 6.24 demonstrate three general properties of chi-square distributions as the degrees of freedom increases: the distribution becomes more symmetric, the center moves to the right, and the variability inflates.

Our principal interest in the chi-square distribution is the calculation of p-values, which (as we have seen before) is related to finding the relevant area in the tail of a distribution. The most common ways to do this are using computer software, using a graphing calculator, or using a table. For folks wanting to use the table option, we provide an outline of how to read the chi-square table in Appendix C.3, which is also where you may find the table. For the examples below, use your preferred approach to confirm you get the same answers.

EXAMPLE 6.25

Figure 6.8(a) shows a chi-square distribution with 3 degrees of freedom and an upper shaded tail starting at 6.25. Find the shaded area.

Using statistical software or a graphing calculator, we can find that the upper tail area for a chi-square distribution with 3 degrees of freedom (df) and a cutoff of 6.25 is 0.1001. That is, the shaded upper tail of Figure 6.8(a) has area 0.1.

EXAMPLE 6.26

Figure 6.8(b) shows the upper tail of a chi-square distribution with 2 degrees of freedom. The bound for this upper tail is at 4.3. Find the tail area.

Using software, we can find that the tail area shaded in Figure 6.8(b) to be 0.1165. If using a table, we would only be able to find a range of values for the tail area: between 0.1 and 0.2.

EXAMPLE 6.27

Figure 6.8(c) shows an upper tail for a chi-square distribution with 5 degrees of freedom and a cutoff of 5.1. Find the tail area.

Using software, we would obtain a tail area of 0.4038. If using the table in Appendix C.3, we would have identified that the tail area is larger than 0.3 but not be able to give the precise value.

GUIDED PRACTICE 6.28

Figure 6.8(d) shows a cutoff of 11.7 on a chi-square distribution with 7 degrees of freedom. Find the area of the upper tail.[34]

GUIDED PRACTICE 6.29

Figure 6.8(e) shows a cutoff of 10 on a chi-square distribution with 4 degrees of freedom. Find the area of the upper tail.[35]

GUIDED PRACTICE 6.30

Figure 6.8(f) shows a cutoff of 9.21 with a chi-square distribution with 3 df. Find the area of the upper tail.[36]

---

[34] The area is 0.1109. If using a table, we would identify that it falls between 0.1 and 0.2.
[35] Precise value: 0.0404. If using the table: between 0.02 and 0.05.
[36] Precise value: 0.0266. If using the table: between 0.02 and 0.05.

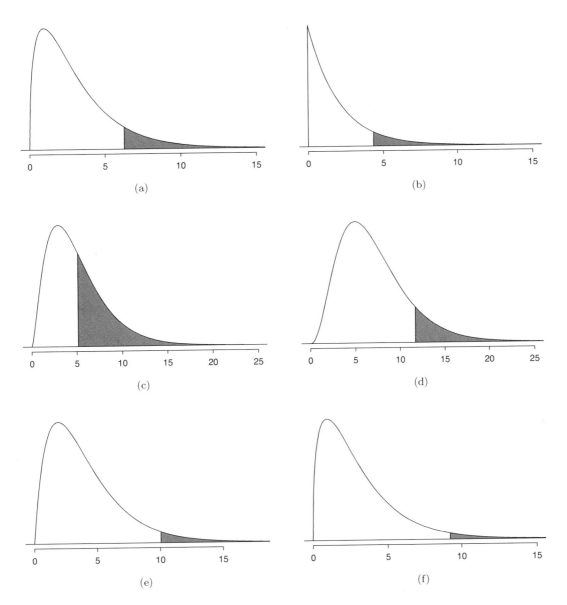

Figure 6.8: **(a)** Chi-square distribution with 3 degrees of freedom, area above 6.25 shaded. **(b)** 2 degrees of freedom, area above 4.3 shaded. **(c)** 5 degrees of freedom, area above 5.1 shaded. **(d)** 7 degrees of freedom, area above 11.7 shaded. **(e)** 4 degrees of freedom, area above 10 shaded. **(f)** 3 degrees of freedom, area above 9.21 shaded.

6.3.4   Finding a p-value for a chi-square distribution

In Section 6.3.2, we identified a new test statistic (X^2) within the context of assessing whether there was evidence of racial bias in how jurors were sampled. The null hypothesis represented the claim that jurors were randomly sampled and there was no racial bias. The alternative hypothesis was that there was racial bias in how the jurors were sampled.

We determined that a large X^2 value would suggest strong evidence favoring the alternative hypothesis: that there was racial bias. However, we could not quantify what the chance was of observing such a large test statistic ($X^2 = 5.89$) if the null hypothesis actually was true. This is where the chi-square distribution becomes useful. If the null hypothesis was true and there was no racial bias, then X^2 would follow a chi-square distribution, with three degrees of freedom in this case. Under certain conditions, the statistic X^2 follows a chi-square distribution with $k - 1$ degrees of freedom, where k is the number of bins.

EXAMPLE 6.31

How many categories were there in the juror example? How many degrees of freedom should be associated with the chi-square distribution used for X^2?

In the jurors example, there were $k = 4$ categories: white, black, Hispanic, and other. According to the rule above, the test statistic X^2 should then follow a chi-square distribution with $k - 1 = 3$ degrees of freedom if H_0 is true.

Just like we checked sample size conditions to use a normal distribution in earlier sections, we must also check a sample size condition to safely apply the chi-square distribution for X^2. Each expected count must be at least 5. In the juror example, the expected counts were 198, 19.25, 33, and 24.75, all easily above 5, so we can apply the chi-square model to the test statistic, $X^2 = 5.89$.

EXAMPLE 6.32

If the null hypothesis is true, the test statistic $X^2 = 5.89$ would be closely associated with a chi-square distribution with three degrees of freedom. Using this distribution and test statistic, identify the p-value.

The chi-square distribution and p-value are shown in Figure 6.9. Because larger chi-square values correspond to stronger evidence against the null hypothesis, we shade the upper tail to represent the p-value. Using statistical software (or the table in Appendix C.3), we can determine that the area is 0.1171. Generally we do not reject the null hypothesis with such a large p-value. In other words, the data do not provide convincing evidence of racial bias in the juror selection.

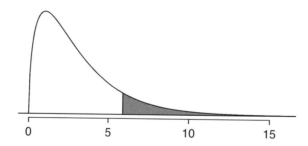

Figure 6.9: The p-value for the juror hypothesis test is shaded in the chi-square distribution with $df = 3$.

CHI-SQUARE TEST FOR ONE-WAY TABLE

Suppose we are to evaluate whether there is convincing evidence that a set of observed counts $O_1, O_2, ..., O_k$ in k categories are unusually different from what might be expected under a null hypothesis. Call the *expected counts* that are based on the null hypothesis $E_1, E_2, ..., E_k$. If each expected count is at least 5 and the null hypothesis is true, then the test statistic below follows a chi-square distribution with $k - 1$ degrees of freedom:

$$X^2 = \frac{(O_1 - E_1)^2}{E_1} + \frac{(O_2 - E_2)^2}{E_2} + \cdots + \frac{(O_k - E_k)^2}{E_k}$$

The p-value for this test statistic is found by looking at the upper tail of this chi-square distribution. We consider the upper tail because larger values of X^2 would provide greater evidence against the null hypothesis.

CONDITIONS FOR THE CHI-SQUARE TEST

There are two conditions that must be checked before performing a chi-square test:

Independence. Each case that contributes a count to the table must be independent of all the other cases in the table.

Sample size / distribution. Each particular scenario (i.e. cell count) must have at least 5 expected cases.

Failing to check conditions may affect the test's error rates.

When examining a table with just two bins, pick a single bin and use the one-proportion methods introduced in Section 6.1.

6.3.5 Evaluating goodness of fit for a distribution

Section 4.2 would be useful background reading for this example, but it is not a prerequisite.

We can apply the chi-square testing framework to the second problem in this section: evaluating whether a certain statistical model fits a data set. Daily stock returns from the S&P500 for 10 can be used to assess whether stock activity each day is independent of the stock's behavior on previous days. This sounds like a very complex question, and it is, but a chi-square test can be used to study the problem. We will label each day as Up or Down (D) depending on whether the market was up or down that day. For example, consider the following changes in price, their new labels of up and down, and then the number of days that must be observed before each Up day:

Change in price	2.52	-1.46	0.51	-4.07	3.36	1.10	-5.46	-1.03	-2.99	1.71
Outcome	Up	D	Up	D	Up	Up	D	D	D	Up
Days to Up	1	-	2	-	2	1	-	-	-	4

If the days really are independent, then the number of days until a positive trading day should follow a geometric distribution. The geometric distribution describes the probability of waiting for the kth trial to observe the first success. Here each up day (Up) represents a success, and down (D) days represent failures. In the data above, it took only one day until the market was up, so the first wait time was 1 day. It took two more days before we observed our next Up trading day, and two more for the third Up day. We would like to determine if these counts (1, 2, 2, 1, 4, and so on) follow the geometric distribution. Figure 6.10 shows the number of waiting days for a positive trading day during 10 years for the S&P500.

Days	1	2	3	4	5	6	7+	Total
Observed	717	369	155	69	28	14	10	1362

Figure 6.10: Observed distribution of the waiting time until a positive trading day for the S&P500.

We consider how many days one must wait until observing an Up day on the S&P500 stock index. If the stock activity was independent from one day to the next and the probability of a positive trading day was constant, then we would expect this waiting time to follow a *geometric distribution*. We can organize this into a hypothesis framework:

H_0: The stock market being up or down on a given day is independent from all other days. We will consider the number of days that pass until an Up day is observed. Under this hypothesis, the number of days until an Up day should follow a geometric distribution.

H_A: The stock market being up or down on a given day is not independent from all other days. Since we know the number of days until an Up day would follow a geometric distribution under the null, we look for deviations from the geometric distribution, which would support the alternative hypothesis.

There are important implications in our result for stock traders: if information from past trading days is useful in telling what will happen today, that information may provide an advantage over other traders.

We consider data for the S&P500 and summarize the waiting times in Figure 6.11 and Figure 6.12. The S&P500 was positive on 54.5% of those days.

Because applying the chi-square framework requires expected counts to be at least 5, we have *binned* together all the cases where the waiting time was at least 7 days to ensure each expected count is well above this minimum. The actual data, shown in the *Observed* row in Figure 6.11, can be compared to the expected counts from the *Geometric Model* row. The method for computing expected counts is discussed in Figure 6.11. In general, the expected counts are determined by (1) identifying the null proportion associated with each bin, then (2) multiplying each null proportion by the total count to obtain the expected counts. That is, this strategy identifies what proportion of the total count we would expect to be in each bin.

Days	1	2	3	4	5	6	7+	Total
Observed	717	369	155	69	28	14	10	1362
Geometric Model	743	338	154	70	32	14	12	1362

Figure 6.11: Distribution of the waiting time until a positive trading day. The expected counts based on the geometric model are shown in the last row. To find each expected count, we identify the probability of waiting D days based on the geometric model $(P(D) = (1 - 0.545)^{D-1}(0.545))$ and multiply by the total number of streaks, 1362. For example, waiting for three days occurs under the geometric model about $0.455^2 \times 0.545 = 11.28\%$ of the time, which corresponds to $0.1128 \times 1362 = 154$ streaks.

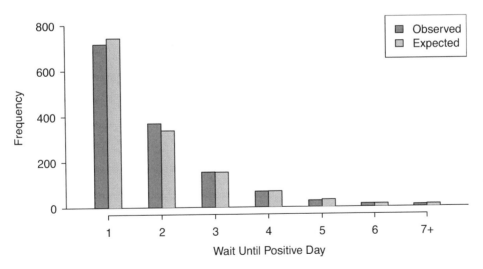

Figure 6.12: Side-by-side bar plot of the observed and expected counts for each waiting time.

EXAMPLE 6.33

Do you notice any unusually large deviations in the graph? Can you tell if these deviations are due to chance just by looking?

It is not obvious whether differences in the observed counts and the expected counts from the geometric distribution are significantly different. That is, it is not clear whether these deviations might be due to chance or whether they are so strong that the data provide convincing evidence against the null hypothesis. However, we can perform a chi-square test using the counts in Figure 6.11.

GUIDED PRACTICE 6.34

Figure 6.11 provides a set of count data for waiting times ($O_1 = 717$, $O_2 = 369$, ...) and expected counts under the geometric distribution ($E_1 = 743$, $E_2 = 338$, ...). Compute the chi-square test statistic, X^2.[37]

GUIDED PRACTICE 6.35

Because the expected counts are all at least 5, we can safely apply the chi-square distribution to X^2. However, how many degrees of freedom should we use?[38]

EXAMPLE 6.36

If the observed counts follow the geometric model, then the chi-square test statistic $X^2 = 4.61$ would closely follow a chi-square distribution with $df = 6$. Using this information, compute a p-value.

Figure 6.13 shows the chi-square distribution, cutoff, and the shaded p-value. Using software, we can find the p-value: 0.5951. Ultimately, we do not have sufficient evidence to reject the notion that the wait times follow a geometric distribution for the last 10 years of data for the S&P500, i.e. we cannot reject the notion that trading days are independent.

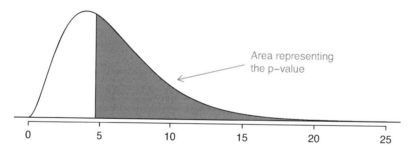

Figure 6.13: Chi-square distribution with 6 degrees of freedom. The p-value for the stock analysis is shaded.

EXAMPLE 6.37

In Example 6.36, we did not reject the null hypothesis that the trading days are independent during the last 10 of data. Why is this so important?

It may be tempting to think the market is "due" for an Up day if there have been several consecutive days where it has been down. However, we haven't found strong evidence that there's any such property where the market is "due" for a correction. At the very least, the analysis suggests any dependence between days is very weak.

---

[37]$X^2 = \frac{(717-743)^2}{743} + \frac{(369-338)^2}{338} + \cdots + \frac{(10-12)^2}{12} = 4.61$
[38]There are $k = 7$ groups, so we use $df = k - 1 = 6$.

Exercises

6.31 True or false, Part I. Determine if the statements below are true or false. For each false statement, suggest an alternative wording to make it a true statement.

(a) The chi-square distribution, just like the normal distribution, has two parameters, mean and standard deviation.

(b) The chi-square distribution is always right skewed, regardless of the value of the degrees of freedom parameter.

(c) The chi-square statistic is always positive.

(d) As the degrees of freedom increases, the shape of the chi-square distribution becomes more skewed.

6.32 True or false, Part II. Determine if the statements below are true or false. For each false statement, suggest an alternative wording to make it a true statement.

(a) As the degrees of freedom increases, the mean of the chi-square distribution increases.

(b) If you found $\chi^2 = 10$ with $df = 5$ you would fail to reject H_0 at the 5% significance level.

(c) When finding the p-value of a chi-square test, we always shade the tail areas in both tails.

(d) As the degrees of freedom increases, the variability of the chi-square distribution decreases.

6.33 Open source textbook. A professor using an open source introductory statistics book predicts that 60% of the students will purchase a hard copy of the book, 25% will print it out from the web, and 15% will read it online. At the end of the semester he asks his students to complete a survey where they indicate what format of the book they used. Of the 126 students, 71 said they bought a hard copy of the book, 30 said they printed it out from the web, and 25 said they read it online.

(a) State the hypotheses for testing if the professor's predictions were inaccurate.

(b) How many students did the professor expect to buy the book, print the book, and read the book exclusively online?

(c) This is an appropriate setting for a chi-square test. List the conditions required for a test and verify they are satisfied.

(d) Calculate the chi-squared statistic, the degrees of freedom associated with it, and the p-value.

(e) Based on the p-value calculated in part (d), what is the conclusion of the hypothesis test? Interpret your conclusion in this context.

6.34 Barking deer. Microhabitat factors associated with forage and bed sites of barking deer in Hainan Island, China were examined. In this region woods make up 4.8% of the land, cultivated grass plot makes up 14.7%, and deciduous forests make up 39.6%. Of the 426 sites where the deer forage, 4 were categorized as woods, 16 as cultivated grassplot, and 61 as deciduous forests. The table below summarizes these data.[39]

Woods	Cultivated grassplot	Deciduous forests	Other	Total
4	16	61	345	426

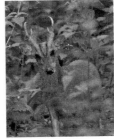

(a) Write the hypotheses for testing if barking deer prefer to forage in certain habitats over others.

(b) What type of test can we use to answer this research question?

(c) Check if the assumptions and conditions required for this test are satisfied.

(d) Do these data provide convincing evidence that barking deer prefer to forage in certain habitats over others? Conduct an appropriate hypothesis test to answer this research question.

Photo by Shrikant Rao
(http://flic.kr/p/4Xjdkk)
CC BY 2.0 license

---

[39]Liwei Teng et al. "Forage and bed sites characteristics of Indian muntjac (Muntiacus muntjak) in Hainan Island, China". In: *Ecological Research* 19.6 (2004), pp. 675–681.

6.4   Testing for independence in two-way tables

We all buy used products – cars, computers, textbooks, and so on – and we sometimes assume the sellers of those products will be forthright about any underlying problems with what they're selling. This is not something we should take for granted. Researchers recruited 219 participants in a study where they would sell a used iPod[40] that was known to have frozen twice in the past. The participants were incentivized to get as much money as they could for the iPod since they would receive a 5% cut of the sale on top of $10 for participating. The researchers wanted to understand what types of questions would elicit the seller to disclose the freezing issue.

Unbeknownst to the participants who were the sellers in the study, the buyers were collaborating with the researchers to evaluate the influence of different questions on the likelihood of getting the sellers to disclose the past issues with the iPod. The scripted buyers started with "Okay, I guess I'm supposed to go first. So you've had the iPod for 2 years ..." and ended with one of three questions:

- General: What can you tell me about it?

- Positive Assumption: It doesn't have any problems, does it?

- Negative Assumption: What problems does it have?

The question is the treatment given to the sellers, and the response is whether the question prompted them to disclose the freezing issue with the iPod. The results are shown in Figure 6.14, and the data suggest that asking the, *What problems does it have?*, was the most effective at getting the seller to disclose the past freezing issues. However, you should also be asking yourself: could we see these results due to chance alone, or is this in fact evidence that some questions are more effective for getting at the truth?

	General	Positive Assumption	Negative Assumption	Total
Disclose Problem	2	23	36	61
Hide Problem	71	50	37	158
Total	73	73	73	219

Figure 6.14: Summary of the iPod study, where a question was posed to the study participant who acted

---

DIFFERENCES OF ONE-WAY TABLES VS TWO-WAY TABLES

A one-way table describes counts for each outcome in a single variable. A two-way table describes counts for *combinations* of outcomes for two variables. When we consider a two-way table, we often would like to know, are these variables related in any way? That is, are they dependent (versus independent)?

The hypothesis test for the iPod experiment is really about assessing whether there is statistically significant evidence that the success each question had on getting the participant to disclose the problem with the iPod. In other words, the goal is to check whether the buyer's question was independent of whether the seller disclosed a problem.

---

[40]For readers not as old as the authors, an iPod is basically an iPhone without any cellular service, assuming it was one of the later generations. Earlier generations were more basic.

6.4.1 Expected counts in two-way tables

Like with one-way tables, we will need to compute estimated counts for each cell in a two-way table.

EXAMPLE 6.38

From the experiment, we can compute the proportion of all sellers who disclosed the freezing problem as $61/219 = 0.2785$. If there really is no difference among the questions and 27.85% of sellers were going to disclose the freezing problem no matter the question that was put to them, how many of the 73 people in the `General` group would we have expected to disclose the freezing problem?

We would predict that $0.2785 \times 73 = 20.33$ sellers would disclose the problem. Obviously we observed fewer than this, though it is not yet clear if that is due to chance variation or whether that is because the questions vary in how effective they are at getting to the truth.

GUIDED PRACTICE 6.39

If the questions were actually equally effective, meaning about 27.85% of respondents would disclose the freezing issue regardless of what question they were asked, about how many sellers would we expect to *hide* the freezing problem from the Positive Assumption group?[41]

We can compute the expected number of sellers who we would expect to disclose or hide the freezing issue for all groups, if the questions had no impact on what they disclosed, using the same strategy employed in Example 6.38 and Guided Practice 6.39. These expected counts were used to construct Figure 6.15, which is the same as Figure 6.14, except now the expected counts have been added in parentheses.

	General	Positive Assumption	Negative Assumption	Total
Disclose Problem	2 **(20.33)**	23 **(20.33)**	36 **(20.33)**	61
Hide Problem	71 **(52.67)**	50 **(52.67)**	37 **(52.67)**	158
Total	73	73	73	219

Figure 6.15: The observed counts and the **(expected counts)**.

The examples and exercises above provided some help in computing expected counts. In general, expected counts for a two-way table may be computed using the row totals, column totals, and the table total. For instance, if there was no difference between the groups, then about 27.85% of each column should be in the first row:

$$0.2785 \times (\text{column 1 total}) = 20.33$$
$$0.2785 \times (\text{column 2 total}) = 20.33$$
$$0.2785 \times (\text{column 3 total}) = 20.33$$

Looking back to how 0.2785 was computed – as the fraction of sellers who disclosed the freezing issue $(158/219)$ – these three expected counts could have been computed as

$$\left(\frac{\text{row 1 total}}{\text{table total}} \right) (\text{column 1 total}) = 20.33$$
$$\left(\frac{\text{row 1 total}}{\text{table total}} \right) (\text{column 2 total}) = 20.33$$
$$\left(\frac{\text{row 1 total}}{\text{table total}} \right) (\text{column 3 total}) = 20.33$$

This leads us to a general formula for computing expected counts in a two-way table when we would like to test whether there is strong evidence of an association between the column variable and row variable.

---

[41]We would expect $(1 - 0.2785) \times 73 = 52.67$. It is okay that this result, like the result from Example 6.38, is a fraction.

COMPUTING EXPECTED COUNTS IN A TWO-WAY TABLE

To identify the expected count for the ith row and j^{th} column, compute

$$\text{Expected Count}_{\text{row } i, \text{ col } j} = \frac{(\text{row } i \text{ total}) \times (\text{column } j \text{ total})}{\text{table total}}$$

6.4.2   The chi-square test for two-way tables

The chi-square test statistic for a two-way table is found the same way it is found for a one-way table. For each table count, compute

General formula	$\dfrac{(\text{observed count } - \text{expected count})^2}{\text{expected count}}$
Row 1, Col 1	$\dfrac{(2 - 20.33)^2}{20.33} = 16.53$
Row 1, Col 2	$\dfrac{(23 - 20.33)^2}{20.33} = 0.35$
$\vdots$	$\vdots$
Row 2, Col 3	$\dfrac{(37 - 52.67)^2}{52.67} = 4.66$

Adding the computed value for each cell gives the chi-square test statistic X^2:

$$X^2 = 16.53 + 0.35 + \cdots + 4.66 = 40.13$$

Just like before, this test statistic follows a chi-square distribution. However, the degrees of freedom are computed a little differently for a two-way table.[42] For two way tables, the degrees of freedom is equal to

$$df = (\text{number of rows minus } 1) \times (\text{number of columns minus } 1)$$

In our example, the degrees of freedom parameter is

$$df = (2 - 1) \times (3 - 1) = 2$$

If the null hypothesis is true (i.e. the questions had no impact on the sellers in the experiment), then the test statistic $X^2 = 40.13$ closely follows a chi-square distribution with 2 degrees of freedom. Using this information, we can compute the p-value for the test, which is depicted in Figure 6.16.

COMPUTING DEGREES OF FREEDOM FOR A TWO-WAY TABLE

When applying the chi-square test to a two-way table, we use

$$df = (R - 1) \times (C - 1)$$

where R is the number of rows in the table and C is the number of columns.

When analyzing 2-by-2 contingency tables, one guideline is to use the two-proportion methods introduced in Section 6.2.

---

[42]Recall: in the one-way table, the degrees of freedom was the number of cells minus 1.

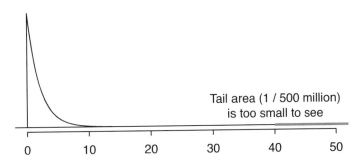

Figure 6.16: Visualization of the p-value for $X^2 = 40.13$ when $df = 2$.

EXAMPLE 6.40

Find the p-value and draw a conclusion about whether the question affects the sellers likelihood of reporting the freezing problem.

Using a computer, we can compute a very precise value for the tail area above $X^2 = 40.13$ for a chi-square distribution with 2 degrees of freedom: 0.000000002. (If using the table in Appendix C.3, we would identify the p-value is smaller than 0.001.) Using a significance level of $\alpha = 0.05$, the null hypothesis is rejected since the p-value is smaller. That is, the data provide convincing evidence that the question asked did affect a seller's likelihood to tell the truth about problems with the iPod.

EXAMPLE 6.41

Figure 6.17 summarizes the results of an experiment evaluating three treatments for Type 2 Diabetes in patients aged 10-17 who were being treated with metformin. The three treatments considered were continued treatment with metformin (met), treatment with metformin combined with rosiglitazone (rosi), or a lifestyle intervention program. Each patient had a primary outcome, which was either lacked glycemic control (failure) or did not lack that control (success). What are appropriate hypotheses for this test?

H_0: There is no difference in the effectiveness of the three treatments.

H_A: There is some difference in effectiveness between the three treatments, e.g. perhaps the rosi treatment performed better than lifestyle.

	Failure	Success	Total
lifestyle	109	125	234
met	120	112	232
rosi	90	143	233
Total	319	380	699

Figure 6.17: Results for the Type 2 Diabetes study.

Ⓖ
GUIDED PRACTICE 6.42

A chi-square test for a two-way table may be used to test the hypotheses in Example 6.41. As a first step, compute the expected values for each of the six table cells.[43]

Ⓖ
GUIDED PRACTICE 6.43

Compute the chi-square test statistic for the data in Figure 6.17.[44]

Ⓖ
GUIDED PRACTICE 6.44

Because there are 3 rows and 2 columns, the degrees of freedom for the test is $df = (3-1) \times (2-1) = 2$. Use $X^2 = 8.16$, $df = 2$, evaluate whether to reject the null hypothesis using a significance level of 0.05.[45]

---

[43]The expected count for row one / column one is found by multiplying the row one total (234) and column one total (319), then dividing by the table total (699): $\frac{234 \times 319}{699} = 106.8$. Similarly for the second column and the first row: $\frac{234 \times 380}{699} = 127.2$. Row 2: 105.9 and 126.1. Row 3: 106.3 and 126.7.

[44]For each cell, compute $\frac{(obs-exp)^2}{exp}$. For instance, the first row and first column: $\frac{(109-106.8)^2}{106.8} = 0.05$. Adding the results of each cell gives the chi-square test statistic: $X^2 = 0.05 + \cdots + 2.11 = 8.16$.

[45] If using a computer, we can identify the p-value as 0.017. That is, we reject the null hypothesis because the p-value is less than 0.05, and we conclude that at least one of the treatments is more or less effective than the others at treating Type 2 Diabetes for glycemic control.

Exercises

6.35 Quitters. Does being part of a support group affect the ability of people to quit smoking? A county health department enrolled 300 smokers in a randomized experiment. 150 participants were assigned to a group that used a nicotine patch and met weekly with a support group; the other 150 received the patch and did not meet with a support group. At the end of the study, 40 of the participants in the patch plus support group had quit smoking while only 30 smokers had quit in the other group.

(a) Create a two-way table presenting the results of this study.

(b) Answer each of the following questions under the null hypothesis that being part of a support group does not affect the ability of people to quit smoking, and indicate whether the expected values are higher or lower than the observed values.

    i. How many subjects in the "patch + support" group would you expect to quit?

    ii. How many subjects in the "patch only" group would you expect to not quit?

6.36 Full body scan, Part II. The table below summarizes a data set we first encountered in Exercise 6.26 regarding views on full-body scans and political affiliation. The differences in each political group may be due to chance. Complete the following computations under the null hypothesis of independence between an individual's party affiliation and his support of full-body scans. It may be useful to first add on an extra column for row totals before proceeding with the computations.

		Party Affiliation		
		Republican	Democrat	Independent
	Should	264	299	351
Answer	Should not	38	55	77
	Don't know/No answer	16	15	22
	Total	318	369	450

(a) How many Republicans would you expect to not support the use of full-body scans?

(b) How many Democrats would you expect to support the use of full- body scans?

(c) How many Independents would you expect to not know or not answer?

6.37 Offshore drilling, Part III. The table below summarizes a data set we first encountered in Exercise 6.23 that examines the responses of a random sample of college graduates and non-graduates on the topic of oil drilling. Complete a chi-square test for these data to check whether there is a statistically significant difference in responses from college graduates and non-graduates.

	College Grad	
	Yes	No
Support	154	132
Oppose	180	126
Do not know	104	131
Total	438	389

6.38 Parasitic worm. Lymphatic filariasis is a disease caused by a parasitic worm. Complications of the disease can lead to extreme swelling and other complications. Here we consider results from a randomized experiment that compared three different drug treatment options to clear people of the this parasite, which people are working to eliminate entirely. The results for the second year of the study are given below:[46]

	Clear at Year 2	Not Clear at Year 2
Three drugs	52	2
Two drugs	31	24
Two drugs annually	42	14

(a) Set up hypotheses for evaluating whether there is any difference in the performance of the treatments, and also check conditions.

(b) Statistical software was used to run a chi-square test, which output:

$$X^2 = 23.7 \qquad\qquad df = 2 \qquad\qquad \text{p-value} = 7.2\text{e-}6$$

Use these results to evaluate the hypotheses from part (a), and provide a conclusion in the context of the problem.

---

[46]Christopher King et al. "A Trial of a Triple-Drug Treatment for Lymphatic Filariasis". In: *New England Journal of Medicine* 379 (2018), pp. 1801–1810.

Chapter exercises

6.39   Active learning. A teacher wanting to increase the active learning component of her course is concerned about student reactions to changes she is planning to make. She conducts a survey in her class, asking students whether they believe more active learning in the classroom (hands on exercises) instead of traditional lecture will helps improve their learning. She does this at the beginning and end of the semester and wants to evaluate whether students' opinions have changed over the semester. Can she used the methods we learned in this chapter for this analysis? Explain your reasoning.

6.40   Website experiment. The OpenIntro website occasionally experiments with design and link placement. We conducted one experiment testing three different placements of a download link for this textbook on the book's main page to see which location, if any, led to the most downloads. The number of site visitors included in the experiment was 701 and is captured in one of the response combinations in the following table:

	Download	No Download
Position 1	13.8%	18.3%
Position 2	14.6%	18.5%
Position 3	12.1%	22.7%

(a) Calculate the actual number of site visitors in each of the six response categories.

(b) Each individual in the experiment had an equal chance of being in any of the three experiment groups. However, we see that there are slightly different totals for the groups. Is there any evidence that the groups were actually imbalanced? Make sure to clearly state hypotheses, check conditions, calculate the appropriate test statistic and the p-value, and make your conclusion in context of the data.

(c) Complete an appropriate hypothesis test to check whether there is evidence that there is a higher rate of site visitors clicking on the textbook link in any of the three groups.

6.41   Shipping holiday gifts. A local news survey asked 500 randomly sampled Los Angeles residents which shipping carrier they prefer to use for shipping holiday gifts. The table below shows the distribution of responses by age group as well as the expected counts for each cell (shown in parentheses).

		Age 18-34		Age 35-54		Age 55+		Total
	USPS	72	(81)	97	(102)	76	(62)	245
	UPS	52	(53)	76	(68)	34	(41)	162
Shipping Method	FedEx	31	(21)	24	(27)	9	(16)	64
	Something else	7	(5)	6	(7)	3	(4)	16
	Not sure	3	(5)	6	(5)	4	(3)	13
	Total	165		209		126		500

(a) State the null and alternative hypotheses for testing for independence of age and preferred shipping method for holiday gifts among Los Angeles residents.

(b) Are the conditions for inference using a chi-square test satisfied?

6.42   The Civil War. A national survey conducted among a simple random sample of 1,507 adults shows that 56% of Americans think the Civil War is still relevant to American politics and political life.[47]

(a) Conduct a hypothesis test to determine if these data provide strong evidence that the majority of the Americans think the Civil War is still relevant.

(b) Interpret the p-value in this context.

(c) Calculate a 90% confidence interval for the proportion of Americans who think the Civil War is still relevant. Interpret the interval in this context, and comment on whether or not the confidence interval agrees with the conclusion of the hypothesis test.

---

[47]Pew Research Center Publications, Civil War at 150: Still Relevant, Still Divisive, data collected between March 30 - April 3, 2011.

6.43 College smokers. We are interested in estimating the proportion of students at a university who smoke. Out of a random sample of 200 students from this university, 40 students smoke.

(a) Calculate a 95% confidence interval for the proportion of students at this university who smoke, and interpret this interval in context. (Reminder: Check conditions.)

(b) If we wanted the margin of error to be no larger than 2% at a 95% confidence level for the proportion of students who smoke, how big of a sample would we need?

6.44 Acetaminophen and liver damage. It is believed that large doses of acetaminophen (the active ingredient in over the counter pain relievers like Tylenol) may cause damage to the liver. A researcher wants to conduct a study to estimate the proportion of acetaminophen users who have liver damage. For participating in this study, he will pay each subject $20 and provide a free medical consultation if the patient has liver damage.

(a) If he wants to limit the margin of error of his 98% confidence interval to 2%, what is the minimum amount of money he needs to set aside to pay his subjects?

(b) The amount you calculated in part (a) is substantially over his budget so he decides to use fewer subjects. How will this affect the width of his confidence interval?

6.45 Life after college. We are interested in estimating the proportion of graduates at a mid-sized university who found a job within one year of completing their undergraduate degree. Suppose we conduct a survey and find out that 348 of the 400 randomly sampled graduates found jobs. The graduating class under consideration included over 4500 students.

(a) Describe the population parameter of interest. What is the value of the point estimate of this parameter?

(b) Check if the conditions for constructing a confidence interval based on these data are met.

(c) Calculate a 95% confidence interval for the proportion of graduates who found a job within one year of completing their undergraduate degree at this university, and interpret it in the context of the data.

(d) What does "95% confidence" mean?

(e) Now calculate a 99% confidence interval for the same parameter and interpret it in the context of the data.

(f) Compare the widths of the 95% and 99% confidence intervals. Which one is wider? Explain.

6.46 Diabetes and unemployment. A Gallup poll surveyed Americans about their employment status and whether or not they have diabetes. The survey results indicate that 1.5% of the 47,774 employed (full or part time) and 2.5% of the 5,855 unemployed 18-29 year olds have diabetes.[48]

(a) Create a two-way table presenting the results of this study.

(b) State appropriate hypotheses to test for difference in proportions of diabetes between employed and unemployed Americans.

(c) The sample difference is about 1%. If we completed the hypothesis test, we would find that the p-value is very small (about 0), meaning the difference is statistically significant. Use this result to explain the difference between statistically significant and practically significant findings.

6.47 Rock-paper-scissors. Rock-paper-scissors is a hand game played by two or more people where players choose to sign either rock, paper, or scissors with their hands. For your statistics class project, you want to evaluate whether players choose between these three options randomly, or if certain options are favored above others. You ask two friends to play rock-paper-scissors and count the times each option is played. The following table summarizes the data:

Rock	Paper	Scissors
43	21	35

Use these data to evaluate whether players choose between these three options randomly, or if certain options are favored above others. Make sure to clearly outline each step of your analysis, and interpret your results in context of the data and the research question.

---

[48]Gallup Wellbeing, Employed Americans in Better Health Than the Unemployed, data collected Jan. 2, 2011 - May 21, 2012.

248   CHAPTER 6. INFERENCE FOR CATEGORICAL DATA

6.48 2010 Healthcare Law. On June 28, 2012 the U.S. Supreme Court upheld the much debated 2010 healthcare law, declaring it constitutional. A Gallup poll released the day after this decision indicates that 46% of 1,012 Americans agree with this decision. At a 95% confidence level, this sample has a 3% margin of error. Based on this information, determine if the following statements are true or false, and explain your reasoning.[49]

(a) We are 95% confident that between 43% and 49% of Americans in this sample support the decision of the U.S. Supreme Court on the 2010 healthcare law.

(b) We are 95% confident that between 43% and 49% of Americans support the decision of the U.S. Supreme Court on the 2010 healthcare law.

(c) If we considered many random samples of 1,012 Americans, and we calculated the sample proportions of those who support the decision of the U.S. Supreme Court, 95% of those sample proportions will be between 43% and 49%.

(d) The margin of error at a 90% confidence level would be higher than 3%.

6.49 Browsing on the mobile device. A survey of 2,254 American adults indicates that 17% of cell phone owners browse the internet exclusively on their phone rather than a computer or other device.[50]

(a) According to an online article, a report from a mobile research company indicates that 38 percent of Chinese mobile web users only access the internet through their cell phones.[51] Conduct a hypothesis test to determine if these data provide strong evidence that the proportion of Americans who only use their cell phones to access the internet is different than the Chinese proportion of 38%.

(b) Interpret the p-value in this context.

(c) Calculate a 95% confidence interval for the proportion of Americans who access the internet on their cell phones, and interpret the interval in this context.

6.50 Coffee and Depression. Researchers conducted a study investigating the relationship between caffeinated coffee consumption and risk of depression in women. They collected data on 50,739 women free of depression symptoms at the start of the study in the year 1996, and these women were followed through 2006. The researchers used questionnaires to collect data on caffeinated coffee consumption, asked each individual about physician- diagnosed depression, and also asked about the use of antidepressants. The table below shows the distribution of incidences of depression by amount of caffeinated coffee consumption.[52]

		Caffeinated coffee consumption					
		≤ 1 cup/week	2-6 cups/week	1 cup/day	2-3 cups/day	≥ 4 cups/day	Total
Clinical	Yes	670	373	905	564	95	2,607
depression	No	11,545	6,244	16,329	11,726	2,288	48,132
	Total	12,215	6,617	17,234	12,290	2,383	50,739

(a) What type of test is appropriate for evaluating if there is an association between coffee intake and depression?

(b) Write the hypotheses for the test you identified in part (a).

(c) Calculate the overall proportion of women who do and do not suffer from depression.

(d) Identify the expected count for the highlighted cell, and calculate the contribution of this cell to the test statistic, i.e. $(Observed - Expected)^2/Expected$.

(e) The test statistic is $\chi^2 = 20.93$. What is the p-value?

(f) What is the conclusion of the hypothesis test?

(g) One of the authors of this study was quoted on the NYTimes as saying it was "too early to recommend that women load up on extra coffee" based on just this study.[53] Do you agree with this statement? Explain your reasoning.

[49]Gallup, Americans Issue Split Decision on Healthcare Ruling, data collected June 28, 2012.
[50]Pew Internet, Cell Internet Use 2012, data collected between March 15 - April 13, 2012.
[51]S. Chang. "The Chinese Love to Use Feature Phone to Access the Internet". In: *M.I.C Gadget* (2012).
[52]M. Lucas et al. "Coffee, caffeine, and risk of depression among women". In: *Archives of internal medicine* 171.17 (2011), p. 1571.
[53]A. O'Connor. "Coffee Drinking Linked to Less Depression in Women". In: *New York Times* (2011).

Chapter 7

Inference for numerical data

Chapter 5 introduced a framework for statistical inference based on confidence intervals and hypotheses using the normal distribution for sample proportions. In this chapter, we encounter several new point estimates and a couple new distributions. In each case, the inference ideas remain the same: determine which point estimate or test statistic is useful, identify an appropriate distribution for the point estimate or test statistic, and apply the ideas of inference.

For videos, slides, and other resources, please visit
www.openintro.org/os

7.1 One-sample means with the t-distribution

Similar to how we can model the behavior of the sample proportion $\hat{p}$ using a normal distribution, the sample mean $\bar{x}$ can also be modeled using a normal distribution when certain conditions are met. However, we'll soon learn that a new distribution, called the t-distribution, tends to be more useful when working with the sample mean. We'll first learn about this new distribution, then we'll use it to construct confidence intervals and conduct hypothesis tests for the mean.

7.1.1 The sampling distribution of $\bar{x}$

The sample mean tends to follow a normal distribution centered at the population mean, μ, when certain conditions are met. Additionally, we can compute a standard error for the sample mean using the population standard deviation σ and the sample size n.

CENTRAL LIMIT THEOREM FOR THE SAMPLE MEAN

When we collect a sufficiently large sample of n independent observations from a population with mean μ and standard deviation σ, the sampling distribution of $\bar{x}$ will be nearly normal with

$$\text{Mean} = \mu \qquad \text{Standard Error } (SE) = \frac{\sigma}{\sqrt{n}}$$

Before diving into confidence intervals and hypothesis tests using $\bar{x}$, we first need to cover two topics:

- When we modeled $\hat{p}$ using the normal distribution, certain conditions had to be satisfied. The conditions for working with $\bar{x}$ are a little more complex, and we'll spend Section 7.1.2 discussing how to check conditions for inference.

- The standard error is dependent on the population standard deviation, σ. However, we rarely know σ, and instead we must estimate it. Because this estimation is itself imperfect, we use a new distribution called the t-distribution to fix this problem, which we discuss in Section 7.1.3.

7.1.2 Evaluating the two conditions required for modeling $\bar{x}$

Two conditions are required to apply the Central Limit Theorem for a sample mean $\bar{x}$:

Independence. The sample observations must be independent, The most common way to satisfy this condition is when the sample is a simple random sample from the population. If the data come from a random process, analogous to rolling a die, this would also satisfy the independence condition.

Normality. When a sample is small, we also require that the sample observations come from a normally distributed population. We can relax this condition more and more for larger and larger sample sizes. This condition is obviously vague, making it difficult to evaluate, so next we introduce a couple rules of thumb to make checking this condition easier.

RULES OF THUMB: HOW TO PERFORM THE NORMALITY CHECK

There is no perfect way to check the normality condition, so instead we use two rules of thumb:

n < 30: If the sample size n is less than 30 and there are no clear outliers in the data, then we typically assume the data come from a nearly normal distribution to satisfy the condition.

n ≥ 30: If the sample size n is at least 30 and there are no *particularly extreme* outliers, then we typically assume the sampling distribution of $\bar{x}$ is nearly normal, even if the underlying distribution of individual observations is not.

In this first course in statistics, you aren't expected to develop perfect judgement on the normality condition. However, you are expected to be able to handle clear cut cases based on the rules of thumb.[1]

EXAMPLE 7.1

Consider the following two plots that come from simple random samples from different populations. Their sample sizes are $n_1 = 15$ and $n_2 = 50$.

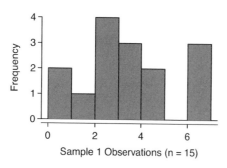

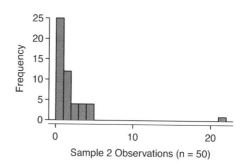

Are the independence and normality conditions met in each case?

Each samples is from a simple random sample of its respective population, so the independence condition is satisfied. Let's next check the normality condition for each using the rule of thumb.

The first sample has fewer than 30 observations, so we are watching for any clear outliers. None are present; while there is a small gap in the histogram on the right, this gap is small and 20% of the observations in this small sample are represented in that far right bar of the histogram, so we can hardly call these clear outliers. With no clear outliers, the normality condition is reasonably met.

The second sample has a sample size greater than 30 and includes an outlier that appears to be roughly 5 times further from the center of the distribution than the next furthest observation. This is an example of a particularly extreme outlier, so the normality condition would not be satisfied.

In practice, it's typical to also do a mental check to evaluate whether we have reason to believe the underlying population would have moderate skew (if $n < 30$) or have particularly extreme outliers ($n \geq 30$) beyond what we observe in the data. For example, consider the number of followers for each individual account on Twitter, and then imagine this distribution. The large majority of accounts have built up a couple thousand followers or fewer, while a relatively tiny fraction have amassed tens of millions of followers, meaning the distribution is extremely skewed. When we know the data come from such an extremely skewed distribution, it takes some effort to understand what sample size is large enough for the normality condition to be satisfied.

7.1.3   Introducing the t-distribution

In practice, we cannot directly calculate the standard error for $\bar{x}$ since we do not know the population standard deviation, σ. We encountered a similar issue when computing the standard error for a sample proportion, which relied on the population proportion, p. Our solution in the proportion context was to use sample value in place of the population value when computing the standard error. We'll employ a similar strategy for computing the standard error of $\bar{x}$, using the sample standard deviation s in place of σ:

$$ SE = \frac{\sigma}{\sqrt{n}} \approx \frac{s}{\sqrt{n}} $$

This strategy tends to work well when we have a lot of data and can estimate σ using s accurately. However, the estimate is less precise with smaller samples, and this leads to problems when using the normal distribution to model $\bar{x}$.

---

[1]More nuanced guidelines would consider further relaxing the *particularly extreme outlier* check when the sample size is very large. However, we'll leave further discussion here to a future course.

We'll find it useful to use a new distribution for inference calculations called the **t-distribution**. A t-distribution, shown as a solid line in Figure 7.1, has a bell shape. However, its tails are thicker than the normal distribution's, meaning observations are more likely to fall beyond two standard deviations from the mean than under the normal distribution. The extra thick tails of the t-distribution are exactly the correction needed to resolve the problem of using s in place of σ in the SE calculation.

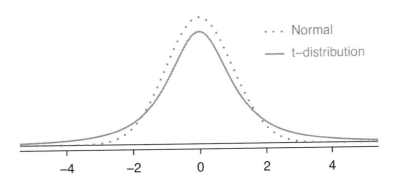

Figure 7.1: Comparison of a t-distribution and a normal distribution.

The t-distribution is always centered at zero and has a single parameter: degrees of freedom. The **degrees of freedom** (**df**) describes the precise form of the bell-shaped t-distribution. Several t-distributions are shown in Figure 7.2 in comparison to the normal distribution.

In general, we'll use a t-distribution with $df = n - 1$ to model the sample mean when the sample size is n. That is, when we have more observations, the degrees of freedom will be larger and the t-distribution will look more like the standard normal distribution; when the degrees of freedom is about 30 or more, the t-distribution is nearly indistinguishable from the normal distribution.

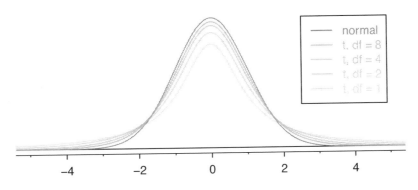

Figure 7.2: The larger the degrees of freedom, the more closely the t-distribution resembles the standard normal distribution.

DEGREES OF FREEDOM (df)

The degrees of freedom describes the shape of the t-distribution. The larger the degrees of freedom, the more closely the distribution approximates the normal model.

When modeling $\bar{x}$ using the t-distribution, use $df = n - 1$.

The t-distribution allows us greater flexibility than the normal distribution when analyzing numerical data. In practice, it's common to use statistical software, such as R, Python, or SAS for these analyses. Alternatively, a graphing calculator or a **t-table** may be used; the t-table is similar to the normal distribution table, and it may be found in Appendix C.2, which includes usage instructions and examples for those who wish to use this option. No matter the approach you choose, apply your method using the examples below to confirm your working understanding of the t-distribution.

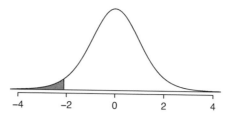

Figure 7.3: The *t*-distribution with 18 degrees of freedom. The area below -2.10 has been shaded.

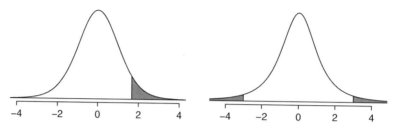

Figure 7.4: Left: The *t*-distribution with 20 degrees of freedom, with the area above 1.65 shaded. Right: The *t*-distribution with 2 degrees of freedom, with the area further than 3 units from 0 shaded.

EXAMPLE 7.2

What proportion of the *t*-distribution with 18 degrees of freedom falls below -2.10?

Just like a normal probability problem, we first draw the picture in Figure 7.3 and shade the area below -2.10. Using statistical software, we can obtain a precise value: 0.0250.

EXAMPLE 7.3

A *t*-distribution with 20 degrees of freedom is shown in the left panel of Figure 7.4. Estimate the proportion of the distribution falling above 1.65.

With a normal distribution, this would correspond to about 0.05, so we should expect the *t*-distribution to give us a value in this neighborhood. Using statistical software: 0.0573.

EXAMPLE 7.4

A *t*-distribution with 2 degrees of freedom is shown in the right panel of Figure 7.4. Estimate the proportion of the distribution falling more than 3 units from the mean (above or below).

With so few degrees of freedom, the *t*-distribution will give a more notably different value than the normal distribution. Under a normal distribution, the area would be about 0.003 using the 68-95-99.7 rule. For a *t*-distribution with $df = 2$, the area in both tails beyond 3 units totals 0.0955. This area is dramatically different than what we obtain from the normal distribution.

GUIDED PRACTICE 7.5

What proportion of the *t*-distribution with 19 degrees of freedom falls above -1.79 units? Use your preferred method for finding tail areas.[2]

---

[2]We want to find the shaded area *above* -1.79 (we leave the picture to you). The lower tail area has an area of 0.0447, so the upper area would have an area of $1 - 0.0447 = 0.9553$.

7.1.4 One sample t-confidence intervals

Let's get our first taste of applying the t-distribution in the context of an example about the mercury content of dolphin muscle. Elevated mercury concentrations are an important problem for both dolphins and other animals, like humans, who occasionally eat them.

Figure 7.5: A Risso's dolphin.

Photo by Mike Baird (www.bairdphotos.com). CC BY 2.0 license.

We will identify a confidence interval for the average mercury content in dolphin muscle using a sample of 19 Risso's dolphins from the Taiji area in Japan. The data are summarized in Figure 7.6. The minimum and maximum observed values can be used to evaluate whether or not there are clear outliers.

n	$\bar{x}$	s	minimum	maximum
19	4.4	2.3	1.7	9.2

Figure 7.6: Summary of mercury content in the muscle of 19 Risso's dolphins from the Taiji area. Measurements are in micrograms of mercury per wet gram of muscle (μg/wet g).

EXAMPLE 7.6

Are the independence and normality conditions satisfied for this data set?

The observations are a simple random sample, therefore independence is reasonable. The summary statistics in Figure 7.6 do not suggest any clear outliers, since all observations are within 2.5 standard deviations of the mean. Based on this evidence, the normality condition seems reasonable.

In the normal model, we used $z^\star$ and the standard error to determine the width of a confidence interval. We revise the confidence interval formula slightly when using the t-distribution:

$$\text{point estimate} \pm t_{df}^\star \times SE \qquad \rightarrow \qquad \bar{x} \pm t_{df}^\star \times \frac{s}{\sqrt{n}}$$

EXAMPLE 7.7

Using the summary statistics in Figure 7.6, compute the standard error for the average mercury content in the $n = 19$ dolphins.

We plug in s and n into the formula: $SE = s/\sqrt{n} = 2.3/\sqrt{19} = 0.528$.

The value $t_{df}^{\star}$ is a cutoff we obtain based on the confidence level and the t-distribution with df degrees of freedom. That cutoff is found in the same way as with a normal distribution: we find $t_{df}^{\star}$ such that the fraction of the t-distribution with df degrees of freedom within a distance $t_{df}^{\star}$ of 0 matches the confidence level of interest.

EXAMPLE 7.8

When $n = 19$, what is the appropriate degrees of freedom? Find $t_{df}^{\star}$ for this degrees of freedom and the confidence level of 95%

The degrees of freedom is easy to calculate: $df = n - 1 = 18$.

Using statistical software, we find the cutoff where the upper tail is equal to 2.5%: $t_{18}^{\star} = 2.10$. The area below -2.10 will also be equal to 2.5%. That is, 95% of the t-distribution with $df = 18$ lies within 2.10 units of 0.

EXAMPLE 7.9

Compute and interpret the 95% confidence interval for the average mercury content in Risso's dolphins.

We can construct the confidence interval as

$$\bar{x} \pm t_{18}^{\star} \times SE \quad \rightarrow \quad 4.4 \pm 2.10 \times 0.528 \quad \rightarrow \quad (3.29, 5.51)$$

We are 95% confident the average mercury content of muscles in Risso's dolphins is between 3.29 and 5.51 μg/wet gram, which is considered extremely high.

FINDING A t-CONFIDENCE INTERVAL FOR THE MEAN

Based on a sample of n independent and nearly normal observations, a confidence interval for the population mean is

$$\text{point estimate} \pm t_{df}^{\star} \times SE \qquad \rightarrow \qquad \bar{x} \pm t_{df}^{\star} \times \frac{s}{\sqrt{n}}$$

where $\bar{x}$ is the sample mean, $t_{df}^{\star}$ corresponds to the confidence level and degrees of freedom df, and SE is the standard error as estimated by the sample.

GUIDED PRACTICE 7.10

The FDA's webpage provides some data on mercury content of fish. Based on a sample of 15 croaker white fish (Pacific), a sample mean and standard deviation were computed as 0.287 and 0.069 ppm (parts per million), respectively. The 15 observations ranged from 0.18 to 0.41 ppm. We will assume these observations are independent. Based on the summary statistics of the data, do you have any objections to the normality condition of the individual observations?[3]

EXAMPLE 7.11

Estimate the standard error of $\bar{x} = 0.287$ ppm using the data summaries in Guided Practice 7.10. If we are to use the t-distribution to create a 90% confidence interval for the actual mean of the mercury content, identify the degrees of freedom and $t_{df}^{\star}$.

The standard error: $SE = \frac{0.069}{\sqrt{15}} = 0.0178$.

Degrees of freedom: $df = n - 1 = 14$.

Since the goal is a 90% confidence interval, we choose $t_{14}^{\star}$ so that the two-tail area is 0.1: $t_{14}^{\star} = 1.76$.

---

[3]The sample size is under 30, so we check for obvious outliers: since all observations are within 2 standard deviations of the mean, there are no such clear outliers.

CONFIDENCE INTERVAL FOR A SINGLE MEAN

Once you've determined a one-mean confidence interval would be helpful for an application, there are four steps to constructing the interval:

Prepare. Identify $\bar{x}$, s, n, and determine what confidence level you wish to use.

Check. Verify the conditions to ensure $\bar{x}$ is nearly normal.

Calculate. If the conditions hold, compute SE, find $t_{df}^{\star}$, and construct the interval.

Conclude. Interpret the confidence interval in the context of the problem.

GUIDED PRACTICE 7.12

Using the information and results of Guided Practice 7.10 and Example 7.11, compute a 90% confidence interval for the average mercury content of croaker white fish (Pacific).[4]

GUIDED PRACTICE 7.13

The 90% confidence interval from Guided Practice 7.12 is 0.256 ppm to 0.318 ppm. Can we say that 90% of croaker white fish (Pacific) have mercury levels between 0.256 and 0.318 ppm?[5]

7.1.5 One sample *t*-tests

Is the typical US runner getting faster or slower over time? We consider this question in the context of the Cherry Blossom Race, which is a 10-mile race in Washington, DC each spring.

The average time for all runners who finished the Cherry Blossom Race in 2006 was 93.29 minutes (93 minutes and about 17 seconds). We want to determine using data from 100 participants in the 2017 Cherry Blossom Race whether runners in this race are getting faster or slower, versus the other possibility that there has been no change.

GUIDED PRACTICE 7.14

What are appropriate hypotheses for this context?[6]

GUIDED PRACTICE 7.15

The data come from a simple random sample of all participants, so the observations are independent. However, should we be worried about the normality condition? See Figure 7.7 for a histogram of the differences and evaluate if we can move forward.[7]

When completing a hypothesis test for the one-sample mean, the process is nearly identical to completing a hypothesis test for a single proportion. First, we find the Z-score using the observed value, null value, and standard error; however, we call it a **T-score** since we use a *t*-distribution for calculating the tail area. Then we find the p-value using the same ideas we used previously: find the one-tail area under the sampling distribution, and double it.

---

[4] $\bar{x} \pm t_{14}^{\star} \times SE \rightarrow 0.287 \pm 1.76 \times 0.0178 \rightarrow (0.256, 0.318)$. We are 90% confident that the average mercury content of croaker white fish (Pacific) is between 0.256 and 0.318 ppm.

[5] No, a confidence interval only provides a range of plausible values for a population parameter, in this case the population mean. It does not describe what we might observe for individual observations.

[6] H_0: The average 10-mile run time was the same for 2006 and 2017. $\mu = 93.29$ minutes. H_A: The average 10-mile run time for 2017 was *different* than that of 2006. $\mu \neq 93.29$ minutes.

[7] With a sample of 100, we should only be concerned if there is are particularly extreme outliers. The histogram of the data doesn't show any outliers of concern (and arguably, no outliers at all).

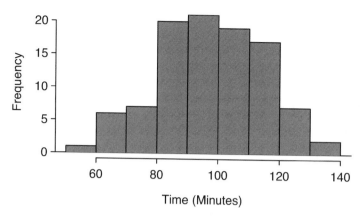

Figure 7.7: A histogram of `time` for the sample Cherry Blossom Race data.

EXAMPLE 7.16

With both the independence and normality conditions satisfied, we can proceed with a hypothesis test using the t-distribution. The sample mean and sample standard deviation of the sample of 100 runners from the 2017 Cherry Blossom Race are 97.32 and 16.98 minutes, respectively. Recall that the sample size is 100 and the average run time in 2006 was 93.29 minutes. Find the test statistic and p-value. What is your conclusion?

To find the test statistic (T-score), we first must determine the standard error:

$$SE = 16.98/\sqrt{100} = 1.70$$

Now we can compute the *T-score* using the sample mean (97.32), null value (93.29), and SE:

$$T = \frac{97.32 - 93.29}{1.70} = 2.37$$

For $df = 100 - 1 = 99$, we can determine using statistical software (or a t-table) that the one-tail area is 0.01, which we double to get the p-value: 0.02.

Because the p-value is smaller than 0.05, we reject the null hypothesis. That is, the data provide strong evidence that the average run time for the Cherry Blossom Run in 2017 is different than the 2006 average. Since the observed value is above the null value and we have rejected the null hypothesis, we would conclude that runners in the race were slower on average in 2017 than in 2006.

HYPOTHESIS TESTING FOR A SINGLE MEAN

Once you've determined a one-mean hypothesis test is the correct procedure, there are four steps to completing the test:

Prepare. Identify the parameter of interest, list out hypotheses, identify the significance level, and identify $\bar{x}$, s, and n.

Check. Verify conditions to ensure $\bar{x}$ is nearly normal.

Calculate. If the conditions hold, compute SE, compute the T-score, and identify the p-value.

Conclude. Evaluate the hypothesis test by comparing the p-value to α, and provide a conclusion in the context of the problem.

Exercises

7.1   Identify the critical t. An independent random sample is selected from an approximately normal population with unknown standard deviation. Find the degrees of freedom and the critical t-value (t^*) for the given sample size and confidence level.

(a) $n = 6$, CL = 90%

(b) $n = 21$, CL = 98%

(c) $n = 29$, CL = 95%

(d) $n = 12$, CL = 99%

7.2 t-distribution. The figure on the right shows three unimodal and symmetric curves: the standard normal (z) distribution, the t-distribution with 5 degrees of freedom, and the t-distribution with 1 degree of freedom. Determine which is which, and explain your reasoning.

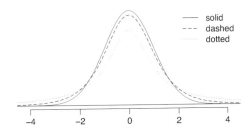

7.3   Find the p-value, Part I. An independent random sample is selected from an approximately normal population with an unknown standard deviation. Find the p-value for the given sample size and test statistic. Also determine if the null hypothesis would be rejected at $\alpha = 0.05$.

(a) $n = 11$, $T = 1.91$

(b) $n = 17$, $T = -3.45$

(c) $n = 7$, $T = 0.83$

(d) $n = 28$, $T = 2.13$

7.4   Find the p-value, Part II. An independent random sample is selected from an approximately normal population with an unknown standard deviation. Find the p-value for the given sample size and test statistic. Also determine if the null hypothesis would be rejected at $\alpha = 0.01$.

(a) $n = 26$, $T = 2.485$

(b) $n = 18$, $T = 0.5$

7.5   Working backwards, Part I. A 95% confidence interval for a population mean, μ, is given as (18.985, 21.015). This confidence interval is based on a simple random sample of 36 observations. Calculate the sample mean and standard deviation. Assume that all conditions necessary for inference are satisfied. Use the t-distribution in any calculations.

7.6   Working backwards, Part II. A 90% confidence interval for a population mean is (65, 77). The population distribution is approximately normal and the population standard deviation is unknown. This confidence interval is based on a simple random sample of 25 observations. Calculate the sample mean, the margin of error, and the sample standard deviation.

7.7 Sleep habits of New Yorkers. New York is known as "the city that never sleeps". A random sample of 25 New Yorkers were asked how much sleep they get per night. Statistical summaries of these data are shown below. The point estimate suggests New Yorkers sleep less than 8 hours a night on average. Is the result statistically significant?

n	$\bar{x}$	s	min	max
25	7.73	0.77	6.17	9.78

(a) Write the hypotheses in symbols and in words.

(b) Check conditions, then calculate the test statistic, T, and the associated degrees of freedom.

(c) Find and interpret the p-value in this context. Drawing a picture may be helpful.

(d) What is the conclusion of the hypothesis test?

(e) If you were to construct a 90% confidence interval that corresponded to this hypothesis test, would you expect 8 hours to be in the interval?

7.8 Heights of adults. Researchers studying anthropometry collected body girth measurements and skeletal diameter measurements, as well as age, weight, height and gender, for 507 physically active individuals. The histogram below shows the sample distribution of heights in centimeters.[8]

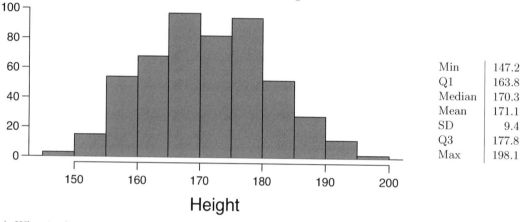

Min	147.2
Q1	163.8
Median	170.3
Mean	171.1
SD	9.4
Q3	177.8
Max	198.1

(a) What is the point estimate for the average height of active individuals? What about the median?

(b) What is the point estimate for the standard deviation of the heights of active individuals? What about the IQR?

(c) Is a person who is 1m 80cm (180 cm) tall considered unusually tall? And is a person who is 1m 55cm (155cm) considered unusually short? Explain your reasoning.

(d) The researchers take another random sample of physically active individuals. Would you expect the mean and the standard deviation of this new sample to be the ones given above? Explain your reasoning.

(e) The sample means obtained are point estimates for the mean height of all active individuals, if the sample of individuals is equivalent to a simple random sample. What measure do we use to quantify the variability of such an estimate? Compute this quantity using the data from the original sample under the condition that the data are a simple random sample.

7.9 Find the mean. You are given the following hypotheses:

$$H_0 : \mu = 60$$
$$H_A : \mu \neq 60$$

We know that the sample standard deviation is 8 and the sample size is 20. For what sample mean would the p-value be equal to 0.05? Assume that all conditions necessary for inference are satisfied.

---

[8]G. Heinz et al. "Exploring relationships in body dimensions". In: *Journal of Statistics Education* 11.2 (2003).

7.10 t^* **vs.** z^*. For a given confidence level, t^*_{df} is larger than z^*. Explain how t^*_{df} being slightly larger than z^* affects the width of the confidence interval.

7.11 Play the piano. Georgianna claims that in a small city renowned for its music school, the average child takes less than 5 years of piano lessons. We have a random sample of 20 children from the city, with a mean of 4.6 years of piano lessons and a standard deviation of 2.2 years.

(a) Evaluate Georgianna's claim (or that the opposite might be true) using a hypothesis test.

(b) Construct a 95% confidence interval for the number of years students in this city take piano lessons, and interpret it in context of the data.

(c) Do your results from the hypothesis test and the confidence interval agree? Explain your reasoning.

7.12 Auto exhaust and lead exposure. Researchers interested in lead exposure due to car exhaust sampled the blood of 52 police officers subjected to constant inhalation of automobile exhaust fumes while working traffic enforcement in a primarily urban environment. The blood samples of these officers had an average lead concentration of 124.32 μg/l and a SD of 37.74 μg/l; a previous study of individuals from a nearby suburb, with no history of exposure, found an average blood level concentration of 35 μg/l.[9]

(a) Write down the hypotheses that would be appropriate for testing if the police officers appear to have been exposed to a different concentration of lead.

(b) Explicitly state and check all conditions necessary for inference on these data.

(c) Regardless of your answers in part (b), test the hypothesis that the downtown police officers have a higher lead exposure than the group in the previous study. Interpret your results in context.

7.13 Car insurance savings. A market researcher wants to evaluate car insurance savings at a competing company. Based on past studies he is assuming that the standard deviation of savings is $100. He wants to collect data such that he can get a margin of error of no more than $10 at a 95% confidence level. How large of a sample should he collect?

7.14 SAT scores. The standard deviation of SAT scores for students at a particular Ivy League college is 250 points. Two statistics students, Raina and Luke, want to estimate the average SAT score of students at this college as part of a class project. They want their margin of error to be no more than 25 points.

(a) Raina wants to use a 90% confidence interval. How large a sample should she collect?

(b) Luke wants to use a 99% confidence interval. Without calculating the actual sample size, determine whether his sample should be larger or smaller than Raina's, and explain your reasoning.

(c) Calculate the minimum required sample size for Luke.

---

[9]WI Mortada et al. "Study of lead exposure from automobile exhaust as a risk for nephrotoxicity among traffic policemen." In: *American journal of nephrology* 21.4 (2000), pp. 274–279.

7.2 Paired data

In an earlier edition of this textbook, we found that Amazon prices were, on average, lower than those of the UCLA Bookstore for UCLA courses in 2010. It's been several years, and many stores have adapted to the online market, so we wondered, how is the UCLA Bookstore doing today?

We sampled 201 UCLA courses. Of those, 68 required books could be found on Amazon. A portion of the data set from these courses is shown in Figure 7.8, where prices are in US dollars.

	subject	course_number	bookstore	amazon	price_difference
1	American Indian Studies	M10	47.97	47.45	0.52
2	Anthropology	2	14.26	13.55	0.71
3	Arts and Architecture	10	13.50	12.53	0.97
⋮	⋮	⋮	⋮	⋮	⋮
68	Jewish Studies	M10	35.96	32.40	3.56

Figure 7.8: Four cases of the `textbooks` data set.

7.2.1 Paired observations

Each textbook has two corresponding prices in the data set: one for the UCLA Bookstore and one for Amazon. When two sets of observations have this special correspondence, they are said to be **paired**.

> **PAIRED DATA**
>
> Two sets of observations are *paired* if each observation in one set has a special correspondence or connection with exactly one observation in the other data set.

To analyze paired data, it is often useful to look at the difference in outcomes of each pair of observations. In the textbook data, we look at the differences in prices, which is represented as the `price_difference` variable in the data set. Here the differences are taken as

UCLA Bookstore price − Amazon price

It is important that we always subtract using a consistent order; here Amazon prices are always subtracted from UCLA prices. The first difference shown in Figure 7.8 is computed as $47.97 - 47.45 = 0.52$. Similarly, the second difference is computed as $14.26 - 13.55 = 0.71$, and the third is $13.50 - 12.53 = 0.97$. A histogram of the differences is shown in Figure 7.9. Using differences between paired observations is a common and useful way to analyze paired data.

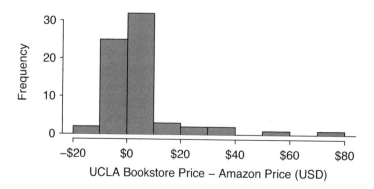

Figure 7.9: Histogram of the difference in price for each book sampled.

7.2.2 Inference for paired data

To analyze a paired data set, we simply analyze the differences. We can use the same t-distribution techniques we applied in Section 7.1.

n_{diff}	$\bar{x}_{diff}$	s_{diff}
68	3.58	13.42

Figure 7.10: Summary statistics for the 68 price differences.

EXAMPLE 7.17

Set up a hypothesis test to determine whether, on average, there is a difference between Amazon's price for a book and the UCLA bookstore's price. Also, check the conditions for whether we can move forward with the test using the t-distribution.

We are considering two scenarios: there is no difference or there is some difference in average prices.

H_0: $\mu_{diff} = 0$. There is no difference in the average textbook price.

H_A: $\mu_{diff} \neq 0$. There is a difference in average prices.

Next, we check the independence and normality conditions. The observations are based on a simple random sample, so independence is reasonable. While there are some outliers, $n = 68$ and none of the outliers are particularly extreme, so the normality of $\bar{x}$ is satisfied. With these conditions satisfied, we can move forward with the t-distribution.

EXAMPLE 7.18

Complete the hypothesis test started in Example 7.17.

To compute the test compute the standard error associated with $\bar{x}_{diff}$ using the standard deviation of the differences ($s_{diff} = 13.42$) and the number of differences ($n_{diff} = 68$):

$$SE_{\bar{x}_{diff}} = \frac{s_{diff}}{\sqrt{n_{diff}}} = \frac{13.42}{\sqrt{68}} = 1.63$$

The test statistic is the T-score of $\bar{x}_{diff}$ under the null condition that the actual mean difference is 0:

$$T = \frac{\bar{x}_{diff} - 0}{SE_{\bar{x}_{diff}}} = \frac{3.58 - 0}{1.63} = 2.20$$

To visualize the p-value, the sampling distribution of $\bar{x}_{diff}$ is drawn as though H_0 is true, and the p-value is represented by the two shaded tails:

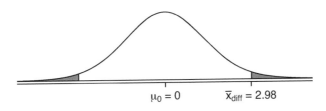

The degrees of freedom is $df = 68 - 1 = 67$. Using statistical software, we find the one-tail area of 0.0156. Doubling this area gives the p-value: 0.0312.

Because the p-value is less than 0.05, we reject the null hypothesis. Amazon prices are, on average, lower than the UCLA Bookstore prices for UCLA courses.

Ⓖ

GUIDED PRACTICE 7.19

Create a 95% confidence interval for the average price difference between books at the UCLA bookstore and books on Amazon.[10]

Ⓖ

GUIDED PRACTICE 7.20

We have strong evidence that Amazon is, on average, less expensive. How should this conclusion affect UCLA student buying habits? Should UCLA students always buy their books on Amazon?[11]

---

[10]Conditions have already verified and the standard error computed in Example 7.17. To find the interval, identify $t_{67}^{\star}$ using statistical software or the t-table ($t_{67}^{\star} = 2.00$), and plug it, the point estimate, and the standard error into the confidence interval formula:

$$\text{point estimate} \pm z^{\star} \times SE \quad \rightarrow \quad 3.58 \pm 2.00 \times 1.63 \quad \rightarrow \quad (0.32, 6.84)$$

We are 95% confident that Amazon is, on average, between $0.32 and $6.84 less expensive than the UCLA Bookstore for UCLA course books.

[11]The average price difference is only mildly useful for this question. Examine the distribution shown in Figure 7.9. There are certainly a handful of cases where Amazon prices are far below the UCLA Bookstore's, which suggests it is worth checking Amazon (and probably other online sites) before purchasing. However, in many cases the Amazon price is above what the UCLA Bookstore charges, and most of the time the price isn't that different. Ultimately, if getting a book immediately from the bookstore is notably more convenient, e.g. to get started on reading or homework, it's likely a good idea to go with the UCLA Bookstore unless the price difference on a specific book happens to be quite large.

For reference, this is a very different result from what we (the authors) had seen in a similar data set from 2010. At that time, Amazon prices were almost uniformly lower than those of the UCLA Bookstore's and by a large margin, making the case to use Amazon over the UCLA Bookstore quite compelling at that time. Now we frequently check multiple websites to find the best price.

Exercises

7.15   Air quality. Air quality measurements were collected in a random sample of 25 country capitals in 2013, and then again in the same cities in 2014. We would like to use these data to compare average air quality between the two years. Should we use a paired or non-paired test? Explain your reasoning.

7.16   True / False: paired. Determine if the following statements are true or false. If false, explain.

(a) In a paired analysis we first take the difference of each pair of observations, and then we do inference on these differences.

(b) Two data sets of different sizes cannot be analyzed as paired data.

(c) Consider two sets of data that are paired with each other. Each observation in one data set has a natural correspondence with exactly one observation from the other data set.

(d) Consider two sets of data that are paired with each other. Each observation in one data set is subtracted from the average of the other data set's observations.

7.17   Paired or not? Part I. In each of the following scenarios, determine if the data are paired.

(a) Compare pre- (beginning of semester) and post-test (end of semester) scores of students.

(b) Assess gender-related salary gap by comparing salaries of randomly sampled men and women.

(c) Compare artery thicknesses at the beginning of a study and after 2 years of taking Vitamin E for the same group of patients.

(d) Assess effectiveness of a diet regimen by comparing the before and after weights of subjects.

7.18   Paired or not? Part II. In each of the following scenarios, determine if the data are paired.

(a) We would like to know if Intel's stock and Southwest Airlines' stock have similar rates of return. To find out, we take a random sample of 50 days, and record Intel's and Southwest's stock on those same days.

(b) We randomly sample 50 items from Target stores and note the price for each. Then we visit Walmart and collect the price for each of those same 50 items.

(c) A school board would like to determine whether there is a difference in average SAT scores for students at one high school versus another high school in the district. To check, they take a simple random sample of 100 students from each high school.

7.19   Global warming, Part I. Let's consider a limited set of climate data, examining temperature differences in 1948 vs 2018. We sampled 197 locations from the National Oceanic and Atmospheric Administration's (NOAA) historical data, where the data was available for both years of interest. We want to know: were there more days with temperatures exceeding 90°F in 2018 or in 1948?[12] The difference in number of days exceeding 90°F (number of days in 2018 - number of days in 1948) was calculated for each of the 197 locations. The average of these differences was 2.9 days with a standard deviation of 17.2 days. We are interested in determining whether these data provide strong evidence that there were more days in 2018 that exceeded 90°F from NOAA's weather stations.

(a) Is there a relationship between the observations collected in 1948 and 2018? Or are the observations in the two groups independent? Explain.

(b) Write hypotheses for this research in symbols and in words.

(c) Check the conditions required to complete this test. A histogram of the differences is given to the right.

(d) Calculate the test statistic and find the p-value.

(e) Use $\alpha = 0.05$ to evaluate the test, and interpret your conclusion in context.

(f) What type of error might we have made? Explain in context what the error means.

(g) Based on the results of this hypothesis test, would you expect a confidence interval for the average difference between the number of days exceeding 90°F from 1948 and 2018 to include 0? Explain your reasoning.

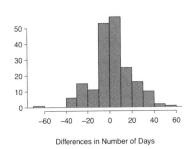

Differences in Number of Days

7.20 High School and Beyond, Part I. The National Center of Education Statistics conducted a survey of high school seniors, collecting test data on reading, writing, and several other subjects. Here we examine a simple random sample of 200 students from this survey. Side-by-side box plots of reading and writing scores as well as a histogram of the differences in scores are shown below.

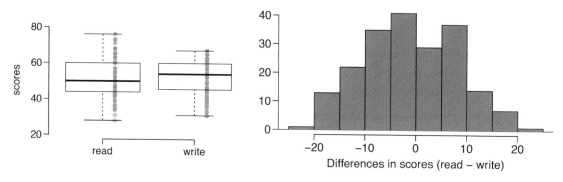

(a) Is there a clear difference in the average reading and writing scores?

(b) Are the reading and writing scores of each student independent of each other?

(c) Create hypotheses appropriate for the following research question: is there an evident difference in the average scores of students in the reading and writing exam?

(d) Check the conditions required to complete this test.

(e) The average observed difference in scores is $\bar{x}_{read-write} = -0.545$, and the standard deviation of the differences is 8.887 points. Do these data provide convincing evidence of a difference between the average scores on the two exams?

(f) What type of error might we have made? Explain what the error means in the context of the application.

(g) Based on the results of this hypothesis test, would you expect a confidence interval for the average difference between the reading and writing scores to include 0? Explain your reasoning.

7.21 Global warming, Part II. We considered the change in the number of days exceeding 90°F from 1948 and 2018 at 197 randomly sampled locations from the NOAA database in Exercise 7.19. The mean and standard deviation of the reported differences are 2.9 days and 17.2 days.

(a) Calculate a 90% confidence interval for the average difference between number of days exceeding 90°F between 1948 and 2018. We've already checked the conditions for you.

(b) Interpret the interval in context.

(c) Does the confidence interval provide convincing evidence that there were more days exceeding 90°F in 2018 than in 1948 at NOAA stations? Explain.

7.22 High school and beyond, Part II. We considered the differences between the reading and writing scores of a random sample of 200 students who took the High School and Beyond Survey in Exercise 7.20. The mean and standard deviation of the differences are $\bar{x}_{read-write} = -0.545$ and 8.887 points.

(a) Calculate a 95% confidence interval for the average difference between the reading and writing scores of all students.

(b) Interpret this interval in context.

(c) Does the confidence interval provide convincing evidence that there is a real difference in the average scores? Explain.

7.3 Difference of two means

In this section we consider a difference in two population means, $\mu_1 - \mu_2$, under the condition that the data are not paired. Just as with a single sample, we identify conditions to ensure we can use the t-distribution with a point estimate of the difference, $\bar{x}_1 - \bar{x}_2$, and a new standard error formula. Other than these two differences, the details are almost identical to the one-mean procedures.

We apply these methods in three contexts: determining whether stem cells can improve heart function, exploring the relationship between pregnant womens' smoking habits and birth weights of newborns, and exploring whether there is statistically significant evidence that one variation of an exam is harder than another variation. This section is motivated by questions like "Is there convincing evidence that newborns from mothers who smoke have a different average birth weight than newborns from mothers who don't smoke?"

7.3.1 Confidence interval for a difference of means

Does treatment using embryonic stem cells (ESCs) help improve heart function following a heart attack? Figure 7.11 contains summary statistics for an experiment to test ESCs in sheep that had a heart attack. Each of these sheep was randomly assigned to the ESC or control group, and the change in their hearts' pumping capacity was measured in the study. Figure 7.12 provides histograms of the two data sets. A positive value corresponds to increased pumping capacity, which generally suggests a stronger recovery. Our goal will be to identify a 95% confidence interval for the effect of ESCs on the change in heart pumping capacity relative to the control group.

	n	$\bar{x}$	s
ESCs	9	3.50	5.17
control	9	-4.33	2.76

Figure 7.11: Summary statistics of the embryonic stem cell study.

The point estimate of the difference in the heart pumping variable is straightforward to find: it is the difference in the sample means.

$$\bar{x}_{esc} - \bar{x}_{control} = 3.50 - (-4.33) = 7.83$$

For the question of whether we can model this difference using a t-distribution, we'll need to check new conditions. Like the 2-proportion cases, we will require a more robust version of independence so we are confident the two groups are also independent. Secondly, we also check for normality in each group separately, which in practice is a check for outliers.

USING THE t-DISTRIBUTION FOR A DIFFERENCE IN MEANS

The t-distribution can be used for inference when working with the standardized difference of two means if

- *Independence, extended.* The data are independent within and between the two groups, e.g. the data come from independent random samples or from a randomized experiment.
- *Normality.* We check the outliers rules of thumb for each group separately.

The standard error may be computed as

$$SE = \sqrt{\frac{\sigma_1^2}{n_1} + \frac{\sigma_2^2}{n_2}}$$

The official formula for the degrees of freedom is quite complex and is generally computed using software, so instead you may use the smaller of $n_1 - 1$ and $n_2 - 1$ for the degrees of freedom if software isn't readily available.

EXAMPLE 7.21

Can the t-distribution be used to make inference using the point estimate, $\bar{x}_{esc} - \bar{x}_{control} = 7.83$?

First, we check for independence. Because the sheep were randomized into the groups, independence within and between groups is satisfied.

Figure 7.12 does not reveal any clear outliers in either group. (The ESC group does look a bit more variability, but this is not the same as having clear outliers.)

With both conditions met, we can use the t-distribution to model the difference of sample means.

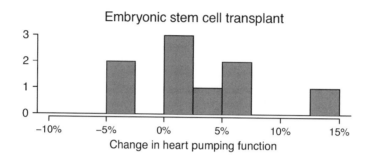

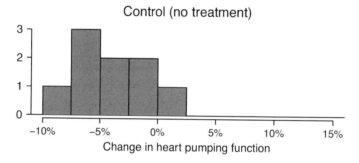

Figure 7.12: Histograms for both the embryonic stem cell and control group.

As with the one-sample case, we always compute the standard error using sample standard deviations rather than population standard deviations:

$$SE = \sqrt{\frac{s_{esc}^2}{n_{esc}} + \frac{s_{control}^2}{n_{control}}} = \sqrt{\frac{5.17^2}{9} + \frac{2.76^2}{9}} = 1.95$$

Generally, we use statistical software to find the appropriate degrees of freedom, or if software isn't available, we can use the smaller of $n_1 - 1$ and $n_2 - 1$ for the degrees of freedom, e.g. if using a t-table to find tail areas. For transparency in the Examples and Guided Practice, we'll use the latter approach for finding df; in the case of the ESC example, this means we'll use $df = 8$.

EXAMPLE 7.22

Calculate a 95% confidence interval for the effect of ESCs on the change in heart pumping capacity of sheep after they've suffered a heart attack.

We will use the sample difference and the standard error that we computed earlier calculations:

$$\bar{x}_{esc} - \bar{x}_{control} = 7.83 \qquad\qquad SE = \sqrt{\frac{5.17^2}{9} + \frac{2.76^2}{9}} = 1.95$$

Using $df = 8$, we can identify the critical value of $t_8^\star = 2.31$ for a 95% confidence interval. Finally, we can enter the values into the confidence interval formula:

$$\text{point estimate} \pm t^\star \times SE \quad\rightarrow\quad 7.83 \pm 2.31 \times 1.95 \quad\rightarrow\quad (3.32, 12.34)$$

We are 95% confident that embryonic stem cells improve the heart's pumping function in sheep that have suffered a heart attack by 3.32% to 12.34%.

As with past statistical inference applications, there is a well-trodden procedure.

Prepare. Retrieve critical contextual information, and if appropriate, set up hypotheses.

Check. Ensure the required conditions are reasonably satisfied.

Calculate. Find the standard error, and then construct a confidence interval, or if conducting a hypothesis test, find a test statistic and p-value.

Conclude. Interpret the results in the context of the application.

The details change a little from one setting to the next, but this general approach remain the same.

7.3.2 Hypothesis tests for the difference of two means

A data set called `ncbirths` represents a random sample of 150 cases of mothers and their newborns in North Carolina over a year. Four cases from this data set are represented in Figure 7.13. We are particularly interested in two variables: `weight` and `smoke`. The `weight` variable represents the weights of the newborns and the `smoke` variable describes which mothers smoked during pregnancy. We would like to know, is there convincing evidence that newborns from mothers who smoke have a different average birth weight than newborns from mothers who don't smoke? We will use the North Carolina sample to try to answer this question. The smoking group includes 50 cases and the nonsmoking group contains 100 cases.

	fage	mage	weeks	weight	sex	smoke
1	NA	13	37	5.00	female	nonsmoker
2	NA	14	36	5.88	female	nonsmoker
3	19	15	41	8.13	male	smoker
⋮	⋮	⋮	⋮	⋮	⋮	
150	45	50	36	9.25	female	nonsmoker

Figure 7.13: Four cases from the `ncbirths` data set. The value "NA", shown for the first two entries of the first variable, indicates that piece of data is missing.

EXAMPLE 7.23

Set up appropriate hypotheses to evaluate whether there is a relationship between a mother smoking and average birth weight.

The null hypothesis represents the case of no difference between the groups.

H_0: There is no difference in average birth weight for newborns from mothers who did and did not smoke. In statistical notation: $\mu_n - \mu_s = 0$, where μ_n represents non-smoking mothers and μ_s represents mothers who smoked.

H_A: There is some difference in average newborn weights from mothers who did and did not smoke ($\mu_n - \mu_s \neq 0$).

We check the two conditions necessary to model the difference in sample means using the t-distribution.

- Because the data come from a simple random sample, the observations are independent, both within and between samples.

- With both data sets over 30 observations, we inspect the data in Figure 7.14 for any particularly extreme outliers and find none.

Since both conditions are satisfied, the difference in sample means may be modeled using a t-distribution.

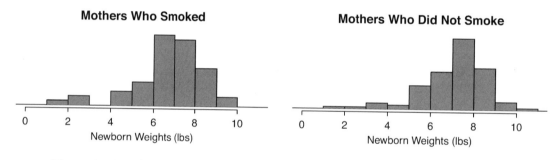

Figure 7.14: The left panel represents birth weights for infants whose mothers smoked. The right panel represents the birth weights for infants whose mothers who did not smoke.

GUIDED PRACTICE 7.24

The summary statistics in Figure 7.15 may be useful for this Guided Practice.[13]

(a) What is the point estimate of the population difference, $\mu_n - \mu_s$?

(b) Compute the standard error of the point estimate from part (a).

	smoker	nonsmoker
mean	6.78	7.18
st. dev.	1.43	1.60
samp. size	50	100

Figure 7.15: Summary statistics for the `ncbirths` data set.

---

[13](a) The difference in sample means is an appropriate point estimate: $\bar{x}_n - \bar{x}_s = 0.40$. (b) The standard error of the estimate can be calculated using the standard error formula:

$$SE = \sqrt{\frac{\sigma_n^2}{n_n} + \frac{\sigma_s^2}{n_s}} \approx \sqrt{\frac{s_n^2}{n_n} + \frac{s_s^2}{n_s}} = \sqrt{\frac{1.60^2}{100} + \frac{1.43^2}{50}} = 0.26$$

EXAMPLE 7.25

Complete the hypothesis test started in Example 7.23 and Guided Practice 7.24. Use a significance level of $\alpha = 0.05$. For reference, $\bar{x}_n - \bar{x}_s = 0.40$, $SE = 0.26$, and the sample sizes were $n_n = 100$ and $n_s = 50$.

We can find the test statistic for this test using the values from Guided Practice 7.24:

$$T = \frac{0.40 - 0}{0.26} = 1.54$$

The p-value is represented by the two shaded tails in the following plot:

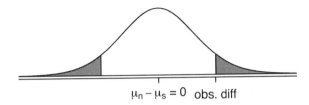

$\mu_n - \mu_s = 0$ obs. diff

We find the single tail area using software (or the t-table in Appendix C.2). We'll use the smaller of $n_n - 1 = 99$ and $n_s - 1 = 49$ as the degrees of freedom: $df = 49$. The one tail area is 0.065; doubling this value gives the two-tail area and p-value, 0.135.

The p-value is larger than the significance value, 0.05, so we do not reject the null hypothesis. There is insufficient evidence to say there is a difference in average birth weight of newborns from North Carolina mothers who did smoke during pregnancy and newborns from North Carolina mothers who did not smoke during pregnancy.

GUIDED PRACTICE 7.26

We've seen much research suggesting smoking is harmful during pregnancy, so how could we fail to reject the null hypothesis in Example 7.25? [14]

GUIDED PRACTICE 7.27

If we made a Type 2 Error and there is a difference, what could we have done differently in data collection to be more likely to detect the difference?[15]

Public service announcement: while we have used this relatively small data set as an example, larger data sets show that women who smoke tend to have smaller newborns. In fact, some in the tobacco industry actually had the audacity to tout that as a *benefit* of smoking:

> *It's true. The babies born from women who smoke are smaller, but they're just as healthy as the babies born from women who do not smoke. And some women would prefer having smaller babies.*

> - Joseph Cullman, Philip Morris' Chairman of the Board
> on CBS' *Face the Nation*, Jan 3, 1971

Fact check: the babies from women who smoke are not actually as healthy as the babies from women who do not smoke.[16]

---

[14]It is possible that there is a difference but we did not detect it. If there is a difference, we made a Type 2 Error.
[15]We could have collected more data. If the sample sizes are larger, we tend to have a better shot at finding a difference if one exists. In fact, this is exactly what we would find if we examined a larger data set!
[16]You can watch an episode of John Oliver on *Last Week Tonight* to explore the present day offenses of the tobacco industry. Please be aware that there is some adult language: youtu.be/6UsHHOCH4q8.

7.3.3 Case study: two versions of a course exam

An instructor decided to run two slight variations of the same exam. Prior to passing out the exams, she shuffled the exams together to ensure each student received a random version. Summary statistics for how students performed on these two exams are shown in Figure 7.16. Anticipating complaints from students who took Version B, she would like to evaluate whether the difference observed in the groups is so large that it provides convincing evidence that Version B was more difficult (on average) than Version A.

Version	n	$\bar{x}$	s	min	max
A	30	79.4	14	45	100
B	27	74.1	20	32	100

Figure 7.16: Summary statistics of scores for each exam version.

GUIDED PRACTICE 7.28

Construct hypotheses to evaluate whether the observed difference in sample means, $\bar{x}_A - \bar{x}_B = 5.3$, is due to chance. We will later evaluate these hypotheses using $\alpha = 0.01$.[17]

GUIDED PRACTICE 7.29

To evaluate the hypotheses in Guided Practice 7.28 using the t-distribution, we must first verify conditions.[18]

(a) Does it seem reasonable that the scores are independent?

(b) Any concerns about outliers?

After verifying the conditions for each sample and confirming the samples are independent of each other, we are ready to conduct the test using the t-distribution. In this case, we are estimating the true difference in average test scores using the sample data, so the point estimate is $\bar{x}_A - \bar{x}_B = 5.3$. The standard error of the estimate can be calculated as

$$SE = \sqrt{\frac{s_A^2}{n_A} + \frac{s_B^2}{n_B}} = \sqrt{\frac{14^2}{30} + \frac{20^2}{27}} = 4.62$$

Finally, we construct the test statistic:

$$T = \frac{\text{point estimate} - \text{null value}}{SE} = \frac{(79.4 - 74.1) - 0}{4.62} = 1.15$$

If we have a computer handy, we can identify the degrees of freedom as 45.97. Otherwise we use the smaller of $n_1 - 1$ and $n_2 - 1$: $df = 26$.

---

[17]H_0: the exams are equally difficult, on average. $\mu_A - \mu_B = 0$. H_A: one exam was more difficult than the other, on average. $\mu_A - \mu_B \neq 0$.

[18](a) Since the exams were shuffled, the "treatment" in this case was randomly assigned, so independence within and between groups is satisfied. (b) The summary statistics suggest the data are roughly symmetric about the mean, and the min/max values don't suggest any outliers of concern.

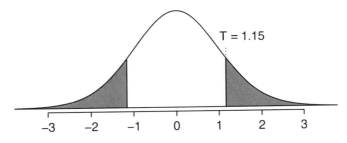

Figure 7.17: The t-distribution with 26 degrees of freedom and the p-value from exam example represented as the shaded areas.

EXAMPLE 7.30

Identify the p-value depicted in Figure 7.17 using $df = 26$, and provide a conclusion in the context of the case study.

Using software, we can find the one-tail area (0.13) and then double this value to get the two-tail area, which is the p-value: 0.26. (Alternatively, we could use the t-table in Appendix C.2.)

In Guided Practice 7.28, we specified that we would use $\alpha = 0.01$. Since the p-value is larger than α, we do not reject the null hypothesis. That is, the data do not convincingly show that one exam version is more difficult than the other, and the teacher should not be convinced that she should add points to the Version B exam scores.

7.3.4 Pooled standard deviation estimate (special topic)

Occasionally, two populations will have standard deviations that are so similar that they can be treated as identical. For example, historical data or a well-understood biological mechanism may justify this strong assumption. In such cases, we can make the t-distribution approach slightly more precise by using a pooled standard deviation.

The **pooled standard deviation** of two groups is a way to use data from both samples to better estimate the standard deviation and standard error. If s_1 and s_2 are the standard deviations of groups 1 and 2 and there are very good reasons to believe that the population standard deviations are equal, then we can obtain an improved estimate of the group variances by pooling their data:

$$s_{pooled}^2 = \frac{s_1^2 \times (n_1 - 1) + s_2^2 \times (n_2 - 1)}{n_1 + n_2 - 2}$$

where n_1 and n_2 are the sample sizes, as before. To use this new statistic, we substitute s_{pooled}^2 in place of s_1^2 and s_2^2 in the standard error formula, and we use an updated formula for the degrees of freedom:

$$df = n_1 + n_2 - 2$$

The benefits of pooling the standard deviation are realized through obtaining a better estimate of the standard deviation for each group and using a larger degrees of freedom parameter for the t-distribution. Both of these changes may permit a more accurate model of the sampling distribution of $\bar{x}_1 - \bar{x}_2$, if the standard deviations of the two groups are indeed equal.

POOL STANDARD DEVIATIONS ONLY AFTER CAREFUL CONSIDERATION

A pooled standard deviation is only appropriate when background research indicates the population standard deviations are nearly equal. When the sample size is large and the condition may be adequately checked with data, the benefits of pooling the standard deviations greatly diminishes.

Exercises

7.23 Friday the 13th, Part I. In the early 1990's, researchers in the UK collected data on traffic flow, number of shoppers, and traffic accident related emergency room admissions on Friday the 13th and the previous Friday, Friday the 6th. The histograms below show the distribution of number of cars passing by a specific intersection on Friday the 6th and Friday the 13th for many such date pairs. Also given are some sample statistics, where the difference is the number of cars on the 6th minus the number of cars on the 13th.[19]

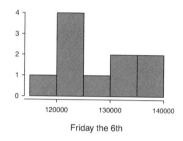
Friday the 6th

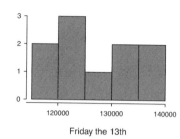
Friday the 13th

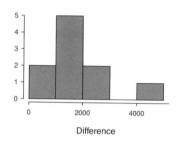
Difference

	6th	13th	Diff.
$\bar{x}$	128,385	126,550	1,835
s	7,259	7,664	1,176
n	10	10	10

(a) Are there any underlying structures in these data that should be considered in an analysis? Explain.

(b) What are the hypotheses for evaluating whether the number of people out on Friday the 6th is different than the number out on Friday the 13th?

(c) Check conditions to carry out the hypothesis test from part (b).

(d) Calculate the test statistic and the p-value.

(e) What is the conclusion of the hypothesis test?

(f) Interpret the p-value in this context.

(g) What type of error might have been made in the conclusion of your test? Explain.

7.24 Diamonds, Part I. Prices of diamonds are determined by what is known as the 4 Cs: cut, clarity, color, and carat weight. The prices of diamonds go up as the carat weight increases, but the increase is not smooth. For example, the difference between the size of a 0.99 carat diamond and a 1 carat diamond is undetectable to the naked human eye, but the price of a 1 carat diamond tends to be much higher than the price of a 0.99 diamond. In this question we use two random samples of diamonds, 0.99 carats and 1 carat, each sample of size 23, and compare the average prices of the diamonds. In order to be able to compare equivalent units, we first divide the price for each diamond by 100 times its weight in carats. That is, for a 0.99 carat diamond, we divide the price by 99. For a 1 carat diamond, we divide the price by 100. The distributions and some sample statistics are shown below.[20]

Conduct a hypothesis test to evaluate if there is a difference between the average standardized prices of 0.99 and 1 carat diamonds. Make sure to state your hypotheses clearly, check relevant conditions, and interpret your results in context of the data.

	0.99 carats	1 carat
Mean	$44.51	$56.81
SD	$13.32	$16.13
n	23	23

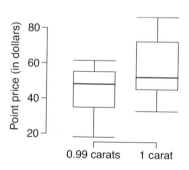

[19]T.J. Scanlon et al. "Is Friday the 13th Bad For Your Health?" In: *BMJ* 307 (1993), pp. 1584–1586.
[20]H. Wickham. *ggplot2: elegant graphics for data analysis*. Springer New York, 2009.

7.25 Friday the 13th, Part II. The Friday the 13th study reported in Exercise 7.23 also provides data on traffic accident related emergency room admissions. The distributions of these counts from Friday the 6th and Friday the 13th are shown below for six such paired dates along with summary statistics. You may assume that conditions for inference are met.

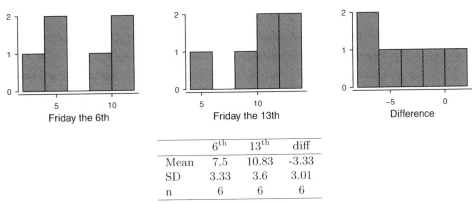

	6th	13th	diff
Mean	7.5	10.83	-3.33
SD	3.33	3.6	3.01
n	6	6	6

(a) Conduct a hypothesis test to evaluate if there is a difference between the average numbers of traffic accident related emergency room admissions between Friday the 6th and Friday the 13th.

(b) Calculate a 95% confidence interval for the difference between the average numbers of traffic accident related emergency room admissions between Friday the 6th and Friday the 13th.

(c) The conclusion of the original study states, "Friday 13th is unlucky for some. The risk of hospital admission as a result of a transport accident may be increased by as much as 52%. Staying at home is recommended." Do you agree with this statement? Explain your reasoning.

7.26 Diamonds, Part II. In Exercise 7.24, we discussed diamond prices (standardized by weight) for diamonds with weights 0. 99 carats and 1 carat. See the table for summary statistics, and then construct a 95% confidence interval for the average difference between the standardized prices of 0.99 and 1 carat diamonds. You may assume the conditions for inference are met.

	0.99 carats	1 carat
Mean	$44.51	$56.81
SD	$13.32	$16.13
n	23	23

7.27 Chicken diet and weight, Part I. Chicken farming is a multi-billion dollar industry, and any methods that increase the growth rate of young chicks can reduce consumer costs while increasing company profits, possibly by millions of dollars. An experiment was conducted to measure and compare the effectiveness of various feed supplements on the growth rate of chickens. Newly hatched chicks were randomly allocated into six groups, and each group was given a different feed supplement. Below are some summary statistics from this data set along with box plots showing the distribution of weights by feed type.[21]

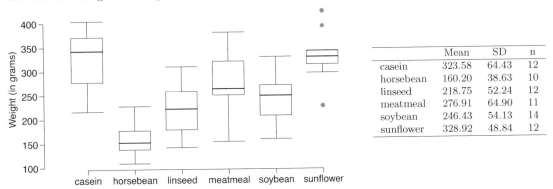

	Mean	SD	n
casein	323.58	64.43	12
horsebean	160.20	38.63	10
linseed	218.75	52.24	12
meatmeal	276.91	64.90	11
soybean	246.43	54.13	14
sunflower	328.92	48.84	12

(a) Describe the distributions of weights of chickens that were fed linseed and horsebean.

(b) Do these data provide strong evidence that the average weights of chickens that were fed linseed and horsebean are different? Use a 5% significance level.

(c) What type of error might we have committed? Explain.

(d) Would your conclusion change if we used $\alpha = 0.01$?

---

[21]Chicken Weights by Feed Type, from the `datasets` package in R..

7.28   Fuel efficiency of manual and automatic cars, Part I. Each year the US Environmental Protection Agency (EPA) releases fuel economy data on cars manufactured in that year. Below are summary statistics on fuel efficiency (in miles/gallon) from random samples of cars with manual and automatic transmissions. Do these data provide strong evidence of a difference between the average fuel efficiency of cars with manual and automatic transmissions in terms of their average city mileage? Assume that conditions for inference are satisfied.[22]

	City MPG	
	Automatic	Manual
Mean	16.12	19.85
SD	3.58	4.51
n	26	26

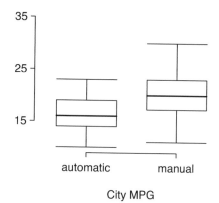

City MPG

7.29   Chicken diet and weight, Part II. Casein is a common weight gain supplement for humans. Does it have an effect on chickens? Using data provided in Exercise 7.27, test the hypothesis that the average weight of chickens that were fed casein is different than the average weight of chickens that were fed soybean. If your hypothesis test yields a statistically significant result, discuss whether or not the higher average weight of chickens can be attributed to the casein diet. Assume that conditions for inference are satisfied.

7.30   Fuel efficiency of manual and automatic cars, Part II. The table provides summary statistics on highway fuel economy of the same 52 cars from Exercise 7.28. Use these statistics to calculate a 98% confidence interval for the difference between average highway mileage of manual and automatic cars, and interpret this interval in the context of the data.[23]

	Hwy MPG	
	Automatic	Manual
Mean	22.92	27.88
SD	5.29	5.01
n	26	26

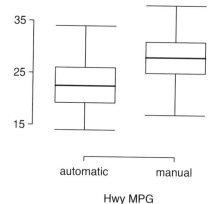

Hwy MPG

---

[22]U.S. Department of Energy, Fuel Economy Data, 2012 Datafile.
[23]U.S. Department of Energy, Fuel Economy Data, 2012 Datafile.

7.31 Prison isolation experiment, Part I. Subjects from Central Prison in Raleigh, NC, volunteered for an experiment involving an "isolation" experience. The goal of the experiment was to find a treatment that reduces subjects' psychopathic deviant T scores. This score measures a person's need for control or their rebellion against control, and it is part of a commonly used mental health test called the Minnesota Multiphasic Personality Inventory (MMPI) test. The experiment had three treatment groups:

(1) Four hours of sensory restriction plus a 15 minute "therapeutic" tape advising that professional help is available.

(2) Four hours of sensory restriction plus a 15 minute "emotionally neutral" tape on training hunting dogs.

(3) Four hours of sensory restriction but no taped message.

Forty-two subjects were randomly assigned to these treatment groups, and an MMPI test was administered before and after the treatment. Distributions of the differences between pre and post treatment scores (pre - post) are shown below, along with some sample statistics. Use this information to independently test the effectiveness of each treatment. Make sure to clearly state your hypotheses, check conditions, and interpret results in the context of the data.[24]

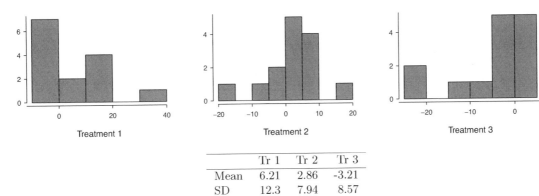

	Tr 1	Tr 2	Tr 3
Mean	6.21	2.86	-3.21
SD	12.3	7.94	8.57
n	14	14	14

7.32 True / False: comparing means. Determine if the following statements are true or false, and explain your reasoning for statements you identify as false.

(a) When comparing means of two samples where $n_1 = 20$ and $n_2 = 40$, we can use the normal model for the difference in means since $n_2 \geq 30$.

(b) As the degrees of freedom increases, the t-distribution approaches normality.

(c) We use a pooled standard error for calculating the standard error of the difference between means when sample sizes of groups are equal to each other.

---

[24]Prison isolation experiment, stat.duke.edu/resources/datasets/prison-isolation.

7.4 Power calculations for a difference of means

Often times in experiment planning, there are two competing considerations:

- We want to collect enough data that we can detect important effects.
- Collecting data can be expensive, and in experiments involving people, there may be some risk to patients.

In this section, we focus on the context of a clinical trial, which is a health-related experiment where the subject are people, and we will determine an appropriate sample size where we can be 80% sure that we would detect any practically important effects.[25]

7.4.1 Going through the motions of a test

We're going to go through the motions of a hypothesis test. This will help us frame our calculations for determining an appropriate sample size for the study.

EXAMPLE 7.31

Suppose a pharmaceutical company has developed a new drug for lowering blood pressure, and they are preparing a clinical trial (experiment) to test the drug's effectiveness. They recruit people who are taking a particular standard blood pressure medication. People in the control group will continue to take their current medication through generic-looking pills to ensure blinding. Write down the hypotheses for a two-sided hypothesis test in this context.

Generally, clinical trials use a two-sided alternative hypothesis, so below are suitable hypotheses for this context:

H_0: The new drug performs exactly as well as the standard medication.
   $\mu_{trmt} - \mu_{ctrl} = 0$.
H_A: The new drug's performance differs from the standard medication.
   $\mu_{trmt} - \mu_{ctrl} \neq 0$.

EXAMPLE 7.32

The researchers would like to run the clinical trial on patients with systolic blood pressures between 140 and 180 mmHg. Suppose previously published studies suggest that the standard deviation of the patients' blood pressures will be about 12 mmHg and the distribution of patient blood pressures will be approximately symmetric.[26] If we had 100 patients per group, what would be the approximate standard error for $\bar{x}_{trmt} - \bar{x}_{ctrl}$?

The standard error is calculated as follows:

$$SE_{\bar{x}_{trmt} - \bar{x}_{ctrl}} = \sqrt{\frac{s^2_{trmt}}{n_{trmt}} + \frac{s^2_{ctrl}}{n_{ctrl}}} = \sqrt{\frac{12^2}{100} + \frac{12^2}{100}} = 1.70$$

This may be an imperfect estimate of $SE_{\bar{x}_{trmt} - \bar{x}_{ctrl}}$, since the standard deviation estimate we used may not be perfectly correct for this group of patients. However, it is sufficient for our purposes.

---

[25]Even though we don't cover it explicitly, similar sample size planning is also helpful for observational studies.
[26]In this particular study, we'd generally measure each patient's blood pressure at the beginning and end of the study, and then the outcome measurement for the study would be the average change in blood pressure. That is, both μ_{trmt} and μ_{ctrl} would represent average differences. This is what you might think of as a 2-sample paired testing structure, and we'd analyze it exactly just like a hypothesis test for a difference in the average change for patients. In the calculations we perform here, we'll suppose that 12 mmHg is the predicted standard deviation of a patient's blood pressure difference over the course of the study.

EXAMPLE 7.33

What does the null distribution of $\bar{x}_{trmt} - \bar{x}_{ctrl}$ look like?

The degrees of freedom are greater than 30, so the distribution of $\bar{x}_{trmt} - \bar{x}_{ctrl}$ will be approximately normal. The standard deviation of this distribution (the standard error) would be about 1.70, and under the null hypothesis, its mean would be 0.

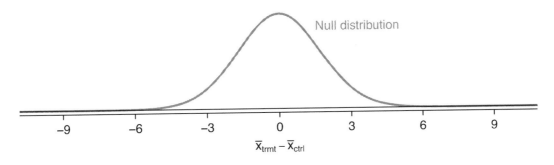

EXAMPLE 7.34

For what values of $\bar{x}_{trmt} - \bar{x}_{ctrl}$ would we reject the null hypothesis?

For $\alpha = 0.05$, we would reject H_0 if the difference is in the lower 2.5% or upper 2.5% tail:

Lower 2.5%: For the normal model, this is 1.96 standard errors below 0, so any difference smaller than $-1.96 \times 1.70 = -3.332$ mmHg.

Upper 2.5%: For the normal model, this is 1.96 standard errors above 0, so any difference larger than $1.96 \times 1.70 = 3.332$ mmHg.

The boundaries of these **rejection regions** are shown below:

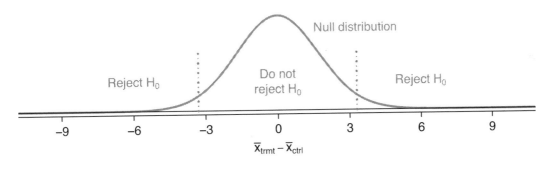

Next, we'll perform some hypothetical calculations to determine the probability we reject the null hypothesis, if the alternative hypothesis were actually true.

7.4.2 Computing the power for a 2-sample test

When planning a study, we want to know how likely we are to detect an effect we care about. In other words, if there is a real effect, and that effect is large enough that it has practical value, then what's the probability that we detect that effect? This probability is called the **power**, and we can compute it for different sample sizes or for different *effect sizes*.

We first determine what is a practically significant result. Suppose that the company researchers care about finding any effect on blood pressure that is 3 mmHg or larger vs the standard medication. Here, 3 mmHg is the minimum **effect size** of interest, and we want to know how likely we are to detect this size of an effect in the study.

EXAMPLE 7.35

Suppose we decided to move forward with 100 patients per treatment group and the new drug reduces blood pressure by an additional 3 mmHg relative to the standard medication. What is the probability that we detect a drop?

Before we even do any calculations, notice that if $\bar{x}_{trmt} - \bar{x}_{ctrl} = -3$ mmHg, there wouldn't even be sufficient evidence to reject H_0. That's not a good sign.

To calculate the probability that we will reject H_0, we need to determine a few things:

- The sampling distribution for $\bar{x}_{trmt} - \bar{x}_{ctrl}$ when the true difference is -3 mmHg. This is the same as the null distribution, except it is shifted to the left by 3:

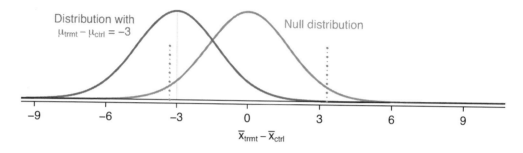

- The rejection regions, which are outside of the dotted lines above.
- The fraction of the distribution that falls in the rejection region.

In short, we need to calculate the probability that $x < -3.332$ for a normal distribution with mean -3 and standard deviation 1.7. To do so, we first shade the area we want to calculate:

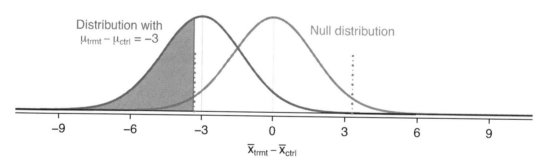

We'll use a normal approximation, which is good approximation when the degrees of freedom is about 30 or more. We'll start by calculating the Z-score and find the tail area using either statistical software or the probability table:

$$Z = \frac{-3.332 - (-3)}{1.7} = -0.20 \qquad \rightarrow \qquad 0.42$$

The power for the test is about 42% when $\mu_{trmt} - \mu_{ctrl} = -3$ and each group has a sample size of 100.

In Example 7.35, we ignored the upper rejection region in the calculation, which was in the opposite direction of the hypothetical truth, i.e. -3. The reasoning? There wouldn't be any value in rejecting the null hypothesis and concluding there was an increase when in fact there was a decrease.

We've also used a normal distribution instead of the t-distribution. This is a convenience, and if the sample size is too small, we'd need to revert back to using the t-distribution. We'll discuss this a bit further at the end of this section.

7.4.3 Determining a proper sample size

In the last example, we found that if we have a sample size of 100 in each group, we can only detect an effect size of 3 mmHg with a probability of about 0.42. Suppose the researchers moved forward and only used 100 patients per group, and the data did not support the alternative hypothesis, i.e. the researchers did not reject H_0. This is a very bad situation to be in for a few reasons:

- In the back of the researchers' minds, they'd all be wondering, *maybe there is a real and meaningful difference, but we weren't able to detect it with such a small sample.*

- The company probably invested hundreds of millions of dollars in developing the new drug, so now they are left with great uncertainty about its potential since the experiment didn't have a great shot at detecting effects that could still be important.

- Patients were subjected to the drug, and we can't even say with much certainty that the drug doesn't help (or harm) patients.

- Another clinical trial may need to be run to get a more conclusive answer as to whether the drug does hold any practical value, and conducting a second clinical trial may take years and many millions of dollars.

We want to avoid this situation, so we need to determine an appropriate sample size to ensure we can be pretty confident that we'll detect any effects that are practically important. As mentioned earlier, a change of 3 mmHg was deemed to be the minimum difference that was practically important. As a first step, we could calculate power for several different sample sizes. For instance, let's try 500 patients per group.

GUIDED PRACTICE 7.36

Calculate the power to detect a change of -3 mmHg when using a sample size of 500 per group.[27]

(a) Determine the standard error (recall that the standard deviation for patients was expected to be about 12 mmHg).

(b) Identify the null distribution and rejection regions.

(c) Identify the alternative distribution when $\mu_{trmt} - \mu_{ctrl} = -3$.

(d) Compute the probability we reject the null hypothesis.

The researchers decided 3 mmHg was the minimum difference that was practically important, and with a sample size of 500, we can be very certain (97.7% or better) that we will detect any such difference. We now have moved to another extreme where we are exposing an unnecessary number of patients to the new drug in the clinical trial. Not only is this ethically questionable, but it would also cost a lot more money than is necessary to be quite sure we'd detect any important effects.

The most common practice is to identify the sample size where the power is around 80%, and sometimes 90%. Other values may be reasonable for a specific context, but 80% and 90% are most commonly targeted as a good balance between high power and not exposing too many patients to a new treatment (or wasting too much money).

We could compute the power of the test at several other possible sample sizes until we find one that's close to 80%, but there's a better way. We should solve the problem backwards.

---

[27](a) The standard error is given as $SE = \sqrt{\frac{12^2}{500} + \frac{12^2}{500}} = 0.76$.

(b) & (c) The null distribution, rejection boundaries, and alternative distribution are shown below:

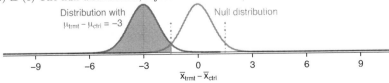

The rejection regions are the areas on the outside of the two dotted lines and are at $\pm 0.76 \times 1.96 = \pm 1.49$.

(d) The area of the alternative distribution where $\mu_{trmt} - \mu_{ctrl} = -3$ has been shaded. We compute the Z-score and find the tail area: $Z = \frac{-1.49 - (-3)}{0.76} = 1.99 \rightarrow 0.977$. With 500 patients per group, we would be about 97.7% sure (or more) that we'd detect any effects that are at least 3 mmHg in size.

EXAMPLE 7.37

What sample size will lead to a power of 80%?

We'll assume we have a large enough sample that the normal distribution is a good approximation for the test statistic, since the normal distribution and the t-distribution look almost identical when the degrees of freedom are moderately large (e.g. $df \geq 30$). If that doesn't turn out to be true, then we'd need to make a correction.

We start by identifying the Z-score that would give us a lower tail of 80%. For a moderately large sample size per group, the Z-score for a lower tail of 80% would be about $Z = 0.84$.

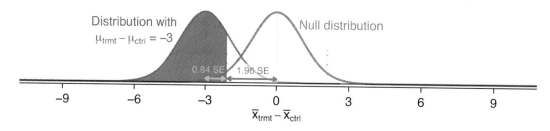

Additionally, the rejection region extends $1.96 \times SE$ from the center of the null distribution for $\alpha = 0.05$. This allows us to calculate the target distance between the center of the null and alternative distributions in terms of the standard error:

$$0.84 \times SE + 1.96 \times SE = 2.8 \times SE$$

In our example, we want the distance between the null and alternative distributions' centers to equal the minimum effect size of interest, 3 mmHg, which allows us to set up an equation between this difference and the standard error:

$$3 = 2.8 \times SE$$

$$3 = 2.8 \times \sqrt{\frac{12^2}{n} + \frac{12^2}{n}}$$

$$n = \frac{2.8^2}{3^2} \times \left(12^2 + 12^2\right) = 250.88$$

We should target 251 patients per group in order to achieve 80% power at the 0.05 significance level for this context.

The standard error difference of $2.8 \times SE$ is specific to a context where the targeted power is 80% and the significance level is $\alpha = 0.05$. If the targeted power is 90% or if we use a different significance level, then we'll use something a little different than $2.8 \times SE$.

Had the suggested sample size been relatively small – roughly 30 or smaller – it would have been a good idea to rework the calculations using the degrees of fredom for the smaller sample size under that initial sample size. That is, we would have revised the 0.84 and 1.96 values based on degrees of freedom implied by the initial sample size. The revised sample size target would generally have then been a little larger.

GUIDED PRACTICE 7.38

Suppose the targeted power was 90% and we were using $\alpha = 0.01$. How many standard errors should separate the centers of the null and alternative distribution, where the alternative distribution is centered at the minimum effect size of interest?[28]

GUIDED PRACTICE 7.39

What are some considerations that are important in determining what the power should be for an experiment?[29]

Figure 7.18 shows the power for sample sizes from 20 patients to 5,000 patients when $\alpha = 0.05$ and the true difference is -3. This curve was constructed by writing a program to compute the power for many different sample sizes.

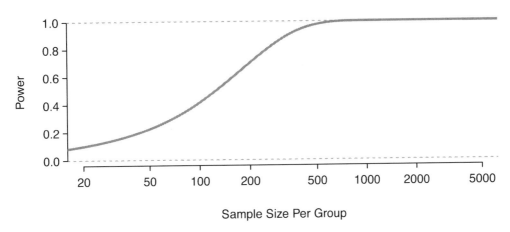

Figure 7.18: The curve shows the power for different sample sizes in the context of the blood pressure example when the true difference is -3. Having more than about 250 to 350 observations doesn't provide much additional value in detecting an effect when $\alpha = 0.05$.

Power calculations for expensive or risky experiments are critical. However, what about experiments that are inexpensive and where the ethical considerations are minimal? For example, if we are doing final testing on a new feature on a popular website, how would our sample size considerations change? As before, we'd want to make sure the sample is big enough. However, suppose the feature has undergone some testing and is known to perform well (e.g. the website's users seem to enjoy the feature). Then it may be reasonable to run a larger experiment if there's value from having a more precise estimate of the feature's effect, such as helping guide the development of the next useful feature.

---

[28]First, find the Z-score such that 90% of the distribution is below it: $Z = 1.28$. Next, find the cutoffs for the rejection regions: ± 2.58. Then the difference in centers should be about $1.28 \times SE + 2.58 \times SE = 3.86 \times SE$.

[29]Answers will vary, but here are a few important considerations:

- Whether there is any risk to patients in the study.
- The cost of enrolling more patients.
- The potential downside of not detecting an effect of interest.

Exercises

7.33 Increasing corn yield. A large farm wants to try out a new type of fertilizer to evaluate whether it will improve the farm's corn production. The land is broken into plots that produce an average of 1,215 pounds of corn with a standard deviation of 94 pounds per plot. The owner is interested in detecting any average difference of at least 40 pounds per plot. How many plots of land would be needed for the experiment if the desired power level is 90%? Assume each plot of land gets treated with either the current fertilizer or the new fertilizer.

7.34 Email outreach efforts. A medical research group is recruiting people to complete short surveys about their medical history. For example, one survey asks for information on a person's family history in regards to cancer. Another survey asks about what topics were discussed during the person's last visit to a hospital. So far, as people sign up, they complete an average of just 4 surveys, and the standard deviation of the number of surveys is about 2.2. The research group wants to try a new interface that they think will encourage new enrollees to complete more surveys, where they will randomize each enrollee to either get the new interface or the current interface. How many new enrollees do they need for each interface to detect an effect size of 0.5 surveys per enrollee, if the desired power level is 80%?

7.5 Comparing many means with ANOVA

Sometimes we want to compare means across many groups. We might initially think to do pairwise comparisons. For example, if there were three groups, we might be tempted to compare the first mean with the second, then with the third, and then finally compare the second and third means for a total of three comparisons. However, this strategy can be treacherous. If we have many groups and do many comparisons, it is likely that we will eventually find a difference just by chance, even if there is no difference in the populations. Instead, we should apply a holistic test to check whether there is evidence that at least one pair groups are in fact different, and this is where *ANOVA* saves the day.

7.5.1 Core ideas of ANOVA

In this section, we will learn a new method called **analysis of variance (ANOVA)** and a new test statistic called F. ANOVA uses a single hypothesis test to check whether the means across many groups are equal:

H_0: The mean outcome is the same across all groups. In statistical notation, $\mu_1 = \mu_2 = \cdots = \mu_k$ where μ_i represents the mean of the outcome for observations in category i.

H_A: At least one mean is different.

Generally we must check three conditions on the data before performing ANOVA:

- the observations are independent within and across groups,
- the data within each group are nearly normal, and
- the variability across the groups is about equal.

When these three conditions are met, we may perform an ANOVA to determine whether the data provide strong evidence against the null hypothesis that all the μ_i are equal.

EXAMPLE 7.40

College departments commonly run multiple lectures of the same introductory course each semester because of high demand. Consider a statistics department that runs three lectures of an introductory statistics course. We might like to determine whether there are statistically significant differences in first exam scores in these three classes (A, B, and C). Describe appropriate hypotheses to determine whether there are any differences between the three classes.

The hypotheses may be written in the following form:

H_0: The average score is identical in all lectures. Any observed difference is due to chance. Notationally, we write $\mu_A = \mu_B = \mu_C$.

H_A: The average score varies by class. We would reject the null hypothesis in favor of the alternative hypothesis if there were larger differences among the class averages than what we might expect from chance alone.

Strong evidence favoring the alternative hypothesis in ANOVA is described by unusually large differences among the group means. We will soon learn that assessing the variability of the group means relative to the variability among individual observations within each group is key to ANOVA's success.

EXAMPLE 7.41

Examine Figure 7.19. Compare groups I, II, and III. Can you visually determine if the differences in the group centers is due to chance or not? Now compare groups IV, V, and VI. Do these differences appear to be due to chance?

Any real difference in the means of groups I, II, and III is difficult to discern, because the data within each group are very volatile relative to any differences in the average outcome. On the other hand, it appears there are differences in the centers of groups IV, V, and VI. For instance, group V appears to have a higher mean than that of the other two groups. Investigating groups IV, V, and VI, we see the differences in the groups' centers are noticeable because those differences are large *relative to the variability in the individual observations within each group*.

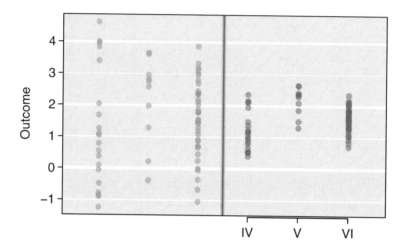

Figure 7.19: Side-by-side dot plot for the outcomes for six groups.

7.5.2   Is batting performance related to player position in MLB?

We would like to discern whether there are real differences between the batting performance of baseball players according to their position: outfielder (OF), infielder (IF), and catcher (C). We will use a data set called bat18, which includes batting records of 429 Major League Baseball (MLB) players from the 2018 season who had at least 100 at bats. Six of the 429 cases represented in bat18 are shown in Figure 7.20, and descriptions for each variable are provided in Figure 7.21. The measure we will use for the player batting performance (the outcome variable) is on-base percentage (OBP). The on-base percentage roughly represents the fraction of the time a player successfully gets on base or hits a home run.

	name	team	position	AB	H	HR	RBI	AVG	OBP
1	Abreu, J	CWS	IF	499	132	22	78	0.265	0.325
2	Acuna Jr., R	ATL	OF	433	127	26	64	0.293	0.366
3	Adames, W	TB	IF	288	80	10	34	0.278	0.348
⋮	⋮	⋮	⋮	⋮	⋮	⋮	⋮		
427	Zimmerman, R	WSH	IF	288	76	13	51	0.264	0.337
428	Zobrist, B	CHC	IF	455	139	9	58	0.305	0.378
429	Zunino, M	SEA	C	373	75	20	44	0.201	0.259

Figure 7.20: Six cases from the bat18 data matrix.

variable	description
name	Player name
team	The abbreviated name of the player's team
position	The player's primary field position (OF, IF, C)
AB	Number of opportunities at bat
H	Number of hits
HR	Number of home runs
RBI	Number of runs batted in
AVG	Batting average, which is equal to H/AB
OBP	On-base percentage, which is roughly equal to the fraction of times a player gets on base or hits a home run

Figure 7.21: Variables and their descriptions for the bat18 data set.

GUIDED PRACTICE 7.42

The null hypothesis under consideration is the following: $\mu_{OF} = \mu_{IF} = \mu_C$. Write the null and corresponding alternative hypotheses in plain language.[30]

EXAMPLE 7.43

The player positions have been divided into three groups: outfield (OF), infield (IF), and catcher (C). What would be an appropriate point estimate of the on-base percentage by outfielders, μ_{OF}?

A good estimate of the on-base percentage by outfielders would be the sample average of OBP for just those players whose position is outfield: $\bar{x}_{OF} = 0.320$.

Figure 7.22 provides summary statistics for each group. A side-by-side box plot for the on-base percentage is shown in Figure 7.23. Notice that the variability appears to be approximately constant across groups; nearly constant variance across groups is an important assumption that must be satisfied before we consider the ANOVA approach.

	OF	IF	C
Sample size (n_i)	160	205	64
Sample mean ($\bar{x}_i$)	0.320	0.318	0.302
Sample SD (s_i)	0.043	0.038	0.038

Figure 7.22: Summary statistics of on-base percentage, split by player position.

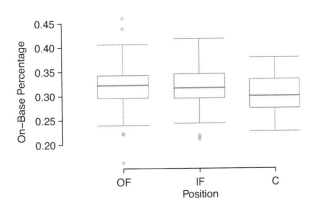

Figure 7.23: Side-by-side box plot of the on-base percentage for 429 players across three groups. With over a hundred players in both the infield and outfield groups, the apparent outliers are not a concern.

---

[30] H_0: The average on-base percentage is equal across the three positions. H_A: The average on-base percentage varies across some (or all) groups.

EXAMPLE 7.44

The largest difference between the sample means is between the catcher and the outfielder positions. Consider again the original hypotheses:

H_0: $\mu_{OF} = \mu_{IF} = \mu_C$

H_A: The average on-base percentage (μ_i) varies across some (or all) groups.

Why might it be inappropriate to run the test by simply estimating whether the difference of μ_C and μ_{OF} is statistically significant at a 0.05 significance level?

The primary issue here is that we are inspecting the data before picking the groups that will be compared. It is inappropriate to examine all data by eye (informal testing) and only afterwards decide which parts to formally test. This is called **data snooping** or **data fishing**. Naturally, we would pick the groups with the large differences for the formal test, and this would leading to an inflation in the Type 1 Error rate. To understand this better, let's consider a slightly different problem.

Suppose we are to measure the aptitude for students in 20 classes in a large elementary school at the beginning of the year. In this school, all students are randomly assigned to classrooms, so any differences we observe between the classes at the start of the year are completely due to chance. However, with so many groups, we will probably observe a few groups that look rather different from each other. If we select only these classes that look so different and then perform a formal test, we will probably make the wrong conclusion that the assignment wasn't random. While we might only formally test differences for a few pairs of classes, we informally evaluated the other classes by eye before choosing the most extreme cases for a comparison.

For additional information on the ideas expressed in Example 7.44, we recommend reading about the **prosecutor's fallacy**.[31]

In the next section we will learn how to use the F statistic and ANOVA to test whether observed differences in sample means could have happened just by chance even if there was no difference in the respective population means.

---

[31]See, for example, statmodeling.stat.columbia.edu/2007/05/18/the_prosecutors.

7.5.3 Analysis of variance (ANOVA) and the F-test

The method of analysis of variance in this context focuses on answering one question: is the variability in the sample means so large that it seems unlikely to be from chance alone? This question is different from earlier testing procedures since we will *simultaneously* consider many groups, and evaluate whether their sample means differ more than we would expect from natural variation. We call this variability the **mean square between groups** (MSG), and it has an associated degrees of freedom, $df_G = k - 1$ when there are k groups. The MSG can be thought of as a scaled variance formula for means. If the null hypothesis is true, any variation in the sample means is due to chance and shouldn't be too large. Details of MSG calculations are provided in the footnote.[32] However, we typically use software for these computations.

The mean square between the groups is, on its own, quite useless in a hypothesis test. We need a benchmark value for how much variability should be expected among the sample means if the null hypothesis is true. To this end, we compute a pooled variance estimate, often abbreviated as the **mean square error** (MSE), which has an associated degrees of freedom value $df_E = n - k$. It is helpful to think of MSE as a measure of the variability within the groups. Details of the computations of the MSE and a link to an extra online section for ANOVA calculations are provided in the footnote[33] for interested readers.

When the null hypothesis is true, any differences among the sample means are only due to chance, and the MSG and MSE should be about equal. As a test statistic for ANOVA, we examine the fraction of MSG and MSE:

$$F = \frac{MSG}{MSE}$$

The MSG represents a measure of the between-group variability, and MSE measures the variability within each of the groups.

GUIDED PRACTICE 7.45

For the baseball data, $MSG = 0.00803$ and $MSE = 0.00158$. Identify the degrees of freedom associated with MSG and MSE and verify the F statistic is approximately 5.077.[34]

We can use the F statistic to evaluate the hypotheses in what is called an **F-test**. A p-value can be computed from the F statistic using an F distribution, which has two associated parameters: df_1 and df_2. For the F statistic in ANOVA, $df_1 = df_G$ and $df_2 = df_E$. An F distribution with 2 and 426 degrees of freedom, corresponding to the F statistic for the baseball hypothesis test, is shown in Figure 7.24.

---

[32]Let $\bar{x}$ represent the mean of outcomes across all groups. Then the mean square between groups is computed as

$$MSG = \frac{1}{df_G} SSG = \frac{1}{k-1} \sum_{i=1}^{k} n_i (\bar{x}_i - \bar{x})^2$$

where SSG is called the **sum of squares between groups** and n_i is the sample size of group i.

[33]Let $\bar{x}$ represent the mean of outcomes across all groups. Then the **sum of squares total** (SST) is computed as

$$SST = \sum_{i=1}^{n} (x_i - \bar{x})^2$$

where the sum is over all observations in the data set. Then we compute the **sum of squared errors** (SSE) in one of two equivalent ways:

$$SSE = SST - SSG$$
$$= (n_1 - 1)s_1^2 + (n_2 - 1)s_2^2 + \cdots + (n_k - 1)s_k^2$$

where s_i^2 is the sample variance (square of the standard deviation) of the residuals in group i. Then the MSE is the standardized form of SSE: $MSE = \frac{1}{df_E} SSE$.

For additional details on ANOVA calculations, see www.openintro.org/d?file=stat_extra_anova_calculations

[34]There are $k = 3$ groups, so $df_G = k - 1 = 2$. There are $n = n_1 + n_2 + n_3 = 429$ total observations, so $df_E = n - k = 426$. Then the F statistic is computed as the ratio of MSG and MSE: $F = \frac{MSG}{MSE} = \frac{0.00803}{0.00158} = 5.082 \approx 5.077$. ($F = 5.077$ was computed by using values for MSG and MSE that were not rounded.)

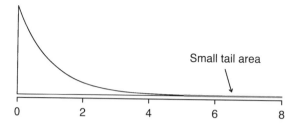

Figure 7.24: An F distribution with $df_1 = 2$ and $df_2 = 426$.

The larger the observed variability in the sample means (MSG) relative to the within-group observations (MSE), the larger F will be and the stronger the evidence against the null hypothesis. Because larger values of F represent stronger evidence against the null hypothesis, we use the upper tail of the distribution to compute a p-value.

THE F STATISTIC AND THE F-TEST

Analysis of variance (ANOVA) is used to test whether the mean outcome differs across 2 or more groups. ANOVA uses a test statistic F, which represents a standardized ratio of variability in the sample means relative to the variability within the groups. If H_0 is true and the model conditions are satisfied, the statistic F follows an F distribution with parameters $df_1 = k - 1$ and $df_2 = n - k$. The upper tail of the F distribution is used to represent the p-value.

EXAMPLE 7.46

The p-value corresponding to the shaded area in Figure 7.24 is equal to about 0.0066. Does this provide strong evidence against the null hypothesis?

The p-value is smaller than 0.05, indicating the evidence is strong enough to reject the null hypothesis at a significance level of 0.05. That is, the data provide strong evidence that the average on-base percentage varies by player's primary field position.

7.5.4   Reading an ANOVA table from software

The calculations required to perform an ANOVA by hand are tedious and prone to human error. For these reasons, it is common to use statistical software to calculate the F statistic and p-value.

An ANOVA can be summarized in a table very similar to that of a regression summary, which we will see in Chapters 8 and 9. Figure 7.25 shows an ANOVA summary to test whether the mean of on-base percentage varies by player positions in the MLB. Many of these values should look familiar; in particular, the F-test statistic and p-value can be retrieved from the last two columns.

	Df	Sum Sq	Mean Sq	F value	Pr(>F)
position	2	0.0161	0.0080	5.0766	0.0066
Residuals	426	0.6740	0.0016		

$s_{pooled} = 0.040$ on $df = 423$

Figure 7.25: ANOVA summary for testing whether the average on-base percentage differs across player positions.

7.5.5 Graphical diagnostics for an ANOVA analysis

There are three conditions we must check for an ANOVA analysis: all observations must be independent, the data in each group must be nearly normal, and the variance within each group must be approximately equal.

Independence. If the data are a simple random sample, this condition is satisfied. For processes and experiments, carefully consider whether the data may be independent (e.g. no pairing). For example, in the MLB data, the data were not sampled. However, there are not obvious reasons why independence would not hold for most or all observations.

Approximately normal. As with one- and two-sample testing for means, the normality assumption is especially important when the sample size is quite small when it is ironically difficult to check for non-normality. A histogram of the observations from each group is shown in Figure 7.26. Since each of the groups we're considering have relatively large sample sizes, what we're looking for are major outliers. None are apparent, so this conditions is reasonably met.

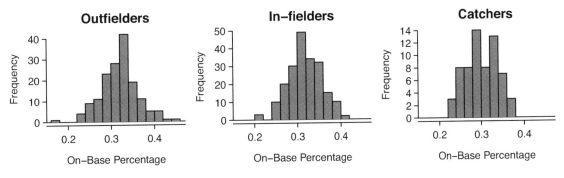

Figure 7.26: Histograms of OBP for each field position.

Constant variance. The last assumption is that the variance in the groups is about equal from one group to the next. This assumption can be checked by examining a side-by-side box plot of the outcomes across the groups, as in Figure 7.23 on page 287. In this case, the variability is similar in the four groups but not identical. We see in Table 7.22 on page 287 that the standard deviation doesn't vary much from one group to the next.

DIAGNOSTICS FOR AN ANOVA ANALYSIS

Independence is always important to an ANOVA analysis. The normality condition is very important when the sample sizes for each group are relatively small. The constant variance condition is especially important when the sample sizes differ between groups.

7.5.6 Multiple comparisons and controlling Type 1 Error rate

When we reject the null hypothesis in an ANOVA analysis, we might wonder, which of these groups have different means? To answer this question, we compare the means of each possible pair of groups. For instance, if there are three groups and there is strong evidence that there are some differences in the group means, there are three comparisons to make: group 1 to group 2, group 1 to group 3, and group 2 to group 3. These comparisons can be accomplished using a two-sample t-test, but we use a modified significance level and a pooled estimate of the standard deviation across groups. Usually this pooled standard deviation can be found in the ANOVA table, e.g. along the bottom of Figure 7.25.

EXAMPLE 7.47

Example 7.40 on page 285 discussed three statistics lectures, all taught during the same semester. Figure 7.27 shows summary statistics for these three courses, and a side-by-side box plot of the data is shown in Figure 7.28. We would like to conduct an ANOVA for these data. Do you see any deviations from the three conditions for ANOVA?

In this case (like many others) it is difficult to check independence in a rigorous way. Instead, the best we can do is use common sense to consider reasons the assumption of independence may not hold. For instance, the independence assumption may not be reasonable if there is a star teaching assistant that only half of the students may access; such a scenario would divide a class into two subgroups. No such situations were evident for these particular data, and we believe that independence is acceptable.

The distributions in the side-by-side box plot appear to be roughly symmetric and show no noticeable outliers.

The box plots show approximately equal variability, which can be verified in Figure 7.27, supporting the constant variance assumption.

Class i	A	B	C
n_i	58	55	51
$\bar{x}_i$	75.1	72.0	78.9
s_i	13.9	13.8	13.1

Figure 7.27: Summary statistics for the first midterm scores in three different lectures of the same course.

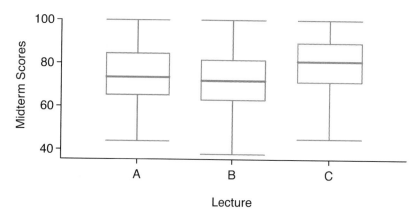

Figure 7.28: Side-by-side box plot for the first midterm scores in three different lectures of the same course.

GUIDED PRACTICE 7.48

ANOVA was conducted for the midterm data, and summary results are shown in Figure 7.29. What should we conclude?[35]

	Df	Sum Sq	Mean Sq	F value	Pr(>F)
lecture	2	1290.11	645.06	3.48	0.0330
Residuals	161	29810.13	185.16		

$$s_{pooled} = 13.61 \text{ on } df = 161$$

Figure 7.29: ANOVA summary table for the midterm data.

There is strong evidence that the different means in each of the three classes is not simply due to chance. We might wonder, which of the classes are actually different? As discussed in earlier chapters, a two-sample t-test could be used to test for differences in each possible pair of groups. However, one pitfall was discussed in Example 7.44 on page 288: when we run so many tests, the Type 1 Error rate increases. This issue is resolved by using a modified significance level.

MULTIPLE COMPARISONS AND THE BONFERRONI CORRECTION FOR α

The scenario of testing many pairs of groups is called **multiple comparisons**. The **Bonferroni correction** suggests that a more stringent significance level is more appropriate for these tests:

$$\alpha^{\star} = \alpha/K$$

where K is the number of comparisons being considered (formally or informally). If there are k groups, then usually all possible pairs are compared and $K = \frac{k(k-1)}{2}$.

EXAMPLE 7.49

In Guided Practice 7.48, you found strong evidence of differences in the average midterm grades between the three lectures. Complete the three possible pairwise comparisons using the Bonferroni correction and report any differences.

We use a modified significance level of $\alpha^{\star} = 0.05/3 = 0.0167$. Additionally, we use the pooled estimate of the standard deviation: $s_{pooled} = 13.61$ on $df = 161$, which is provided in the ANOVA summary table.

Lecture A versus Lecture B: The estimated difference and standard error are, respectively,

$$\bar{x}_A - \bar{x}_B = 75.1 - 72 = 3.1 \qquad SE = \sqrt{\frac{13.61^2}{58} + \frac{13.61^2}{55}} = 2.56$$

(See Section 7.3.4 on page 273 for additional details.) This results in a T-score of 1.21 on $df = 161$ (we use the df associated with s_{pooled}). Statistical software was used to precisely identify the two-sided p-value since the modified significance level of 0.0167 is not found in the t-table. The p-value (0.228) is larger than $\alpha^* = 0.0167$, so there is not strong evidence of a difference in the means of lectures A and B.

Lecture A versus Lecture C: The estimated difference and standard error are 3.8 and 2.61, respectively. This results in a T score of 1.46 on $df = 161$ and a two-sided p-value of 0.1462. This p-value is larger than α^*, so there is not strong evidence of a difference in the means of lectures A and C.

Lecture B versus Lecture C: The estimated difference and standard error are 6.9 and 2.65, respectively. This results in a T score of 2.60 on $df = 161$ and a two-sided p-value of 0.0102. This p-value is smaller than α^*. Here we find strong evidence of a difference in the means of lectures B and C.

---

[35]The p-value of the test is 0.0330, less than the default significance level of 0.05. Therefore, we reject the null hypothesis and conclude that the difference in the average midterm scores are not due to chance.

We might summarize the findings of the analysis from Example 7.49 using the following notation:

$$\mu_A \overset{?}{=} \mu_B \qquad\qquad \mu_A \overset{?}{=} \mu_C \qquad\qquad \mu_B \neq \mu_C$$

The midterm mean in lecture A is not statistically distinguishable from those of lectures B or C. However, there is strong evidence that lectures B and C are different. In the first two pairwise comparisons, we did not have sufficient evidence to reject the null hypothesis. Recall that failing to reject H_0 does not imply H_0 is true.

REJECT H_0 WITH ANOVA BUT FIND NO DIFFERENCES IN GROUP MEANS

It is possible to reject the null hypothesis using ANOVA and then to not subsequently identify differences in the pairwise comparisons. However, *this does not invalidate the ANOVA conclusion*. It only means we have not been able to successfully identify which specific groups differ in their means.

The ANOVA procedure examines the big picture: it considers all groups simultaneously to decipher whether there is evidence that some difference exists. Even if the test indicates that there is strong evidence of differences in group means, identifying with high confidence a specific difference as statistically significant is more difficult.

Consider the following analogy: we observe a Wall Street firm that makes large quantities of money based on predicting mergers. Mergers are generally difficult to predict, and if the prediction success rate is extremely high, that may be considered sufficiently strong evidence to warrant investigation by the Securities and Exchange Commission (SEC). While the SEC may be quite certain that there is insider trading taking place at the firm, the evidence against any single trader may not be very strong. It is only when the SEC considers all the data that they identify the pattern. This is effectively the strategy of ANOVA: stand back and consider all the groups simultaneously.

Exercises

7.35 Fill in the blank. When doing an ANOVA, you observe large differences in means between groups. Within the ANOVA framework, this would most likely be interpreted as evidence strongly favoring the _____ hypothesis.

7.36 Which test? We would like to test if students who are in the social sciences, natural sciences, arts and humanities, and other fields spend the same amount of time studying for this course. What type of test should we use? Explain your reasoning.

7.37 Chicken diet and weight, Part III. In Exercises 7.27 and 7.29 we compared the effects of two types of feed at a time. A better analysis would first consider all feed types at once: casein, horsebean, linseed, meat meal, soybean, and sunflower. The ANOVA output below can be used to test for differences between the average weights of chicks on different diets.

	Df	Sum Sq	Mean Sq	F value	Pr(>F)
feed	5	231,129.16	46,225.83	15.36	0.0000
Residuals	65	195,556.02	3,008.55		

Conduct a hypothesis test to determine if these data provide convincing evidence that the average weight of chicks varies across some (or all) groups. Make sure to check relevant conditions. Figures and summary statistics are shown below.

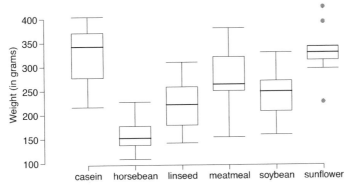

	Mean	SD	n
casein	323.58	64.43	12
horsebean	160.20	38.63	10
linseed	218.75	52.24	12
meatmeal	276.91	64.90	11
soybean	246.43	54.13	14
sunflower	328.92	48.84	12

7.38 Teaching descriptive statistics. A study compared five different methods for teaching descriptive statistics. The five methods were traditional lecture and discussion, programmed textbook instruction, programmed text with lectures, computer instruction, and computer instruction with lectures. 45 students were randomly assigned, 9 to each method. After completing the course, students took a 1-hour exam.

(a) What are the hypotheses for evaluating if the average test scores are different for the different teaching methods?

(b) What are the degrees of freedom associated with the F-test for evaluating these hypotheses?

(c) Suppose the p-value for this test is 0.0168. What is the conclusion?

7.39 Coffee, depression, and physical activity. Caffeine is the world's most widely used stimulant, with approximately 80% consumed in the form of coffee. Participants in a study investigating the relationship between coffee consumption and exercise were asked to report the number of hours they spent per week on moderate (e.g., brisk walking) and vigorous (e.g., strenuous sports and jogging) exercise. Based on these data the researchers estimated the total hours of metabolic equivalent tasks (MET) per week, a value always greater than 0. The table below gives summary statistics of MET for women in this study based on the amount of coffee consumed.[36]

	$\leq$ 1 cup/week	2-6 cups/week	1 cup/day	2-3 cups/day	$\geq$ 4 cups/day	Total
Mean	18.7	19.6	19.3	18.9	17.5	
SD	21.1	25.5	22.5	22.0	22.0	
n	12,215	6,617	17,234	12,290	2,383	50,739

Caffeinated coffee consumption

(a) Write the hypotheses for evaluating if the average physical activity level varies among the different levels of coffee consumption.

(b) Check conditions and describe any assumptions you must make to proceed with the test.

(c) Below is part of the output associated with this test. Fill in the empty cells.

	Df	Sum Sq	Mean Sq	F value	Pr(>F)
coffee					0.0003
Residuals		25,564,819			
Total		25,575,327			

(d) What is the conclusion of the test?

7.40 Student performance across discussion sections. A professor who teaches a large introductory statistics class (197 students) with eight discussion sections would like to test if student performance differs by discussion section, where each discussion section has a different teaching assistant. The summary table below shows the average final exam score for each discussion section as well as the standard deviation of scores and the number of students in each section.

	Sec 1	Sec 2	Sec 3	Sec 4	Sec 5	Sec 6	Sec 7	Sec 8
n_i	33	19	10	29	33	10	32	31
$\bar{x}_i$	92.94	91.11	91.80	92.45	89.30	88.30	90.12	93.35
s_i	4.21	5.58	3.43	5.92	9.32	7.27	6.93	4.57

The ANOVA output below can be used to test for differences between the average scores from the different discussion sections.

	Df	Sum Sq	Mean Sq	F value	Pr(>F)
section	7	525.01	75.00	1.87	0.0767
Residuals	189	7584.11	40.13		

Conduct a hypothesis test to determine if these data provide convincing evidence that the average score varies across some (or all) groups. Check conditions and describe any assumptions you must make to proceed with the test.

[36]M. Lucas et al. "Coffee, caffeine, and risk of depression among women". In: *Archives of internal medicine* 171.17 (2011), p. 1571.

7.41 GPA and major. Undergraduate students taking an introductory statistics course at Duke University conducted a survey about GPA and major. The side-by-side box plots show the distribution of GPA among three groups of majors. Also provided is the ANOVA output.

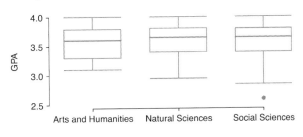

	Df	Sum Sq	Mean Sq	F value	Pr(>F)
major	2	0.03	0.015	0.185	0.8313
Residuals	195	15.77	0.081		

(a) Write the hypotheses for testing for a difference between average GPA across majors.

(b) What is the conclusion of the hypothesis test?

(c) How many students answered these questions on the survey, i.e. what is the sample size?

7.42 Work hours and education. The General Social Survey collects data on demographics, education, and work, among many other characteristics of US residents.[37] Using ANOVA, we can consider educational attainment levels for all 1,172 respondents at once. Below are the distributions of hours worked by educational attainment and relevant summary statistics that will be helpful in carrying out this analysis.

	Educational attainment					
	Less than HS	HS	Jr Coll	Bachelor's	Graduate	Total
Mean	38.67	39.6	41.39	42.55	40.85	40.45
SD	15.81	14.97	18.1	13.62	15.51	15.17
n	121	546	97	253	155	1,172

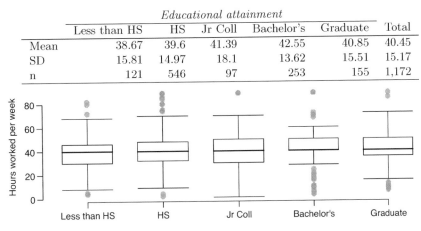

(a) Write hypotheses for evaluating whether the average number of hours worked varies across the five groups.

(b) Check conditions and describe any assumptions you must make to proceed with the test.

(c) Below is part of the output associated with this test. Fill in the empty cells.

	Df	Sum Sq	Mean Sq	F-value	Pr(>F)
degree			501.54		0.0682
Residuals		267,382			
Total					

(d) What is the conclusion of the test?

7.43 True / False: ANOVA, Part I. Determine if the following statements are true or false in ANOVA, and explain your reasoning for statements you identify as false.

(a) As the number of groups increases, the modified significance level for pairwise tests increases as well.

(b) As the total sample size increases, the degrees of freedom for the residuals increases as well.

(c) The constant variance condition can be somewhat relaxed when the sample sizes are relatively consistent across groups.

(d) The independence assumption can be relaxed when the total sample size is large.

---

[37]National Opinion Research Center, General Social Survey, 2018.

7.44 Child care hours. The China Health and Nutrition Survey aims to examine the effects of the health, nutrition, and family planning policies and programs implemented by national and local governments.[38] It, for example, collects information on number of hours Chinese parents spend taking care of their children under age 6. The side-by-side box plots below show the distribution of this variable by educational attainment of the parent. Also provided below is the ANOVA output for comparing average hours across educational attainment categories.

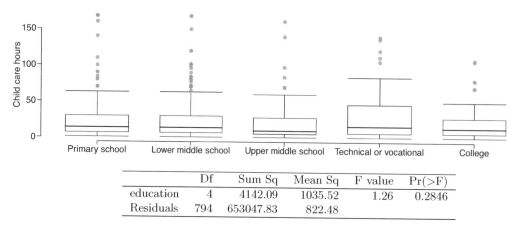

	Df	Sum Sq	Mean Sq	F value	Pr(>F)
education	4	4142.09	1035.52	1.26	0.2846
Residuals	794	653047.83	822.48		

(a) Write the hypotheses for testing for a difference between the average number of hours spent on child care across educational attainment levels.

(b) What is the conclusion of the hypothesis test?

7.45 Prison isolation experiment, Part II. Exercise 7.31 introduced an experiment that was conducted with the goal of identifying a treatment that reduces subjects' psychopathic deviant T scores, where this score measures a person's need for control or his rebellion against control. In Exercise 7.31 you evaluated the success of each treatment individually. An alternative analysis involves comparing the success of treatments. The relevant ANOVA output is given below.

	Df	Sum Sq	Mean Sq	F value	Pr(>F)
treatment	2	639.48	319.74	3.33	0.0461
Residuals	39	3740.43	95.91		

$$s_{pooled} = 9.793 \text{ on } df = 39$$

(a) What are the hypotheses?

(b) What is the conclusion of the test? Use a 5% significance level.

(c) If in part (b) you determined that the test is significant, conduct pairwise tests to determine which groups are different from each other. If you did not reject the null hypothesis in part (b), recheck your answer. Summary statistics for each group are provided below

	Tr 1	Tr 2	Tr 3
Mean	6.21	2.86	-3.21
SD	12.3	7.94	8.57
n	14	14	14

7.46 True / False: ANOVA, Part II. Determine if the following statements are true or false, and explain your reasoning for statements you identify as false.

If the null hypothesis that the means of four groups are all the same is rejected using ANOVA at a 5% significance level, then ...

(a) we can then conclude that all the means are different from one another.

(b) the standardized variability between groups is higher than the standardized variability within groups.

(c) the pairwise analysis will identify at least one pair of means that are significantly different.

(d) the appropriate α to be used in pairwise comparisons is 0.05 / 4 = 0.0125 since there are four groups.

---

[38]UNC Carolina Population Center, China Health and Nutrition Survey, 2006.

Chapter exercises

7.47 Gaming and distracted eating, Part I. A group of researchers are interested in the possible effects of distracting stimuli during eating, such as an increase or decrease in the amount of food consumption. To test this hypothesis, they monitored food intake for a group of 44 patients who were randomized into two equal groups. The treatment group ate lunch while playing solitaire, and the control group ate lunch without any added distractions. Patients in the treatment group ate 52.1 grams of biscuits, with a standard deviation of 45.1 grams, and patients in the control group ate 27.1 grams of biscuits, with a standard deviation of 26.4 grams. Do these data provide convincing evidence that the average food intake (measured in amount of biscuits consumed) is different for the patients in the treatment group? Assume that conditions for inference are satisfied.[39]

7.48 Gaming and distracted eating, Part II. The researchers from Exercise 7.47 also investigated the effects of being distracted by a game on how much people eat. The 22 patients in the treatment group who ate their lunch while playing solitaire were asked to do a serial-order recall of the food lunch items they ate. The average number of items recalled by the patients in this group was 4. 9, with a standard deviation of 1.8. The average number of items recalled by the patients in the control group (no distraction) was 6.1, with a standard deviation of 1.8. Do these data provide strong evidence that the average number of food items recalled by the patients in the treatment and control groups are different?

7.49 Sample size and pairing. Determine if the following statement is true or false, and if false, explain your reasoning: If comparing means of two groups with equal sample sizes, always use a paired test.

7.50 College credits. A college counselor is interested in estimating how many credits a student typically enrolls in each semester. The counselor decides to randomly sample 100 students by using the registrar's database of students. The histogram below shows the distribution of the number of credits taken by these students. Sample statistics for this distribution are also provided.

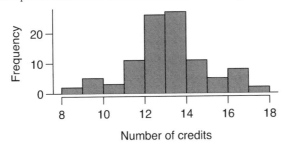

Min	8
Q1	13
Median	14
Mean	13.65
SD	1.91
Q3	15
Max	18

(a) What is the point estimate for the average number of credits taken per semester by students at this college? What about the median?

(b) What is the point estimate for the standard deviation of the number of credits taken per semester by students at this college? What about the IQR?

(c) Is a load of 16 credits unusually high for this college? What about 18 credits? Explain your reasoning.

(d) The college counselor takes another random sample of 100 students and this time finds a sample mean of 14.02 units. Should she be surprised that this sample statistic is slightly different than the one from the original sample? Explain your reasoning.

(e) The sample means given above are point estimates for the mean number of credits taken by all students at that college. What measures do we use to quantify the variability of this estimate? Compute this quantity using the data from the original sample.

[39]R.E. Oldham-Cooper et al. "Playing a computer game during lunch affects fullness, memory for lunch, and later snack intake". In: *The American Journal of Clinical Nutrition* 93.2 (2011), p. 308.

7.51 Hen eggs. The distribution of the number of eggs laid by a certain species of hen during their breeding period has a mean of 35 eggs with a standard deviation of 18.2. Suppose a group of researchers randomly samples 45 hens of this species, counts the number of eggs laid during their breeding period, and records the sample mean. They repeat this 1,000 times, and build a distribution of sample means.

(a) What is this distribution called?

(b) Would you expect the shape of this distribution to be symmetric, right skewed, or left skewed? Explain your reasoning.

(c) Calculate the variability of this distribution and state the appropriate term used to refer to this value.

(d) Suppose the researchers' budget is reduced and they are only able to collect random samples of 10 hens. The sample mean of the number of eggs is recorded, and we repeat this 1,000 times, and build a new distribution of sample means. How will the variability of this new distribution compare to the variability of the original distribution?

7.52 Forest management. Forest rangers wanted to better understand the rate of growth for younger trees in the park. They took measurements of a random sample of 50 young trees in 2009 and again measured those same trees in 2019. The data below summarize their measurements, where the heights are in feet:

	2009	2019	Differences
$\bar{x}$	12.0	24.5	12.5
s	3.5	9.5	7.2
n	50	50	50

Construct a 99% confidence interval for the average growth of (what had been) younger trees in the park over 2009-2019.

7.53 Experiment resizing. At a startup company running a new weather app, an engineering team generally runs experiments where a random sample of 1% of the app's visitors in the control group and another 1% were in the treatment group to test each new feature. The team's core goal is to increase a metric called *daily visitors*, which is essentially the number of visitors to the app each day. They track this metric in each experiment arm and as their core experiment metric. In their most recent experiment, the team tested including a new animation when the app started, and the number of daily visitors in this experiment stabilized at +1.2% with a 95% confidence interval of (-0.2%, +2.6%). This means if this new app start animation was launched, the team thinks they might lose as many as 0.2% of daily visitors or gain as many as 2.6% more daily visitors. Suppose you are consulting as the team's data scientist, and after discussing with the team, you and they agree that they should run another experiment that is bigger. You also agree that this new experiment should be able to detect a gain in the daily visitors metric of 1.0% or more with 80% power. Now they turn to you and ask, "How big of an experiment do we need to run to ensure we can detect this effect?"

(a) How small must the standard error be if the team is to be able to detect an effect of 1.0% with 80% power and a significance level of $\alpha = 0.05$? You may safely assume the percent change in daily visitors metric follows a normal distribution.

(b) Consider the first experiment, where the point estimate was +1.2% and the 95% confidence interval was (-0.2%, +2.6%). If that point estimate followed a normal distribution, what was the standard error of the estimate?

(c) The ratio of the standard error from part (a) vs the standard error from part (b) should be 2.03. How much bigger of an experiment is needed to shrink a standard error by a factor of 2.03?

(d) Using your answer from part (c) and that the original experiment was a 1% vs 1% experiment to recommend an experiment size to the team.

7.54 Torque on a rusty bolt. Project Farm is a YouTube channel that routinely compares different products. In one episode, the channel evaluated different options for loosening rusty bolts.[40] Eight options were evaluated, including a control group where no treatment was given ("none" in the graph), to determine which was most effective. For all treatments, there were four bolts tested, except for a treatment of heat with a blow torch, where only two data points were collected. The results are shown in the figure below:

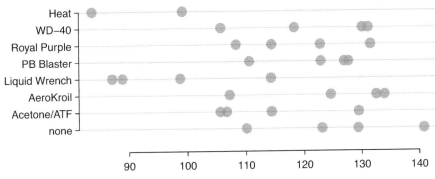

Torque Required to Loosen Rusty Bolt, in Foot–Pounds

(a) Do you think it is reasonable to apply ANOVA in this case?

(b) Regardless of your answer in part (a), describe hypotheses for ANOVA in this context, and use the table below to carry out the test. Give your conclusion in the context of the data.

	Df	Sum Sq	Mean Sq	F value	Pr(>F)
treatment	7	3603.43	514.78	4.03	0.0056
Residuals	22	2812.80	127.85		

(c) The table below are p-values for pairwise t-tests comparing each of the different groups. These p-values have not been corrected for multiple comparisons. Which pair of groups appears most likely to represent a difference?

	AeroKroil	Heat	Liquid Wrench	none	PB Blaster	Royal Purple	WD-40
Acetone/ATF	0.2026	0.0308	0.0476	0.1542	0.3294	0.5222	0.3744
AeroKroil		0.0027	0.0025	0.8723	0.7551	0.5143	0.6883
Heat			0.5580	0.0020	0.0050	0.0096	0.0059
Liquid Wrench				0.0017	0.0053	0.0117	0.0065
none					0.6371	0.4180	0.5751
PB Blaster						0.7318	0.9286
Royal Purple							0.8000

(d) There are 28 p-values shown in the table in part (c). Determine if any of them are statistically significant after correcting for multiple comparisons. If so, which one(s)? Explain your answer.

7.55 Exclusive relationships. A survey conducted on a reasonably random sample of 203 undergraduates asked, among many other questions, about the number of exclusive relationships these students have been in. The histogram below shows the distribution of the data from this sample. The sample average is 3.2 with a standard deviation of 1.97.

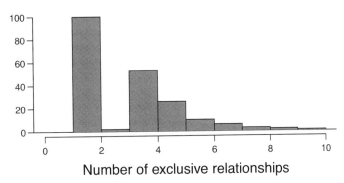

Number of exclusive relationships

Estimate the average number of exclusive relationships Duke students have been in using a 90% confidence interval and interpret this interval in context. Check any conditions required for inference, and note any assumptions you must make as you proceed with your calculations and conclusions.

---

[40]Project Farm on YouTube, youtu.be/xUEob2oAKVs, April 16, 2018.

7.56   Age at first marriage, Part I. The National Survey of Family Growth conducted by the Centers for Disease Control gathers information on family life, marriage and divorce, pregnancy, infertility, use of contraception, and men's and women's health. One of the variables collected on this survey is the age at first marriage. The histogram below shows the distribution of ages at first marriage of 5,534 randomly sampled women between 2006 and 2010. The average age at first marriage among these women is 23.44 with a standard deviation of 4.72.[41]

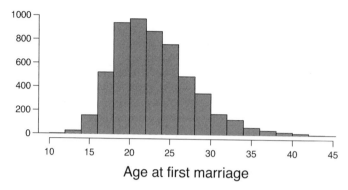

Estimate the average age at first marriage of women using a 95% confidence interval, and interpret this interval in context. Discuss any relevant assumptions.

7.57   Online communication. A study suggests that the average college student spends 10 hours per week communicating with others online. You believe that this is an underestimate and decide to collect your own sample for a hypothesis test. You randomly sample 60 students from your dorm and find that on average they spent 13.5 hours a week communicating with others online. A friend of yours, who offers to help you with the hypothesis test, comes up with the following set of hypotheses. Indicate any errors you see.

$$H_0 : \bar{x} < 10 \ hours$$
$$H_A : \bar{x} > 13.5 \ hours$$

7.58   Age at first marriage, Part II. Exercise 7.56 presents the results of a 2006 - 2010 survey showing that the average age of women at first marriage is 23.44. Suppose a social scientist thinks this value has changed since the survey was taken. Below is how she set up her hypotheses. Indicate any errors you see.

$$H_0 : \bar{x} \neq 23.44 \ years \ old$$
$$H_A : \bar{x} = 23.44 \ years \ old$$

[41] Centers for Disease Control and Prevention, National Survey of Family Growth, 2010.

Chapter 8

Introduction to linear regression

Linear regression is a very powerful statistical technique. Many people have some familiarity with regression just from reading the news, where straight lines are overlaid on scatterplots. Linear models can be used for prediction or to evaluate whether there is a linear relationship between two numerical variables.

For videos, slides, and other resources, please visit
www.openintro.org/os

8.1 Fitting a line, residuals, and correlation

It's helpful to think deeply about the line fitting process. In this section, we define the form of a linear model, explore criteria for what makes a good fit, and introduce a new statistic called *correlation*.

8.1.1 Fitting a line to data

Figure 8.1 shows two variables whose relationship can be modeled perfectly with a straight line. The equation for the line is

$$y = 5 + 64.96x$$

Consider what a perfect linear relationship means: we know the exact value of y just by knowing the value of x. This is unrealistic in almost any natural process. For example, if we took family income (x), this value would provide some useful information about how much financial support a college may offer a prospective student (y). However, the prediction would be far from perfect, since other factors play a role in financial support beyond a family's finances.

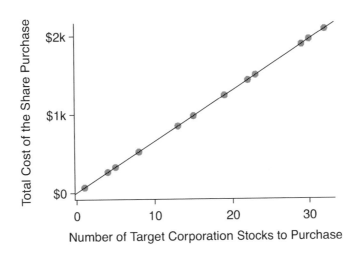

Figure 8.1: Requests from twelve separate buyers were simultaneously placed with a trading company to purchase Target Corporation stock (ticker TGT, December 28th, 2018), and the total cost of the shares were reported. Because the cost is computed using a linear formula, the linear fit is perfect.

Linear regression is the statistical method for fitting a line to data where the relationship between two variables, x and y, can be modeled by a straight line with some error:

$$y = \beta_0 + \beta_1 x + \varepsilon$$

The values β_0 and β_1 represent the model's parameters (β is the Greek letter *beta*), and the error is represented by ε (the Greek letter *epsilon*). The parameters are estimated using data, and we write their point estimates as b_0 and b_1. When we use x to predict y, we usually call x the explanatory or **predictor** variable, and we call y the response; we also often drop the ϵ term when writing down the model since our main focus is often on the prediction of the average outcome.

It is rare for all of the data to fall perfectly on a straight line. Instead, it's more common for data to appear as a *cloud of points*, such as those examples shown in Figure 8.2. In each case, the data fall around a straight line, even if none of the observations fall exactly on the line. The first plot shows a relatively strong downward linear trend, where the remaining variability in the data around the line is minor relative to the strength of the relationship between x and y. The second plot shows an upward trend that, while evident, is not as strong as the first. The last plot shows a

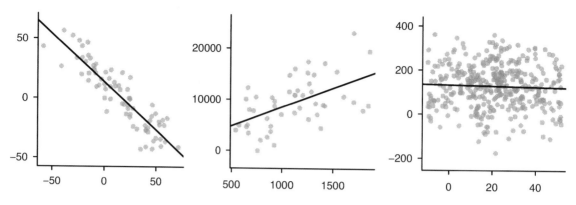

Figure 8.2: Three data sets where a linear model may be useful even though the data do not all fall exactly on the line.

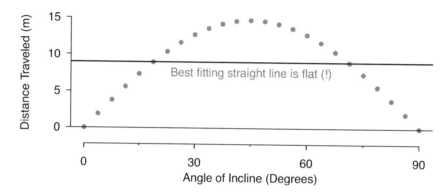

Figure 8.3: A linear model is not useful in this nonlinear case. These data are from an introductory physics experiment.

very weak downward trend in the data, so slight we can hardly notice it. In each of these examples, we will have some uncertainty regarding our estimates of the model parameters, β_0 and β_1. For instance, we might wonder, should we move the line up or down a little, or should we tilt it more or less? As we move forward in this chapter, we will learn about criteria for line-fitting, and we will also learn about the uncertainty associated with estimates of model parameters.

There are also cases where fitting a straight line to the data, even if there is a clear relationship between the variables, is not helpful. One such case is shown in Figure 8.3 where there is a very clear relationship between the variables even though the trend is not linear. We discuss nonlinear trends in this chapter and the next, but details of fitting nonlinear models are saved for a later course.

8.1.2 Using linear regression to predict possum head lengths

Brushtail possums are a marsupial that lives in Australia, and a photo of one is shown in Figure 8.4. Researchers captured 104 of these animals and took body measurements before releasing the animals back into the wild. We consider two of these measurements: the total length of each possum, from head to tail, and the length of each possum's head.

Figure 8.5 shows a scatterplot for the head length and total length of the possums. Each point represents a single possum from the data. The head and total length variables are associated: possums with an above average total length also tend to have above average head lengths. While the relationship is not perfectly linear, it could be helpful to partially explain the connection between these variables with a straight line.

Figure 8.4: The common brushtail possum of Australia.

Photo by Greg Schechter (https://flic.kr/p/9BAFbR). CC BY 2.0 license.

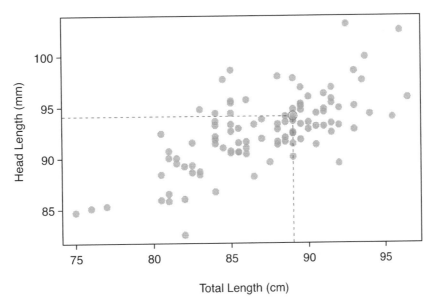

Figure 8.5: A scatterplot showing head length against total length for 104 brushtail possums. A point representing a possum with head length 94.1mm and total length 89cm is highlighted.

We want to describe the relationship between the head length and total length variables in the possum data set using a line. In this example, we will use the total length as the predictor variable, x, to predict a possum's head length, y. We could fit the linear relationship by eye, as in Figure 8.6. The equation for this line is

$$\hat{y} = 41 + 0.59x$$

A "hat" on y is used to signify that this is an estimate. We can use this line to discuss properties of possums. For instance, the equation predicts a possum with a total length of 80 cm will have a head length of

$$\hat{y} = 41 + 0.59 \times 80$$
$$= 88.2$$

The estimate may be viewed as an average: the equation predicts that possums with a total length of 80 cm will have an average head length of 88.2 mm. Absent further information about an 80 cm possum, the prediction for head length that uses the average is a reasonable estimate.

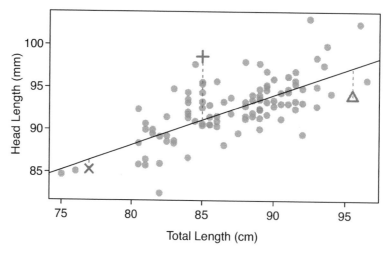

Figure 8.6: A reasonable linear model was fit to represent the relationship between head length and total length.

EXAMPLE 8.1

What other variables might help us predict the head length of a possum besides its length?

Perhaps the relationship would be a little different for male possums than female possums, or perhaps it would differ for possums from one region of Australia versus another region. In Chapter 9, we'll learn about how we can include more than one predictor. Before we get there, we first need to better understand how to best build a simple linear model with one predictor.

8.1.3 Residuals

Residuals are the leftover variation in the data after accounting for the model fit:

$$\text{Data} = \text{Fit} + \text{Residual}$$

Each observation will have a residual, and three of the residuals for the linear model we fit for the possum data is shown in Figure 8.6. If an observation is above the regression line, then its residual, the vertical distance from the observation to the line, is positive. Observations below the line have negative residuals. One goal in picking the right linear model is for these residuals to be as small as possible.

Let's look closer at the three residuals featured in Figure 8.6. The observation marked by an "×" has a small, negative residual of about -1; the observation marked by "+" has a large residual of about +7; and the observation marked by "△" has a moderate residual of about -4. The size of a residual is usually discussed in terms of its absolute value. For example, the residual for "△" is larger than that of "×" because $|-4|$ is larger than $|-1|$.

RESIDUAL: DIFFERENCE BETWEEN OBSERVED AND EXPECTED

The residual of the ith observation (x_i, y_i) is the difference of the observed response (y_i) and the response we would predict based on the model fit $(\hat{y}_i)$:

$$e_i = y_i - \hat{y}_i$$

We typically identify $\hat{y}_i$ by plugging x_i into the model.

EXAMPLE 8.2

The linear fit shown in Figure 8.6 is given as $\hat{y} = 41 + 0.59x$. Based on this line, formally compute the residual of the observation $(77.0, 85.3)$. This observation is denoted by "×" in Figure 8.6. Check it against the earlier visual estimate, -1.

We first compute the predicted value of point "×" based on the model:

$$\hat{y}_\times = 41 + 0.59x_\times = 41 + 0.59 \times 77.0 = 86.4$$

Next we compute the difference of the actual head length and the predicted head length:

$$e_\times = y_\times - \hat{y}_\times = 85.3 - 86.4 = -1.1$$

The model's error is $e_\times = -1.1$mm, which is very close to the visual estimate of -1mm. The negative residual indicates that the linear model overpredicted head length for this particular possum.

GUIDED PRACTICE 8.3

If a model underestimates an observation, will the residual be positive or negative? What about if it overestimates the observation?[1]

GUIDED PRACTICE 8.4

Compute the residuals for the "+" observation $(85.0, 98.6)$ and the "△" observation $(95.5, 94.0)$ in the figure using the linear relationship $\hat{y} = 41 + 0.59x$.[2]

Residuals are helpful in evaluating how well a linear model fits a data set. We often display them in a **residual plot** such as the one shown in Figure 8.7 for the regression line in Figure 8.6. The residuals are plotted at their original horizontal locations but with the vertical coordinate as the residual. For instance, the point $(85.0, 98.6)_+$ had a residual of 7.45, so in the residual plot it is placed at $(85.0, 7.45)$. Creating a residual plot is sort of like tipping the scatterplot over so the regression line is horizontal.

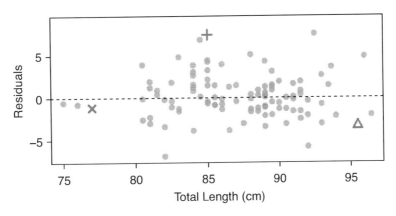

Figure 8.7: Residual plot for the model in Figure 8.6.

---

[1]If a model underestimates an observation, then the model estimate is below the actual. The residual, which is the actual observation value minus the model estimate, must then be positive. The opposite is true when the model overestimates the observation: the residual is negative.

[2](+) First compute the predicted value based on the model:

$$\hat{y}_+ = 41 + 0.59x_+ = 41 + 0.59 \times 85.0 = 91.15$$

Then the residual is given by

$$e_+ = y_+ - \hat{y}_+ = 98.6 - 91.15 = 7.45$$

This was close to the earlier estimate of 7.

(△) $\hat{y}_\triangle = 41 + 0.59x_\triangle = 97.3$. $e_\triangle = y_\triangle - \hat{y}_\triangle = -3.3$, close to the estimate of -4.

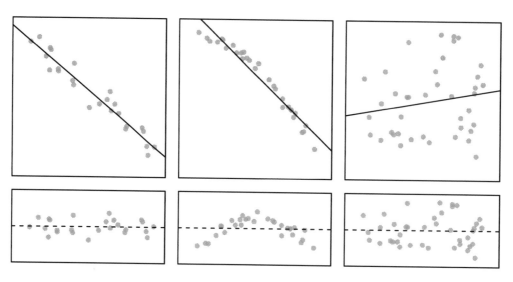

Figure 8.8: Sample data with their best fitting lines (top row) and their corresponding residual plots (bottom row).

EXAMPLE 8.5

One purpose of residual plots is to identify characteristics or patterns still apparent in data after fitting a model. Figure 8.8 shows three scatterplots with linear models in the first row and residual plots in the second row. Can you identify any patterns remaining in the residuals?

In the first data set (first column), the residuals show no obvious patterns. The residuals appear to be scattered randomly around the dashed line that represents 0.

The second data set shows a pattern in the residuals. There is some curvature in the scatterplot, which is more obvious in the residual plot. We should not use a straight line to model these data. Instead, a more advanced technique should be used.

The last plot shows very little upwards trend, and the residuals also show no obvious patterns. It is reasonable to try to fit a linear model to the data. However, it is unclear whether there is statistically significant evidence that the slope parameter is different from zero. The point estimate of the slope parameter, labeled b_1, is not zero, but we might wonder if this could just be due to chance. We will address this sort of scenario in Section 8.4.

8.1.4 Describing linear relationships with correlation

We've seen plots with strong linear relationships and others with very weak linear relationships. It would be useful if we could quantify the strength of these linear relationships with a statistic.

CORRELATION: STRENGTH OF A LINEAR RELATIONSHIP

Correlation, which always takes values between -1 and 1, describes the strength of the linear relationship between two variables. We denote the correlation by R.

We can compute the correlation using a formula, just as we did with the sample mean and standard deviation. This formula is rather complex,[3] and like with other statistics, we generally perform

---

[3]Formally, we can compute the correlation for observations (x_1, y_1), (x_2, y_2), ..., (x_n, y_n) using the formula

$$R = \frac{1}{n-1} \sum_{i=1}^{n} \frac{x_i - \bar{x}}{s_x} \frac{y_i - \bar{y}}{s_y}$$

where $\bar{x}$, $\bar{y}$, s_x, and s_y are the sample means and standard deviations for each variable.

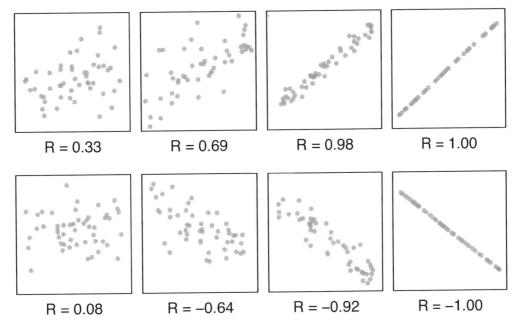

Figure 8.9: Sample scatterplots and their correlations. The first row shows variables with a positive relationship, represented by the trend up and to the right. The second row shows variables with a negative trend, where a large value in one variable is associated with a low value in the other.

the calculations on a computer or calculator. Figure 8.9 shows eight plots and their corresponding correlations. Only when the relationship is perfectly linear is the correlation either -1 or 1. If the relationship is strong and positive, the correlation will be near +1. If it is strong and negative, it will be near -1. If there is no apparent linear relationship between the variables, then the correlation will be near zero.

The correlation is intended to quantify the strength of a linear trend. Nonlinear trends, even when strong, sometimes produce correlations that do not reflect the strength of the relationship; see three such examples in Figure 8.10.

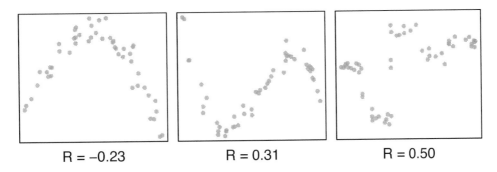

Figure 8.10: Sample scatterplots and their correlations. In each case, there is a strong relationship between the variables. However, because the relationship is nonlinear, the correlation is relatively weak.

GUIDED PRACTICE 8.6

No straight line is a good fit for the data sets represented in Figure 8.10. Try drawing nonlinear curves on each plot. Once you create a curve for each, describe what is important in your fit.[4]

---

[4]We'll leave it to you to draw the lines. In general, the lines you draw should be close to most points and reflect overall trends in the data.

Exercises

8.1 Visualize the residuals. The scatterplots shown below each have a superimposed regression line. If we were to construct a residual plot (residuals versus x) for each, describe what those plots would look like.

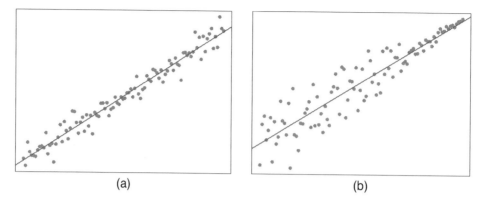

(a)   (b)

8.2 Trends in the residuals. Shown below are two plots of residuals remaining after fitting a linear model to two different sets of data. Describe important features and determine if a linear model would be appropriate for these data. Explain your reasoning.

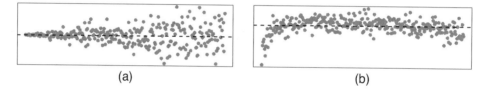

(a)   (b)

8.3 Identify relationships, Part I. For each of the six plots, identify the strength of the relationship (e.g. weak, moderate, or strong) in the data and whether fitting a linear model would be reasonable.

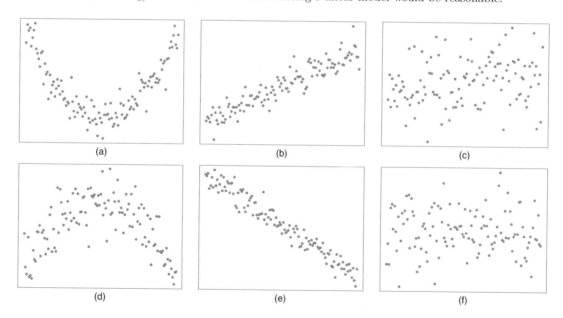

(a)   (b)   (c)

(d)   (e)   (f)

8.4 Identify relationships, Part II. For each of the six plots, identify the strength of the relationship (e.g. weak, moderate, or strong) in the data and whether fitting a linear model would be reasonable.

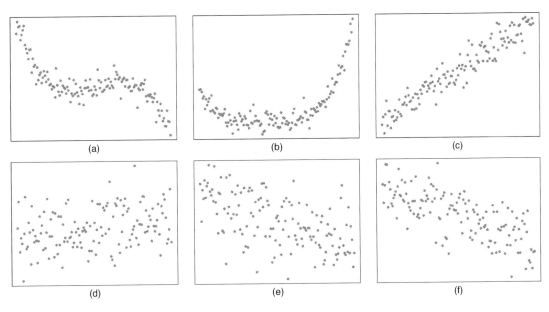

(a)   (b)   (c)

(d)   (e)   (f)

8.5 Exams and grades. The two scatterplots below show the relationship between final and mid-semester exam grades recorded during several years for a Statistics course at a university.

(a) Based on these graphs, which of the two exams has the strongest correlation with the final exam grade? Explain.

(b) Can you think of a reason why the correlation between the exam you chose in part (a) and the final exam is higher?

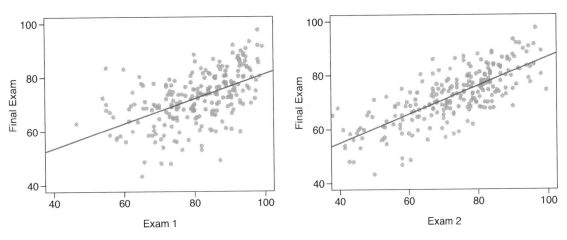

8.6 Husbands and wives, Part I. The Great Britain Office of Population Census and Surveys once collected data on a random sample of 170 married couples in Britain, recording the age (in years) and heights (converted here to inches) of the husbands and wives.[5] The scatterplot on the left shows the wife's age plotted against her husband's age, and the plot on the right shows wife's height plotted against husband's height.

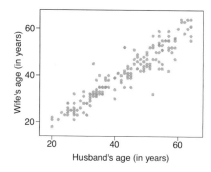 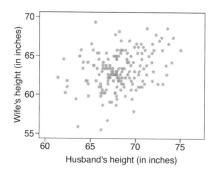

(a) Describe the relationship between husbands' and wives' ages.

(b) Describe the relationship between husbands' and wives' heights.

(c) Which plot shows a stronger correlation? Explain your reasoning.

(d) Data on heights were originally collected in centimeters, and then converted to inches. Does this conversion affect the correlation between husbands' and wives' heights?

8.7 Match the correlation, Part I. Match each correlation to the corresponding scatterplot.

(a) $R = -0.7$
(b) $R = 0.45$
(c) $R = 0.06$
(d) $R = 0.92$

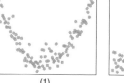 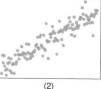 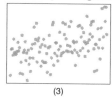 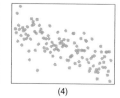

       (1)   (2)   (3)   (4)

8.8 Match the correlation, Part II. Match each correlation to the corresponding scatterplot.

(a) $R = 0.49$
(b) $R = -0.48$
(c) $R = -0.03$
(d) $R = -0.85$

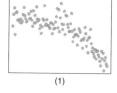 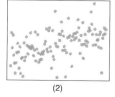 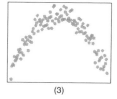 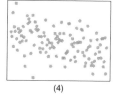

       (1)   (2)   (3)   (4)

8.9 Speed and height. 1,302 UCLA students were asked to fill out a survey where they were asked about their height, fastest speed they have ever driven, and gender. The scatterplot on the left displays the relationship between height and fastest speed, and the scatterplot on the right displays the breakdown by gender in this relationship.

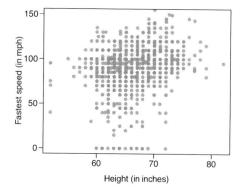 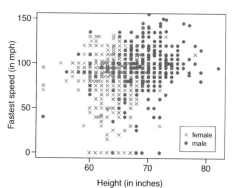

(a) Describe the relationship between height and fastest speed.

(b) Why do you think these variables are positively associated?

(c) What role does gender play in the relationship between height and fastest driving speed?

---

[5]D.J. Hand. *A handbook of small data sets.* Chapman & Hall/CRC, 1994.

8.10 Guess the correlation. Eduardo and Rosie are both collecting data on number of rainy days in a year and the total rainfall for the year. Eduardo records rainfall in inches and Rosie in centimeters. How will their correlation coefficients compare?

8.11 The Coast Starlight, Part I. The Coast Starlight Amtrak train runs from Seattle to Los Angeles. The scatterplot below displays the distance between each stop (in miles) and the amount of time it takes to travel from one stop to another (in minutes).

(a) Describe the relationship between distance and travel time.

(b) How would the relationship change if travel time was instead measured in hours, and distance was instead measured in kilometers?

(c) Correlation between travel time (in miles) and distance (in minutes) is $r = 0.636$. What is the correlation between travel time (in kilometers) and distance (in hours)?

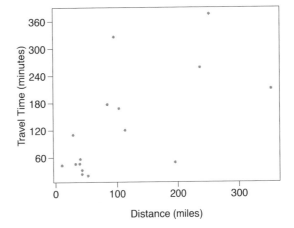

8.12 Crawling babies, Part I. A study conducted at the University of Denver investigated whether babies take longer to learn to crawl in cold months, when they are often bundled in clothes that restrict their movement, than in warmer months.[6] Infants born during the study year were split into twelve groups, one for each birth month. We consider the average crawling age of babies in each group against the average temperature when the babies are six months old (that's when babies often begin trying to crawl). Temperature is measured in degrees Fahrenheit (°F) and age is measured in weeks.

(a) Describe the relationship between temperature and crawling age.

(b) How would the relationship change if temperature was measured in degrees Celsius (°C) and age was measured in months?

(c) The correlation between temperature in °F and age in weeks was $r = -0.70$. If we converted the temperature to °C and age to months, what would the correlation be?

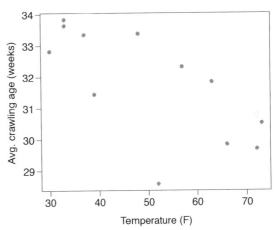

[6] J.B. Benson. "Season of birth and onset of locomotion: Theoretical and methodological implications". In: *Infant behavior and development* 16.1 (1993), pp. 69–81. ISSN: 0163-6383.

8.13 Body measurements, Part I. Researchers studying anthropometry collected body girth measurements and skeletal diameter measurements, as well as age, weight, height and gender for 507 physically active individuals.[7] The scatterplot below shows the relationship between height and shoulder girth (over deltoid muscles), both measured in centimeters.

(a) Describe the relationship between shoulder girth and height.

(b) How would the relationship change if shoulder girth was measured in inches while the units of height remained in centimeters?

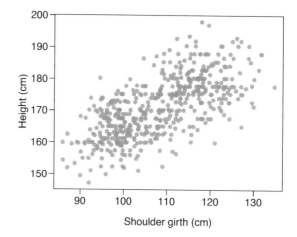

8.14 Body measurements, Part II. The scatterplot below shows the relationship between weight measured in kilograms and hip girth measured in centimeters from the data described in Exercise 8.13.

(a) Describe the relationship between hip girth and weight.

(b) How would the relationship change if weight was measured in pounds while the units for hip girth remained in centimeters?

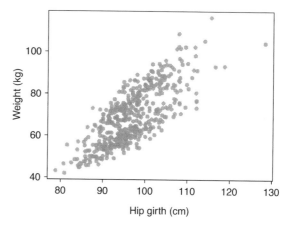

8.15 Correlation, Part I. What would be the correlation between the ages of husbands and wives if men always married woman who were

(a) 3 years younger than themselves?

(b) 2 years older than themselves?

(c) half as old as themselves?

8.16 Correlation, Part II. What would be the correlation between the annual salaries of males and females at a company if for a certain type of position men always made

(a) $5,000 more than women?

(b) 25% more than women?

(c) 15% less than women?

---

[7]G. Heinz et al. "Exploring relationships in body dimensions". In: *Journal of Statistics Education* 11.2 (2003).

8.2 Least squares regression

Fitting linear models by eye is open to criticism since it is based on an individual's preference. In this section, we use *least squares regression* as a more rigorous approach.

8.2.1 Gift aid for freshman at Elmhurst College

This section considers family income and gift aid data from a random sample of fifty students in the freshman class of Elmhurst College in Illinois. Gift aid is financial aid that does not need to be paid back, as opposed to a loan. A scatterplot of the data is shown in Figure 8.11 along with two linear fits. The lines follow a negative trend in the data; students who have higher family incomes tended to have lower gift aid from the university.

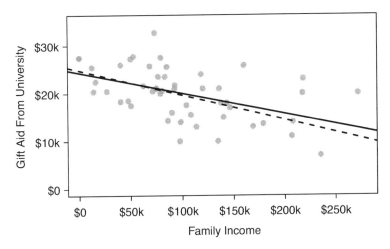

Figure 8.11: Gift aid and family income for a random sample of 50 freshman students from Elmhurst College. Two lines are fit to the data, the solid line being the *least squares line*.

GUIDED PRACTICE 8.7

Is the correlation positive or negative in Figure 8.11?[8]

8.2.2 An objective measure for finding the best line

We begin by thinking about what we mean by "best". Mathematically, we want a line that has small residuals. The first option that may come to mind is to minimize the sum of the residual magnitudes:

$$|e_1| + |e_2| + \cdots + |e_n|$$

which we could accomplish with a computer program. The resulting dashed line shown in Figure 8.11 demonstrates this fit can be quite reasonable. However, a more common practice is to choose the line that minimizes the sum of the squared residuals:

$$e_1^2 + e_2^2 + \cdots + e_n^2$$

---

[8]Larger family incomes are associated with lower amounts of aid, so the correlation will be negative. Using a computer, the correlation can be computed: -0.499.

The line that minimizes this **least squares criterion** is represented as the solid line in Figure 8.11. This is commonly called the **least squares line**. The following are three possible reasons to choose this option instead of trying to minimize the sum of residual magnitudes without any squaring:

1. It is the most commonly used method.

2. Computing the least squares line is widely supported in statistical software.

3. In many applications, a residual twice as large as another residual is more than twice as bad. For example, being off by 4 is usually more than twice as bad as being off by 2. Squaring the residuals accounts for this discrepancy.

The first two reasons are largely for tradition and convenience; the last reason explains why the least squares criterion is typically most helpful.[9]

8.2.3   Conditions for the least squares line

When fitting a least squares line, we generally require

Linearity. The data should show a linear trend. If there is a nonlinear trend (e.g. left panel of Figure 8.12), an advanced regression method from another book or later course should be applied.

Nearly normal residuals. Generally, the residuals must be nearly normal. When this condition is found to be unreasonable, it is usually because of outliers or concerns about influential points, which we'll talk about more in Sections 8.3. An example of a residual that would be a potentially concern is shown in Figure 8.12, where one observation is clearly much further from the regression line than the others.

Constant variability. The variability of points around the least squares line remains roughly constant. An example of non-constant variability is shown in the third panel of Figure 8.12, which represents the most common pattern observed when this condition fails: the variability of y is larger when x is larger.

Independent observations. Be cautious about applying regression to **time series** data, which are sequential observations in time such as a stock price each day. Such data may have an underlying structure that should be considered in a model and analysis. An example of a data set where successive observations are not independent is shown in the fourth panel of Figure 8.12. There are also other instances where correlations within the data are important, which is further discussed in Chapter 9.

GUIDED PRACTICE 8.8

Ⓖ   Should we have concerns about applying least squares regression to the Elmhurst data in Figure 8.11?[10]

---

[9]There are applications where the sum of residual magnitudes may be more useful, and there are plenty of other criteria we might consider. However, this book only applies the least squares criterion.

[10]The trend appears to be linear, the data fall around the line with no obvious outliers, the variance is roughly constant. These are also not time series observations. Least squares regression can be applied to these data.

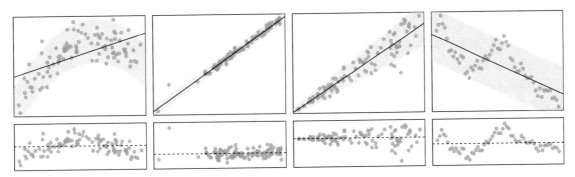

Figure 8.12: Four examples showing when the methods in this chapter are insufficient to apply to the data. First panel: linearity fails. Second panel: there are outliers, most especially one point that is very far away from the line. Third panel: the variability of the errors is related to the value of x. Fourth panel: a time series data set is shown, where successive observations are highly correlated.

8.2.4 Finding the least squares line

For the Elmhurst data, we could write the equation of the least squares regression line as

$$\widehat{aid} = \beta_0 + \beta_1 \times family\_income$$

Here the equation is set up to predict gift aid based on a student's family income, which would be useful to students considering Elmhurst. These two values, β_0 and β_1, are the parameters of the regression line.

As in Chapters 5, 6, and 7, the parameters are estimated using observed data. In practice, this estimation is done using a computer in the same way that other estimates, like a sample mean, can be estimated using a computer or calculator. However, we can also find the parameter estimates by applying two properties of the least squares line:

- The slope of the least squares line can be estimated by

$$b_1 = \frac{s_y}{s_x} R$$

  where R is the correlation between the two variables, and s_x and s_y are the sample standard deviations of the explanatory variable and response, respectively.

- If $\bar{x}$ is the sample mean of the explanatory variable and $\bar{y}$ is the sample mean of the vertical variable, then the point $(\bar{x}, \bar{y})$ is on the least squares line.

  Figure 8.13 shows the sample means for the family income and gift aid as $101,780 and $19,940, respectively. We could plot the point $(101.8, 19.94)$ on Figure 8.11 on page 317 to verify it falls on the least squares line (the solid line).

Next, we formally find the point estimates b_0 and b_1 of the parameters β_0 and β_1.

	Family Income (x)	Gift Aid (y)
mean	$\bar{x} = \$101,780$	$\bar{y} = \$19,940$
sd	$s_x = \$63,200$	$s_y = \$5,460$
		$R = -0.499$

Figure 8.13: Summary statistics for family income and gift aid.

GUIDED PRACTICE 8.9

Using the summary statistics in Figure 8.13, compute the slope for the regression line of gift aid against family income.[11]

You might recall the **point-slope** form of a line from math class, which we can use to find the model fit, including the estimate of b_0. Given the slope of a line and a point on the line, (x_0, y_0), the equation for the line can be written as

$$y - y_0 = slope \times (x - x_0)$$

IDENTIFYING THE LEAST SQUARES LINE FROM SUMMARY STATISTICS

To identify the least squares line from summary statistics:

- Estimate the slope parameter, $b_1 = (s_y/s_x)R$.
- Noting that the point $(\bar{x}, \bar{y})$ is on the least squares line, use $x_0 = \bar{x}$ and $y_0 = \bar{y}$ with the point-slope equation: $y - \bar{y} = b_1(x - \bar{x})$.
- Simplify the equation, which would reveal that $b_0 = \bar{y} - b_1\bar{x}$.

EXAMPLE 8.10

Using the point $(101780, 19940)$ from the sample means and the slope estimate $b_1 = -0.0431$ from Guided Practice 8.9, find the least-squares line for predicting aid based on family income.

Apply the point-slope equation using $(101.78, 19.94)$ and the slope $b_1 = -0.0431$:

$$y - y_0 = b_1(x - x_0)$$
$$y - 19{,}940 = -0.0431(x - 101{,}780)$$

Expanding the right side and then adding 19,940 to each side, the equation simplifies:

$$\widehat{aid} = 24{,}327 - 0.0431 \times family\_income$$

Here we have replaced y with $\widehat{aid}$ and x with *family_income* to put the equation in context. The final equation should always include a "hat" on the variable being predicted, whether it is a generic "y" or a named variable like "*aid*".

A computer is usually used to compute the least squares line, and a summary table generated using software for the Elmhurst regression line is shown in Figure 8.14. The first column of numbers provides estimates for b_0 and b_1, respectively. These results match those from Example 8.10 (with some minor rounding error).

	Estimate	Std. Error	t value	Pr(>|t|)
(Intercept)	24319.3	1291.5	18.83	<0.0001
family_income	-0.0431	0.0108	-3.98	0.0002

Figure 8.14: Summary of least squares fit for the Elmhurst data. Compare the parameter estimates in the first column to the results of Example 8.10.

---

[11]Compute the slope using the summary statistics from Figure 8.13:

$$b_1 = \frac{s_y}{s_x} R = \frac{5{,}460}{63{,}200}(-0.499) = -0.0431$$

EXAMPLE 8.11

Examine the second, third, and fourth columns in Figure 8.14. Can you guess what they represent? (If you have not reviewed any inference chapter yet, skip this example.)

We'll describe the meaning of the columns using the second row, which corresponds to β_1. The first column provides the point estimate for β_1, as we calculated in an earlier example: $b_1 = -0.0431$. The second column is a standard error for this point estimate: $SE_{b_1} = 0.0108$. The third column is a t-test statistic for the null hypothesis that $\beta_1 = 0$: $T = -3.98$. The last column is the p-value for the t-test statistic for the null hypothesis $\beta_1 = 0$ and a two-sided alternative hypothesis: 0.0002. We will get into more of these details in Section 8.4.

EXAMPLE 8.12

Suppose a high school senior is considering Elmhurst College. Can she simply use the linear equation that we have estimated to calculate her financial aid from the university?

She may use it as an estimate, though some qualifiers on this approach are important. First, the data all come from one freshman class, and the way aid is determined by the university may change from year to year. Second, the equation will provide an imperfect estimate. While the linear equation is good at capturing the trend in the data, no individual student's aid will be perfectly predicted.

8.2.5 Interpreting regression model parameter estimates

Interpreting parameters in a regression model is often one of the most important steps in the analysis.

EXAMPLE 8.13

The intercept and slope estimates for the Elmhurst data are $b_0 = 24{,}319$ and $b_1 = -0.0431$. What do these numbers really mean?

Interpreting the slope parameter is helpful in almost any application. For each additional $1,000 of family income, we would expect a student to receive a net difference of $1,000 \times (-0.0431) = -\$43.10$ in aid on average, i.e. $43.10 *less*. Note that a higher family income corresponds to less aid because the coefficient of family income is negative in the model. We must be cautious in this interpretation: while there is a real association, we cannot interpret a causal connection between the variables because these data are observational. That is, increasing a student's family income may not cause the student's aid to drop. (It would be reasonable to contact the college and ask if the relationship is causal, i.e. if Elmhurst College's aid decisions are partially based on students' family income.)

The estimated intercept $b_0 = 24{,}319$ describes the average aid if a student's family had no income. The meaning of the intercept is relevant to this application since the family income for some students at Elmhurst is $0. In other applications, the intercept may have little or no practical value if there are no observations where x is near zero.

INTERPRETING PARAMETERS ESTIMATED BY LEAST SQUARES

The slope describes the estimated difference in the y variable if the explanatory variable x for a case happened to be one unit larger. The intercept describes the average outcome of y if $x = 0$ *and* the linear model is valid all the way to $x = 0$, which in many applications is not the case.

8.2.6 Extrapolation is treacherous

When those blizzards hit the East Coast this winter, it proved to my satisfaction that global warming was a fraud. That snow was freezing cold. But in an alarming trend, temperatures this spring have risen. Consider this: On February 6th it was 10 degrees. Today it hit almost 80. At this rate, by August it will be 220 degrees. So clearly folks the climate debate rages on.

Stephen Colbert
April 6th, 2010[12]

Linear models can be used to approximate the relationship between two variables. However, these models have real limitations. Linear regression is simply a modeling framework. The truth is almost always much more complex than our simple line. For example, we do not know how the data outside of our limited window will behave.

EXAMPLE 8.14

Use the model $\widehat{aid} = 24{,}319 - 0.0431 \times family\_income$ to estimate the aid of another freshman student whose family had income of $1 million.

We want to calculate the aid for $family\_income = 1{,}000{,}000$:

$$24{,}319 - 0.0431 \times family\_income = 24{,}319 - 0.0431 \times 1{,}000{,}000 = -18{,}781$$

The model predicts this student will have -$18,781 in aid (!). However, Elmhurst College does not offer *negative aid* where they select some students to pay extra on top of tuition to attend.

Applying a model estimate to values outside of the realm of the original data is called **extrapolation**. Generally, a linear model is only an approximation of the real relationship between two variables. If we extrapolate, we are making an unreliable bet that the approximate linear relationship will be valid in places where it has not been analyzed.

8.2.7 Using R^2 to describe the strength of a fit

We evaluated the strength of the linear relationship between two variables earlier using the correlation, R. However, it is more common to explain the strength of a linear fit using R^2, called **R-squared**. If provided with a linear model, we might like to describe how closely the data cluster around the linear fit.

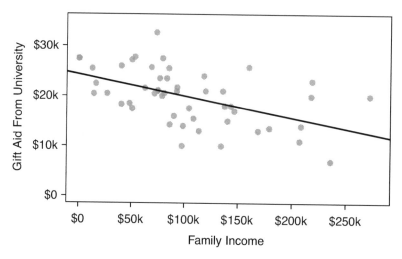

Figure 8.15: Gift aid and family income for a random sample of 50 freshman students from Elmhurst College, shown with the least squares regression line.

[12]www.cc.com/video-clips/l4nkoq

The R^2 of a linear model describes the amount of variation in the response that is explained by the least squares line. For example, consider the Elmhurst data, shown in Figure 8.15. The variance of the response variable, aid received, is about $s_{aid}^2 \approx 29.8$ million. However, if we apply our least squares line, then this model reduces our uncertainty in predicting aid using a student's family income. The variability in the residuals describes how much variation remains after using the model: $s_{RES}^2 \approx 22.4$ million. In short, there was a reduction of

$$\frac{s_{aid}^2 - s_{RES}^2}{s_{aid}^2} = \frac{29{,}800{,}000 - 22{,}400{,}000}{29{,}800{,}000} = \frac{7{,}500{,}000}{29{,}800{,}000} = 0.25$$

or about 25% in the data's variation by using information about family income for predicting aid using a linear model. This corresponds exactly to the R-squared value:

$$R = -0.499 \qquad\qquad R^2 = 0.25$$

GUIDED PRACTICE 8.15

If a linear model has a very strong negative relationship with a correlation of -0.97, how much of the variation in the response is explained by the explanatory variable?[13]

8.2.8 Categorical predictors with two levels

Categorical variables are also useful in predicting outcomes. Here we consider a categorical predictor with two levels (recall that a *level* is the same as a *category*). We'll consider Ebay auctions for a video game, *Mario Kart* for the Nintendo Wii, where both the total price of the auction and the condition of the game were recorded. Here we want to predict total price based on game condition, which takes values used and new. A plot of the auction data is shown in Figure 8.16.

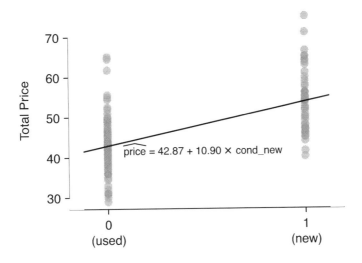

Figure 8.16: Total auction prices for the video game *Mario Kart*, divided into used ($x = 0$) and new ($x = 1$) condition games. The least squares regression line is also shown.

To incorporate the game condition variable into a regression equation, we must convert the categories into a numerical form. We will do so using an **indicator variable** called cond_new, which takes value 1 when the game is new and 0 when the game is used. Using this indicator variable, the linear model may be written as

$$\widehat{price} = \beta_0 + \beta_1 \times \texttt{cond\_new}$$

---

[13] About $R^2 = (-0.97)^2 = 0.94$ or 94% of the variation is explained by the linear model.

	Estimate	Std. Error	t value	Pr(>|t|)
(Intercept)	42.87	0.81	52.67	<0.0001
cond_new	10.90	1.26	8.66	<0.0001

Figure 8.17: Least squares regression summary for the final auction price against the condition of the game.

The parameter estimates are given in Figure 8.17, and the model equation can be summarized as

$$\widehat{price} = 42.87 + 10.90 \times \texttt{cond\_new}$$

For categorical predictors with just two levels, the linearity assumption will always be satisfied. However, we must evaluate whether the residuals in each group are approximately normal and have approximately equal variance. As can be seen in Figure 8.16, both of these conditions are reasonably satisfied by the auction data.

EXAMPLE 8.16

Interpret the two parameters estimated in the model for the price of *Mario Kart* in eBay auctions.

The intercept is the estimated price when `cond_new` takes value 0, i.e. when the game is in used condition. That is, the average selling price of a used version of the game is $42.87.

The slope indicates that, on average, new games sell for about $10.90 more than used games.

INTERPRETING MODEL ESTIMATES FOR CATEGORICAL PREDICTORS

The estimated intercept is the value of the response variable for the first category (i.e. the category corresponding to an indicator value of 0). The estimated slope is the average change in the response variable between the two categories.

We'll elaborate further on this topic in Chapter 9, where we examine the influence of many predictor variables simultaneously using multiple regression.

Exercises

8.17 Units of regression. Consider a regression predicting weight (kg) from height (cm) for a sample of adult males. What are the units of the correlation coefficient, the intercept, and the slope?

8.18 Which is higher? Determine if I or II is higher or if they are equal. Explain your reasoning. For a regression line, the uncertainty associated with the slope estimate, b_1, is higher when

  I. there is a lot of scatter around the regression line or

  II. there is very little scatter around the regression line

8.19 Over-under, Part I. Suppose we fit a regression line to predict the shelf life of an apple based on its weight. For a particular apple, we predict the shelf life to be 4.6 days. The apple's residual is -0.6 days. Did we over or under estimate the shelf-life of the apple? Explain your reasoning.

8.20 Over-under, Part II. Suppose we fit a regression line to predict the number of incidents of skin cancer per 1,000 people from the number of sunny days in a year. For a particular year, we predict the incidence of skin cancer to be 1.5 per 1,000 people, and the residual for this year is 0.5. Did we over or under estimate the incidence of skin cancer? Explain your reasoning.

8.21 Tourism spending. The Association of Turkish Travel Agencies reports the number of foreign tourists visiting Turkey and tourist spending by year.[14] Three plots are provided: scatterplot showing the relationship between these two variables along with the least squares fit, residuals plot, and histogram of residuals.

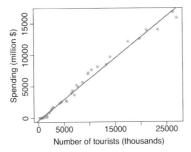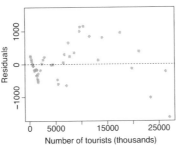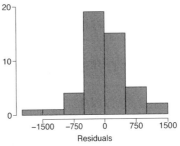

(a) Describe the relationship between number of tourists and spending.

(b) What are the explanatory and response variables?

(c) Why might we want to fit a regression line to these data?

(d) Do the data meet the conditions required for fitting a least squares line? In addition to the scatterplot, use the residual plot and histogram to answer this question.

---

[14]Association of Turkish Travel Agencies, Foreign Visitors Figure & Tourist Spendings By Years.

8.22 Nutrition at Starbucks, Part I. The scatterplot below shows the relationship between the number of calories and amount of carbohydrates (in grams) Starbucks food menu items contain.[15] Since Starbucks only lists the number of calories on the display items, we are interested in predicting the amount of carbs a menu item has based on its calorie content.

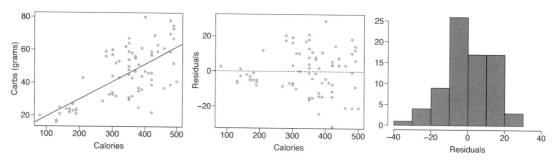

(a) Describe the relationship between number of calories and amount of carbohydrates (in grams) that Starbucks food menu items contain.

(b) In this scenario, what are the explanatory and response variables?

(c) Why might we want to fit a regression line to these data?

(d) Do these data meet the conditions required for fitting a least squares line?

8.23 The Coast Starlight, Part II. Exercise 8.11 introduces data on the Coast Starlight Amtrak train that runs from Seattle to Los Angeles. The mean travel time from one stop to the next on the Coast Starlight is 129 mins, with a standard deviation of 113 minutes. The mean distance traveled from one stop to the next is 108 miles with a standard deviation of 99 miles. The correlation between travel time and distance is 0.636.

(a) Write the equation of the regression line for predicting travel time.

(b) Interpret the slope and the intercept in this context.

(c) Calculate R^2 of the regression line for predicting travel time from distance traveled for the Coast Starlight, and interpret R^2 in the context of the application.

(d) The distance between Santa Barbara and Los Angeles is 103 miles. Use the model to estimate the time it takes for the Starlight to travel between these two cities.

(e) It actually takes the Coast Starlight about 168 mins to travel from Santa Barbara to Los Angeles. Calculate the residual and explain the meaning of this residual value.

(f) Suppose Amtrak is considering adding a stop to the Coast Starlight 500 miles away from Los Angeles. Would it be appropriate to use this linear model to predict the travel time from Los Angeles to this point?

8.24 Body measurements, Part III. Exercise 8.13 introduces data on shoulder girth and height of a group of individuals. The mean shoulder girth is 107.20 cm with a standard deviation of 10.37 cm. The mean height is 171.14 cm with a standard deviation of 9.41 cm. The correlation between height and shoulder girth is 0.67.

(a) Write the equation of the regression line for predicting height.

(b) Interpret the slope and the intercept in this context.

(c) Calculate R^2 of the regression line for predicting height from shoulder girth, and interpret it in the context of the application.

(d) A randomly selected student from your class has a shoulder girth of 100 cm. Predict the height of this student using the model.

(e) The student from part (d) is 160 cm tall. Calculate the residual, and explain what this residual means.

(f) A one year old has a shoulder girth of 56 cm. Would it be appropriate to use this linear model to predict the height of this child?

---

[15]Source: Starbucks.com, collected on March 10, 2011,
www.starbucks.com/menu/nutrition.

8.25 Murders and poverty, Part I. The following regression output is for predicting annual murders per million from percentage living in poverty in a random sample of 20 metropolitan areas.

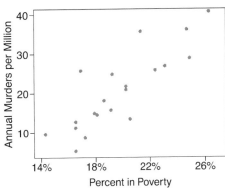

	Estimate	Std. Error	t value	Pr(>|t|)
(Intercept)	-29.901	7.789	-3.839	0.001
poverty%	2.559	0.390	6.562	0.000
$s = 5.512$		$R^2 = 70.52\%$		$R^2_{adj} = 68.89\%$

(a) Write out the linear model.

(b) Interpret the intercept.

(c) Interpret the slope.

(d) Interpret R^2.

(e) Calculate the correlation coefficient.

8.26 Cats, Part I. The following regression output is for predicting the heart weight (in g) of cats from their body weight (in kg). The coefficients are estimated using a dataset of 144 domestic cats.

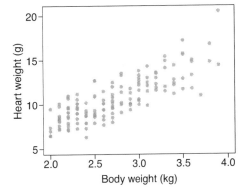

	Estimate	Std. Error	t value	Pr(>|t|)
(Intercept)	-0.357	0.692	-0.515	0.607
body wt	4.034	0.250	16.119	0.000
$s = 1.452$		$R^2 = 64.66\%$		$R^2_{adj} = 64.41\%$

(a) Write out the linear model.

(b) Interpret the intercept.

(c) Interpret the slope.

(d) Interpret R^2.

(e) Calculate the correlation coefficient.

8.3   Types of outliers in linear regression

In this section, we identify criteria for determining which outliers are important and influential. Outliers in regression are observations that fall far from the cloud of points. These points are especially important because they can have a strong influence on the least squares line.

EXAMPLE 8.17

There are six plots shown in Figure 8.18 along with the least squares line and residual plots. For each scatterplot and residual plot pair, identify the outliers and note how they influence the least squares line. Recall that an outlier is any point that doesn't appear to belong with the vast majority of the other points.

(1) There is one outlier far from the other points, though it only appears to slightly influence the line.

(2) There is one outlier on the right, though it is quite close to the least squares line, which suggests it wasn't very influential.

(3) There is one point far away from the cloud, and this outlier appears to pull the least squares line up on the right; examine how the line around the primary cloud doesn't appear to fit very well.

(4) There is a primary cloud and then a small secondary cloud of four outliers. The secondary cloud appears to be influencing the line somewhat strongly, making the least square line fit poorly almost everywhere. There might be an interesting explanation for the dual clouds, which is something that could be investigated.

(5) There is no obvious trend in the main cloud of points and the outlier on the right appears to largely control the slope of the least squares line.

(6) There is one outlier far from the cloud. However, it falls quite close to the least squares line and does not appear to be very influential.

Examine the residual plots in Figure 8.18. You will probably find that there is some trend in the main clouds of (3) and (4). In these cases, the outliers influenced the slope of the least squares lines. In (5), data with no clear trend were assigned a line with a large trend simply due to one outlier (!).

LEVERAGE

Points that fall horizontally away from the center of the cloud tend to pull harder on the line, so we call them points with **high leverage**.

Points that fall horizontally far from the line are points of high leverage; these points can strongly influence the slope of the least squares line. If one of these high leverage points does appear to actually invoke its influence on the slope of the line – as in cases (3), (4), and (5) of Example 8.17 – then we call it an **influential point**. Usually we can say a point is influential if, had we fitted the line without it, the influential point would have been unusually far from the least squares line.

It is tempting to remove outliers. Don't do this without a very good reason. Models that ignore exceptional (and interesting) cases often perform poorly. For instance, if a financial firm ignored the largest market swings – the "outliers" – they would soon go bankrupt by making poorly thought-out investments.

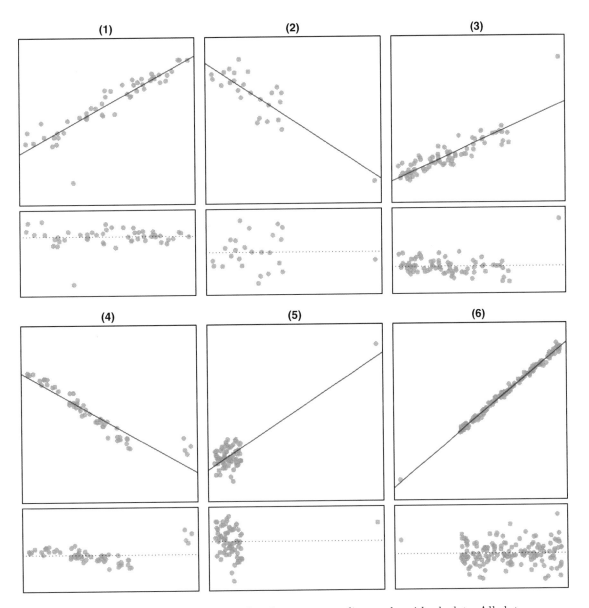

Figure 8.18: Six plots, each with a least squares line and residual plot. All data sets have at least one outlier.

Exercises

8.27 Outliers, Part I. Identify the outliers in the scatterplots shown below, and determine what type of outliers they are. Explain your reasoning.

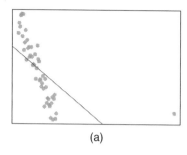   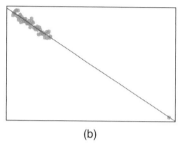   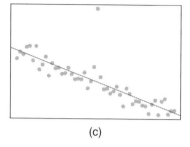

(a)   (b)   (c)

8.28 Outliers, Part II. Identify the outliers in the scatterplots shown below and determine what type of outliers they are. Explain your reasoning.

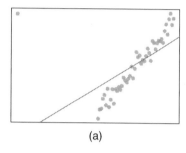   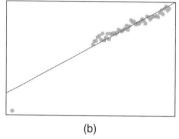   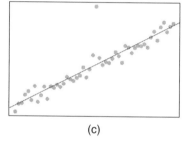

(a)   (b)   (c)

8.29 Urban homeowners, Part I. The scatterplot below shows the percent of families who own their home vs. the percent of the population living in urban areas.[16] There are 52 observations, each corresponding to a state in the US. Puerto Rico and District of Columbia are also included.

(a) Describe the relationship between the percent of families who own their home and the percent of the population living in urban areas.

(b) The outlier at the bottom right corner is District of Columbia, where 100% of the population is considered urban. What type of an outlier is this observation?

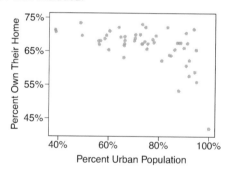

8.30 Crawling babies, Part II. Exercise 8.12 introduces data on the average monthly temperature during the month babies first try to crawl (about 6 months after birth) and the average first crawling age for babies born in a given month. A scatterplot of these two variables reveals a potential outlying month when the average temperature is about 53°F and average crawling age is about 28.5 weeks. Does this point have high leverage? Is it an influential point?

---

[16]United States Census Bureau, 2010 Census Urban and Rural Classification and Urban Area Criteria and Housing Characteristics: 2010.

8.4   Inference for linear regression

In this section, we discuss uncertainty in the estimates of the slope and y-intercept for a regression line. Just as we identified standard errors for point estimates in previous chapters, we first discuss standard errors for these new estimates.

8.4.1   Midterm elections and unemployment

Elections for members of the United States House of Representatives occur every two years, coinciding every four years with the U.S. Presidential election. The set of House elections occurring during the middle of a Presidential term are called midterm elections. In America's two-party system, one political theory suggests the higher the unemployment rate, the worse the President's party will do in the midterm elections.

To assess the validity of this claim, we can compile historical data and look for a connection. We consider every midterm election from 1898 to 2018, with the exception of those elections during the Great Depression. Figure 8.19 shows these data and the least-squares regression line:

$$\% \text{ change in House seats for President's party}$$
$$= -7.36 - 0.89 \times (\text{unemployment rate})$$

We consider the percent change in the number of seats of the President's party (e.g. percent change in the number of seats for Republicans in 2018) against the unemployment rate.

Examining the data, there are no clear deviations from linearity, the constant variance condition, or substantial outliers. While the data are collected sequentially, a separate analysis was used to check for any apparent correlation between successive observations; no such correlation was found.

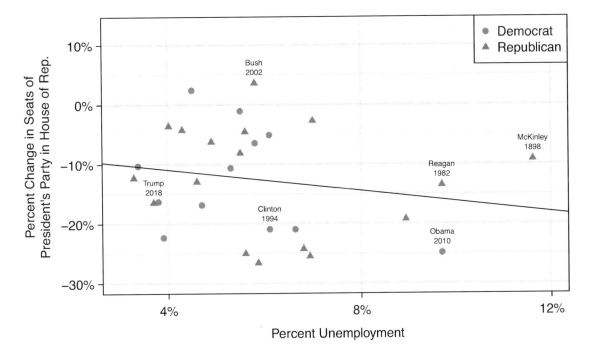

Figure 8.19: The percent change in House seats for the President's party in each election from 1898 to 2010 plotted against the unemployment rate. The two points for the Great Depression have been removed, and a least squares regression line has been fit to the data.

GUIDED PRACTICE 8.18

The data for the Great Depression (1934 and 1938) were removed because the unemployment rate was 21% and 18%, respectively. Do you agree that they should be removed for this investigation? Why or why not?[17]

There is a negative slope in the line shown in Figure 8.19. However, this slope (and the y-intercept) are only estimates of the parameter values. We might wonder, is this convincing evidence that the "true" linear model has a negative slope? That is, do the data provide strong evidence that the political theory is accurate, where the unemployment rate is a useful predictor of the midterm election? We can frame this investigation into a statistical hypothesis test:

H_0: $\beta_1 = 0$. The true linear model has slope zero.

H_A: $\beta_1 \neq 0$. The true linear model has a slope different than zero. The unemployment is predictive of whether the President's party wins or loses seats in the House of Representatives.

We would reject H_0 in favor of H_A if the data provide strong evidence that the true slope parameter is different than zero. To assess the hypotheses, we identify a standard error for the estimate, compute an appropriate test statistic, and identify the p-value.

8.4.2   Understanding regression output from software

Just like other point estimates we have seen before, we can compute a standard error and test statistic for b_1. We will generally label the test statistic using a T, since it follows the t-distribution.

We will rely on statistical software to compute the standard error and leave the explanation of how this standard error is determined to a second or third statistics course. Figure 8.20 shows software output for the least squares regression line in Figure 8.19. The row labeled *unemp* includes the point estimate and other hypothesis test information for the slope, which is the coefficient of the unemployment variable.

	Estimate	Std. Error	t value	Pr($>$|t|)
(Intercept)	-7.3644	5.1553	-1.43	0.1646
unemp	-0.8897	0.8350	-1.07	0.2961
				df = 27

Figure 8.20: Output from statistical software for the regression line modeling the midterm election losses for the President's party as a response to unemployment.

EXAMPLE 8.19

What do the first and second columns of Figure 8.20 represent?

The entries in the first column represent the least squares estimates, b_0 and b_1, and the values in the second column correspond to the standard errors of each estimate. Using the estimates, we could write the equation for the least square regression line as

$$\hat{y} = -7.3644 - 0.8897x$$

where $\hat{y}$ in this case represents the predicted change in the number of seats for the president's party, and x represents the unemployment rate.

---

[17]We will provide two considerations. Each of these points would have very high leverage on any least-squares regression line, and years with such high unemployment may not help us understand what would happen in other years where the unemployment is only modestly high. On the other hand, these are exceptional cases, and we would be discarding important information if we exclude them from a final analysis.

We previously used a t-test statistic for hypothesis testing in the context of numerical data. Regression is very similar. In the hypotheses we consider, the null value for the slope is 0, so we can compute the test statistic using the T (or Z) score formula:

$$T = \frac{\text{estimate} - \text{null value}}{\text{SE}} = \frac{-0.8897 - 0}{0.8350} = -1.07$$

This corresponds to the third column of Figure 8.20.

EXAMPLE 8.20

Use the table in Figure 8.20 to determine the p-value for the hypothesis test.

The last column of the table gives the p-value for the two-sided hypothesis test for the coefficient of the unemployment rate: 0.2961. That is, the data do not provide convincing evidence that a higher unemployment rate has any correspondence with smaller or larger losses for the President's party in the House of Representatives in midterm elections.

INFERENCE FOR REGRESSION

We usually rely on statistical software to identify point estimates, standard errors, test statistics, and p-values in practice. However, be aware that software will not generally check whether the method is appropriate, meaning we must still verify conditions are met.

EXAMPLE 8.21

Examine Figure 8.15 on page 322, which relates the Elmhurst College aid and student family income. How sure are you that the slope is statistically significantly different from zero? That is, do you think a formal hypothesis test would reject the claim that the true slope of the line should be zero?

While the relationship between the variables is not perfect, there is an evident decreasing trend in the data. This suggests the hypothesis test will reject the null claim that the slope is zero.

GUIDED PRACTICE 8.22

Figure 8.21 shows statistical software output from fitting the least squares regression line shown in Figure 8.15. Use this output to formally evaluate the following hypotheses.[18]

H_0: The true coefficient for family income is zero.

H_A: The true coefficient for family income is not zero.

	Estimate	Std. Error	t value	Pr($>$|t|)
(Intercept)	24319.3	1291.5	18.83	$<$0.0001
family_income	-0.0431	0.0108	-3.98	0.0002
				$df = 48$

Figure 8.21: Summary of least squares fit for the Elmhurst College data, where we are predicting the gift aid by the university based on the family income of students.

---

[18]We look in the second row corresponding to the family income variable. We see the point estimate of the slope of the line is -0.0431, the standard error of this estimate is 0.0108, and the t-test statistic is $T = -3.98$. The p-value corresponds exactly to the two-sided test we are interested in: 0.0002. The p-value is so small that we reject the null hypothesis and conclude that family income and financial aid at Elmhurst College for freshman entering in the year 2011 are negatively correlated and the true slope parameter is indeed less than 0, just as we believed in Example 8.21.

8.4.3   Confidence interval for a coefficient

Similar to how we can conduct a hypothesis test for a model coefficient using regression output, we can also construct a confidence interval for that coefficient.

EXAMPLE 8.23

Compute the 95% confidence interval for the `family_income` coefficient using the regression output from Table 8.21.

The point estimate is -0.0431 and the standard error is $SE = 0.0108$. When constructing a confidence interval for a model coefficient, we generally use a t-distribution. The degrees of freedom for the distribution are noted in the regression output, $df = 48$, allowing us to identify $t_{48}^{\star} = 2.01$ for use in the confidence interval.

We can now construct the confidence interval in the usual way:

$$\text{point estimate} \pm t_{48}^{\star} \times SE \quad \rightarrow \quad -0.0431 \pm 2.01 \times 0.0108 \quad \rightarrow \quad (-0.0648, -0.0214)$$

We are 95% confident that with each dollar increase in `family_income`, the university's gift aid is predicted to decrease on average by $0.0214 to $0.0648.

CONFIDENCE INTERVALS FOR COEFFICIENTS

Confidence intervals for model coefficients can be computed using the t-distribution:

$$b_i \; \pm \; t_{df}^{\star} \times SE_{b_i}$$

where $t_{df}^{\star}$ is the appropriate t-value corresponding to the confidence level with the model's degrees of freedom.

On the topic of intervals in this book, we've focused exclusively on confidence intervals for model parameters. However, there are other types of intervals that may be of interest, including prediction intervals for a response value and also confidence intervals for a mean response value in the context of regression. These two interval types are introduced in an online extra that you may download at

www.openintro.org/d?file=stat_extra_linear_regression_supp

Exercises

In the following exercises, visually check the conditions for fitting a least squares regression line. However, you do not need to report these conditions in your solutions.

8.31 Body measurements, Part IV. The scatterplot and least squares summary below show the relationship between weight measured in kilograms and height measured in centimeters of 507 physically active individuals.

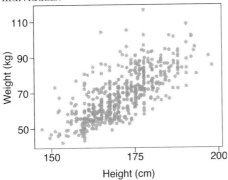

	Estimate	Std. Error	t value	Pr(>|t|)
(Intercept)	-105.0113	7.5394	-13.93	0.0000
height	1.0176	0.0440	23.13	0.0000

(a) Describe the relationship between height and weight.

(b) Write the equation of the regression line. Interpret the slope and intercept in context.

(c) Do the data provide strong evidence that an increase in height is associated with an increase in weight? State the null and alternative hypotheses, report the p-value, and state your conclusion.

(d) The correlation coefficient for height and weight is 0.72. Calculate R^2 and interpret it in context.

8.32 Beer and blood alcohol content. Many people believe that gender, weight, drinking habits, and many other factors are much more important in predicting blood alcohol content (BAC) than simply considering the number of drinks a person consumed. Here we examine data from sixteen student volunteers at Ohio State University who each drank a randomly assigned number of cans of beer. These students were evenly divided between men and women, and they differed in weight and drinking habits. Thirty minutes later, a police officer measured their blood alcohol content (BAC) in grams of alcohol per deciliter of blood.[19] The scatterplot and regression table summarize the findings.

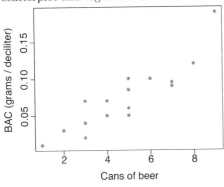

	Estimate	Std. Error	t value	Pr(>|t|)
(Intercept)	-0.0127	0.0126	-1.00	0.3320
beers	0.0180	0.0024	7.48	0.0000

(a) Describe the relationship between the number of cans of beer and BAC.

(b) Write the equation of the regression line. Interpret the slope and intercept in context.

(c) Do the data provide strong evidence that drinking more cans of beer is associated with an increase in blood alcohol? State the null and alternative hypotheses, report the p-value, and state your conclusion.

(d) The correlation coefficient for number of cans of beer and BAC is 0.89. Calculate R^2 and interpret it in context.

(e) Suppose we visit a bar, ask people how many drinks they have had, and also take their BAC. Do you think the relationship between number of drinks and BAC would be as strong as the relationship found in the Ohio State study?

---

[19] J. Malkevitch and L.M. Lesser. *For All Practical Purposes: Mathematical Literacy in Today's World.* WH Freeman & Co, 2008.

8.33 Husbands and wives, Part II. The scatterplot below summarizes husbands' and wives' heights in a random sample of 170 married couples in Britain, where both partners' ages are below 65 years. Summary output of the least squares fit for predicting wife's height from husband's height is also provided in the table.

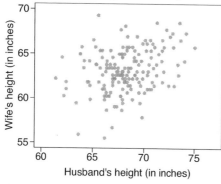

	Estimate	Std. Error	t value	Pr(>|t|)
(Intercept)	43.5755	4.6842	9.30	0.0000
height_husband	0.2863	0.0686	4.17	0.0000

(a) Is there strong evidence that taller men marry taller women? State the hypotheses and include any information used to conduct the test.

(b) Write the equation of the regression line for predicting wife's height from husband's height.

(c) Interpret the slope and intercept in the context of the application.

(d) Given that $R^2 = 0.09$, what is the correlation of heights in this data set?

(e) You meet a married man from Britain who is 5'9" (69 inches). What would you predict his wife's height to be? How reliable is this prediction?

(f) You meet another married man from Britain who is 6'7" (79 inches). Would it be wise to use the same linear model to predict his wife's height? Why or why not?

8.34 Urban homeowners, Part II. Exercise 8.29 gives a scatterplot displaying the relationship between the percent of families that own their home and the percent of the population living in urban areas. Below is a similar scatterplot, excluding District of Columbia, as well as the residuals plot. There were 51 cases.

(a) For these data, $R^2 = 0.28$. What is the correlation? How can you tell if it is positive or negative?

(b) Examine the residual plot. What do you observe? Is a simple least squares fit appropriate for these data?

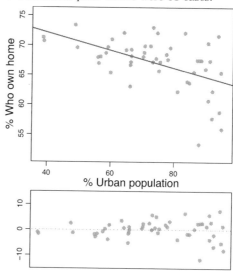

8.35 Murders and poverty, Part II. Exercise 8.25 presents regression output from a model for predicting annual murders per million from percentage living in poverty based on a random sample of 20 metropolitan areas. The model output is also provided below.

	Estimate	Std. Error	t value	Pr(>|t|)
(Intercept)	-29.901	7.789	-3.839	0.001
poverty%	2.559	0.390	6.562	0.000

$$s = 5.512 \qquad R^2 = 70.52\% \qquad R^2_{adj} = 68.89\%$$

(a) What are the hypotheses for evaluating whether poverty percentage is a significant predictor of murder rate?

(b) State the conclusion of the hypothesis test from part (a) in context of the data.

(c) Calculate a 95% confidence interval for the slope of poverty percentage, and interpret it in context of the data.

(d) Do your results from the hypothesis test and the confidence interval agree? Explain.

8.36 Babies. Is the gestational age (time between conception and birth) of a low birth-weight baby useful in predicting head circumference at birth? Twenty-five low birth-weight babies were studied at a Harvard teaching hospital; the investigators calculated the regression of head circumference (measured in centimeters) against gestational age (measured in weeks). The estimated regression line is

$$\widehat{head\ circumference} = 3.91 + 0.78 \times gestational\ age$$

(a) What is the predicted head circumference for a baby whose gestational age is 28 weeks?

(b) The standard error for the coefficient of gestational age is 0. 35, which is associated with $df = 23$. Does the model provide strong evidence that gestational age is significantly associated with head circumference?

Chapter exercises

8.37   True / False. Determine if the following statements are true or false. If false, explain why.

(a) A correlation coefficient of -0.90 indicates a stronger linear relationship than a correlation of 0.5.

(b) Correlation is a measure of the association between any two variables.

8.38   Trees. The scatterplots below show the relationship between height, diameter, and volume of timber in 31 felled black cherry trees. The diameter of the tree is measured 4.5 feet above the ground.[20]

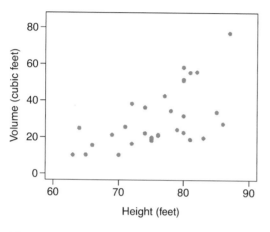
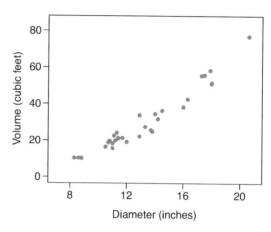

(a) Describe the relationship between volume and height of these trees.

(b) Describe the relationship between volume and diameter of these trees.

(c) Suppose you have height and diameter measurements for another black cherry tree. Which of these variables would be preferable to use to predict the volume of timber in this tree using a simple linear regression model? Explain your reasoning.

8.39   Husbands and wives, Part III. Exercise 8.33 presents a scatterplot displaying the relationship between husbands' and wives' ages in a random sample of 170 married couples in Britain, where both partners' ages are below 65 years. Given below is summary output of the least squares fit for predicting wife's age from husband's age.

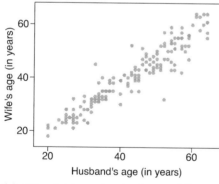

	Estimate	Std. Error	t value	Pr(>|t|)
(Intercept)	1.5740	1.1501	1.37	0.1730
age_husband	0.9112	0.0259	35.25	0.0000
				$df = 168$

(a) We might wonder, is the age difference between husbands and wives consistent across ages? If this were the case, then the slope parameter would be $\beta_1 = 1$. Use the information above to evaluate if there is strong evidence that the difference in husband and wife ages differs for different ages.

(b) Write the equation of the regression line for predicting wife's age from husband's age.

(c) Interpret the slope and intercept in context.

(d) Given that $R^2 = 0.88$, what is the correlation of ages in this data set?

(e) You meet a married man from Britain who is 55 years old. What would you predict his wife's age to be? How reliable is this prediction?

(f) You meet another married man from Britain who is 85 years old. Would it be wise to use the same linear model to predict his wife's age? Explain.

---

[20]Source: R Dataset, stat.ethz.ch/R-manual/R-patched/library/datasets/html/trees.html.

8.40 Cats, Part II. Exercise 8.26 presents regression output from a model for predicting the heart weight (in g) of cats from their body weight (in kg). The coefficients are estimated using a dataset of 144 domestic cat. The model output is also provided below.

	Estimate	Std. Error	t value	$Pr(>|t|)$
(Intercept)	-0.357	0.692	-0.515	0.607
body wt	4.034	0.250	16.119	0.000

$$s = 1.452 \qquad R^2 = 64.66\% \qquad R^2_{adj} = 64.41\%$$

(a) We see that the point estimate for the slope is positive. What are the hypotheses for evaluating whether body weight is positively associated with heart weight in cats?

(b) State the conclusion of the hypothesis test from part (a) in context of the data.

(c) Calculate a 95% confidence interval for the slope of body weight, and interpret it in context of the data.

(d) Do your results from the hypothesis test and the confidence interval agree? Explain.

8.41 Nutrition at Starbucks, Part II. Exercise 8.22 introduced a data set on nutrition information on Starbucks food menu items. Based on the scatterplot and the residual plot provided, describe the relationship between the protein content and calories of these menu items, and determine if a simple linear model is appropriate to predict amount of protein from the number of calories.

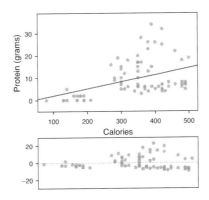

8.42 Helmets and lunches. The scatterplot shows the relationship between socioeconomic status measured as the percentage of children in a neighborhood receiving reduced-fee lunches at school (lunch) and the percentage of bike riders in the neighborhood wearing helmets (helmet). The average percentage of children receiving reduced-fee lunches is 30.8% with a standard deviation of 26.7% and the average percentage of bike riders wearing helmets is 38.8% with a standard deviation of 16.9%.

(a) If the R^2 for the least-squares regression line for these data is 72%, what is the correlation between lunch and helmet?

(b) Calculate the slope and intercept for the least-squares regression line for these data.

(c) Interpret the intercept of the least-squares regression line in the context of the application.

(d) Interpret the slope of the least-squares regression line in the context of the application.

(e) What would the value of the residual be for a neighborhood where 40% of the children receive reduced-fee lunches and 40% of the bike riders wear helmets? Interpret the meaning of this residual in the context of the application.

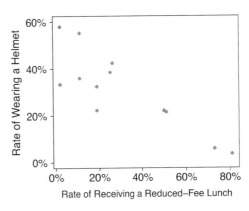

8.43 Match the correlation, Part III. Match each correlation to the corresponding scatterplot.

(a) $r = -0.72$

(b) $r = 0.07$

(c) $r = 0.86$

(d) $r = 0.99$

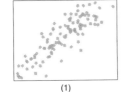
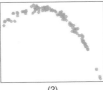
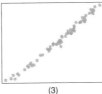
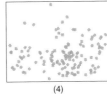

(1)   (2)   (3)   (4)

8.44 Rate my professor. Many college courses conclude by giving students the opportunity to evaluate the course and the instructor anonymously. However, the use of these student evaluations as an indicator of course quality and teaching effectiveness is often criticized because these measures may reflect the influence of non-teaching related characteristics, such as the physical appearance of the instructor. Researchers at University of Texas, Austin collected data on teaching evaluation score (higher score means better) and standardized beauty score (a score of 0 means average, negative score means below average, and a positive score means above average) for a sample of 463 professors.[21] The scatterplot below shows the relationship between these variables, and regression output is provided for predicting teaching evaluation score from beauty score.

	Estimate	Std. Error	t value	Pr(>|t|)
(Intercept)	4.010	0.0255	157.21	0.0000
beauty		0.0322	4.13	0.0000

(a) Given that the average standardized beauty score is -0.0883 and average teaching evaluation score is 3.9983, calculate the slope. Alternatively, the slope may be computed using just the information provided in the model summary table.

(b) Do these data provide convincing evidence that the slope of the relationship between teaching evaluation and beauty is positive? Explain your reasoning.

(c) List the conditions required for linear regression and check if each one is satisfied for this model based on the following diagnostic plots.

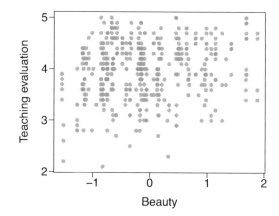

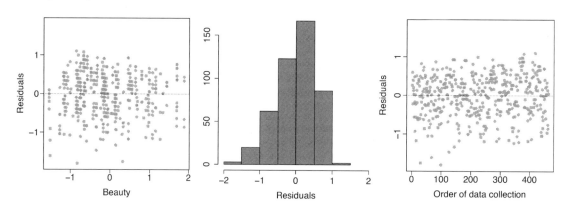

[21]Daniel S Hamermesh and Amy Parker. "Beauty in the classroom: Instructors' pulchritude and putative pedagogical productivity". In: *Economics of Education Review* 24.4 (2005), pp. 369–376.

Chapter 9

Multiple and logistic regression

The principles of simple linear regression lay the foundation for more sophisticated regression models used in a wide range of challenging settings. In Chapter 9, we explore multiple regression, which introduces the possibility of more than one predictor in a linear model, and logistic regression, a technique for predicting categorical outcomes with two levels.

For videos, slides, and other resources, please visit
www.openintro.org/os

9.1 Introduction to multiple regression

Multiple regression extends simple two-variable regression to the case that still has one response but many predictors (denoted x_1, x_2, x_3, ...). The method is motivated by scenarios where many variables may be simultaneously connected to an output.

We will consider data about loans from the peer-to-peer lender, Lending Club, which is a data set we first encountered in Chapters 1 and 2. The loan data includes terms of the loan as well as information about the borrower. The outcome variable we would like to better understand is the interest rate assigned to the loan. For instance, all other characteristics held constant, does it matter how much debt someone already has? Does it matter if their income has been verified? Multiple regression will help us answer these and other questions.

The data set `loans` includes results from 10,000 loans, and we'll be looking at a subset of the available variables, some of which will be new from those we saw in earlier chapters. The first six observations in the data set are shown in Figure 9.1, and descriptions for each variable are shown in Figure 9.2. Notice that the past bankruptcy variable (`bankruptcy`) is an indicator variable, where it takes the value 1 if the borrower had a past bankruptcy in their record and 0 if not. Using an indicator variable in place of a category name allows for these variables to be directly used in regression. Two of the other variables are categorical (`income_ver` and `issued`), each of which can take one of a few different non-numerical values; we'll discuss how these are handled in the model in Section 9.1.1.

	interest_rate	income_ver	debt_to_income	credit_util	bankruptcy	term	issued	credit_checks
1	14.07	verified	18.01	0.55	0	60	Mar2018	6
2	12.61	not	5.04	0.15	1	36	Feb2018	1
3	17.09	source_only	21.15	0.66	0	36	Feb2018	4
4	6.72	not	10.16	0.20	0	36	Jan2018	0
5	14.07	verified	57.96	0.75	0	36	Mar2018	7
6	6.72	not	6.46	0.09	0	36	Jan2018	6
⋮	⋮	⋮	⋮	⋮	⋮	⋮	⋮	⋮

Figure 9.1: First six rows from the `loans` data set.

variable	description
interest_rate	Interest rate for the loan.
income_ver	Categorical variable describing whether the borrower's income source and amount have been verified, with levels verified, source_only, and not.
debt_to_income	Debt-to-income ratio, which is the percentage of total debt of the borrower divided by their total income.
credit_util	Of all the credit available to the borrower, what fraction are they utilizing. For example, the credit utilization on a credit card would be the card's balance divided by the card's credit limit.
bankruptcy	An indicator variable for whether the borrower has a past bankruptcy in her record. This variable takes a value of 1 if the answer is "yes" and 0 if the answer is "no".
term	The length of the loan, in months.
issued	The month and year the loan was issued, which for these loans is always during the first quarter of 2018.
credit_checks	Number of credit checks in the last 12 months. For example, when filing an application for a credit card, it is common for the company receiving the application to run a credit check.

Figure 9.2: Variables and their descriptions for the `loans` data set.

9.1.1 Indicator and categorical variables as predictors

Let's start by fitting a linear regression model for interest rate with a single predictor indicating whether or not a person has a bankruptcy in their record:

$$\widehat{rate} = 12.33 + 0.74 \times bankruptcy$$

Results of this model are shown in Figure 9.3.

	Estimate	Std. Error	t value	Pr(>|t|)
(Intercept)	12.3380	0.0533	231.49	<0.0001
bankruptcy	0.7368	0.1529	4.82	<0.0001
				$df = 9998$

Figure 9.3: Summary of a linear model for predicting interest rate based on whether the borrower has a bankruptcy in their record.

EXAMPLE 9.1

Interpret the coefficient for the past bankruptcy variable in the model. Is this coefficient significantly different from 0?

The `bankruptcy` variable takes one of two values: 1 when the borrower has a bankruptcy in their history and 0 otherwise. A slope of 0.74 means that the model predicts a 0.74% higher interest rate for those borrowers with a bankruptcy in their record. (See Section 8.2.8 for a review of the interpretation for two-level categorical predictor variables.) Examining the regression output in Figure 9.3, we can see that the p-value for `bankruptcy` is very close to zero, indicating there is strong evidence the coefficient is different from zero when using this simple one-predictor model.

Suppose we had fit a model using a 3-level categorical variable, such as `income_ver`. The output from software is shown in Figure 9.4. This regression output provides multiple rows for the `income_ver` variable. Each row represents the relative difference for each level of `income_ver`. However, we are missing one of the levels: `not` (for *not verified*). The missing level is called the **reference level**, and it represents the default level that other levels are measured against.

	Estimate	Std. Error	t value	Pr(>|t|)
(Intercept)	11.0995	0.0809	137.18	<0.0001
income_ver: *source_only*	1.4160	0.1107	12.79	<0.0001
income_ver: *verified*	3.2543	0.1297	25.09	<0.0001
				$df = 9998$

Figure 9.4: Summary of a linear model for predicting interest rate based on whether the borrower's income source and amount has been verified. This predictor has three levels, which results in 2 rows in the regression output.

EXAMPLE 9.2

How would we write an equation for this regression model?

The equation for the regression model may be written as a model with two predictors:

$$\widehat{rate} = 11.10 + 1.42 \times \texttt{income\_ver}_{source\_only} + 3.25 \times \texttt{income\_ver}_{verified}$$

We use the notation $\texttt{variable}_{level}$ to represent indicator variables for when the categorical variable takes a particular value. For example, $\texttt{income\_ver}_{source\_only}$ would take a value of 1 if `income_ver` was `source_only` for a loan, and it would take a value of 0 otherwise. Likewise, $\texttt{income\_ver}_{verified}$ would take a value of 1 if `income_ver` took a value of `verified` and 0 if it took any other value.

The notation used in Example 9.2 may feel a bit confusing. Let's figure out how to use the equation for each level of the `income_ver` variable.

EXAMPLE 9.3

Using the model from Example 9.2, compute the average interest rate for borrowers whose income source and amount are both unverified.

When `income_ver` takes a value of `not`, then both indicator functions in the equation from Example 9.2 are set to zero:

$$\widehat{rate} = 11.10 + 1.42 \times 0 + 3.25 \times 0$$
$$= 11.10$$

The average interest rate for these borrowers is 11.1%. Because the `not` level does not have its own coefficient and it is the reference value, the indicators for the other levels for this variable all drop out.

EXAMPLE 9.4

Using the model from Example 9.2, compute the average interest rate for borrowers whose income source is verified but the amount is not.

When `income_ver` takes a value of `source_only`, then the corresponding variable takes a value of 1 while the other ($income\_ver_{verified}$) is 0:

$$\widehat{rate} = 11.10 + 1.42 \times 1 + 3.25 \times 0$$
$$= 12.52$$

The average interest rate for these borrowers is 12.52%.

GUIDED PRACTICE 9.5

Compute the average interest rate for borrowers whose income source and amount are both verified.[1]

PREDICTORS WITH SEVERAL CATEGORIES

When fitting a regression model with a categorical variable that has k levels where $k > 2$, software will provide a coefficient for $k - 1$ of those levels. For the last level that does not receive a coefficient, this is the **reference level**, and the coefficients listed for the other levels are all considered relative to this reference level.

---

[1] When `income_ver` takes a value of `verified`, then the corresponding variable takes a value of 1 while the other ($income\_ver_{source\_only}$) is 0:

$$\widehat{rate} = 11.10 + 1.42 \times 0 + 3.25 \times 1$$
$$= 14.35$$

The average interest rate for these borrowers is 14.35%.

GUIDED PRACTICE 9.6

Interpret the coefficients in the `income_ver` model.[2]

The higher interest rate for borrowers who have verified their income source or amount is surprising. Intuitively, we'd think that a loan would look *less* risky if the borrower's income has been verified. However, note that the situation may be more complex, and there may be confounding variables that we didn't account for. For example, perhaps lender require borrowers with poor credit to verify their income. That is, verifying income in our data set might be a signal of some concerns about the borrower rather than a reassurance that the borrower will pay back the loan. For this reason, the borrower could be deemed higher risk, resulting in a higher interest rate. (What other confounding variables might explain this counter-intuitive relationship suggested by the model?)

GUIDED PRACTICE 9.7

How much larger of an interest rate would we expect for a borrower who has verified their income source and amount vs a borrower whose income source has only been verified?[3]

9.1.2   Including and assessing many variables in a model

The world is complex, and it can be helpful to consider many factors at once in statistical modeling. For example, we might like to use the full context of borrower to predict the interest rate they receive rather than using a single variable. This is the strategy used in **multiple regression**. While we remain cautious about making any causal interpretations using multiple regression on observational data, such models are a common first step in gaining insights or providing some evidence of a causal connection.

We want to construct a model that accounts for not only for any past bankruptcy or whether the borrower had their income source or amount verified, but simultaneously accounts for all the variables in the data set: `income_ver`, `debt_to_income`, `credit_util`, `bankruptcy`, `term`, `issued`, and `credit_checks`.

$$\widehat{\text{rate}} = \beta_0 + \beta_1 \times \text{income\_ver}_{\text{source\_only}} + \beta_2 \times \text{income\_ver}_{\text{verified}} + \beta_3 \times \text{debt\_to\_income}$$
$$+ \beta_4 \times \text{credit\_util} + \beta_5 \times \text{bankruptcy} + \beta_6 \times \text{term}$$
$$+ \beta_7 \times \text{issued}_{\text{Jan2018}} + \beta_8 \times \text{issued}_{\text{Mar2018}} + \beta_9 \times \text{credit\_checks}$$

This equation represents a holistic approach for modeling all of the variables simultaneously. Notice that there are two coefficients for `income_ver` and also two coefficients for `issued`, since both are 3-level categorical variables.

We estimate the parameters β_0, β_1, β_2, ..., β_9 in the same way as we did in the case of a single predictor. We select b_0, b_1, b_2, ..., b_9 that minimize the sum of the squared residuals:

$$SSE = e_1^2 + e_2^2 + \cdots + e_{10000}^2 = \sum_{i=1}^{10000} e_i^2 = \sum_{i=1}^{10000} (y_i - \hat{y}_i)^2 \tag{9.8}$$

where y_i and $\hat{y}_i$ represent the observed interest rates and their estimated values according to the model, respectively. 10,000 residuals are calculated, one for each observation. We typically use a computer to minimize the sum of squares and compute point estimates, as shown in the sample output in Figure 9.5. Using this output, we identify the point estimates b_i of each β_i, just as we did in the one-predictor case.

---

[2]Each of the coefficients gives the incremental interest rate for the corresponding level relative to the `not` level, which is the reference level. For example, for a borrower whose income source and amount have been verified, the model predicts that they will have a 3.25% higher interest rate than a borrower who has not had their income source or amount verified.

[3]Relative to the `not` category, the `verified` category has an interest rate of 3.25% higher, while the `source_only` category is only 1.42% higher. Thus, `verified` borrowers will tend to get an interest rate about 3.25%−1.42% = 1.83% higher than `source_only` borrowers.

	Estimate	Std. Error	t value	Pr(>|t|)
(Intercept)	1.9251	0.2102	9.16	<0.0001
income_ver: *source_only*	0.9750	0.0991	9.83	<0.0001
income_ver: *verified*	2.5374	0.1172	21.65	<0.0001
debt_to_income	0.0211	0.0029	7.18	<0.0001
credit_util	4.8959	0.1619	30.24	<0.0001
bankruptcy	0.3864	0.1324	2.92	0.0035
term	0.1537	0.0039	38.96	<0.0001
issued: *Jan2018*	0.0276	0.1081	0.26	0.7981
issued: *Mar2018*	-0.0397	0.1065	-0.37	0.7093
credit_checks	0.2282	0.0182	12.51	<0.0001

$$df = 9990$$

Figure 9.5: Output for the regression model, where `interest_rate` is the outcome and the variables listed are the predictors.

MULTIPLE REGRESSION MODEL

A multiple regression model is a linear model with many predictors. In general, we write the model as

$$\hat{y} = \beta_0 + \beta_1 x_1 + \beta_2 x_2 + \cdots + \beta_k x_k$$

when there are k predictors. We always estimate the β_i parameters using statistical software.

EXAMPLE 9.9

Write out the regression model using the point estimates from Figure 9.5. How many predictors are there in this model?

The fitted model for the interest rate is given by:

$$\widehat{\text{rate}} = 1.925 + 0.975 \times \text{income\_ver}_{source\_only} + 2.537 \times \text{income\_ver}_{verified} + 0.021 \times \text{debt\_to\_income}$$
$$+ 4.896 \times \text{credit\_util} + 0.386 \times \text{bankruptcy} + 0.154 \times \text{term}$$
$$+ 0.028 \times \text{issued}_{Jan2018} - 0.040 \times \text{issued}_{Mar2018} + 0.228 \times \text{credit\_checks}$$

If we count up the number of predictor coefficients, we get the *effective* number of predictors in the model: $k = 9$. Notice that the `issued` categorical predictor counts as two, once for the two levels shown in the model. In general, a categorical predictor with p different levels will be represented by $p - 1$ terms in a multiple regression model.

GUIDED PRACTICE 9.10

What does β_4, the coefficient of variable `credit_util`, represent? What is the point estimate of β_4?[4]

---

[4] β_4 represents the change in interest rate we would expect if someone's credit utilization was 0 and went to 1, all other factors held even. The point estimate is $b_4 = 4.90\%$.

EXAMPLE 9.11

Compute the residual of the first observation in Figure 9.1 on page 343 using the equation identified in Guided Practice 9.9.

To compute the residual, we first need the predicted value, which we compute by plugging values into the equation from Example 9.9. For example, $\texttt{income\_ver}_{\texttt{source\_only}}$ takes a value of 0, $\texttt{income\_ver}_{\texttt{verified}}$ takes a value of 1 (since the borrower's income source and amount were verified), $\texttt{debt\_to\_income}$ was 18.01, and so on. This leads to a prediction of $\widehat{rate}_1 = 18.09$. The observed interest rate was 14.07%, which leads to a residual of $e_1 = 14.07 - 18.09 = -4.02$.

EXAMPLE 9.12

We estimated a coefficient for $\texttt{bankruptcy}$ in Section 9.1.1 of $b_4 = 0.74$ with a standard error of $SE_{b_1} = 0.15$ when using simple linear regression. Why is there a difference between that estimate and the estimated coefficient of 0.39 in the multiple regression setting?

If we examined the data carefully, we would see that some predictors are correlated. For instance, when we estimated the connection of the outcome $\texttt{interest\_rate}$ and predictor $\texttt{bankruptcy}$ using simple linear regression, we were unable to control for other variables like whether the borrower had her income verified, the borrower's debt-to-income ratio, and other variables. That original model was constructed in a vacuum and did not consider the full context. When we include all of the variables, underlying and unintentional bias that was missed by these other variables is reduced or eliminated. Of course, bias can still exist from other confounding variables.

Example 9.12 describes a common issue in multiple regression: correlation among predictor variables. We say the two predictor variables are **collinear** (pronounced as *co-linear*) when they are correlated, and this collinearity complicates model estimation. While it is impossible to prevent collinearity from arising in observational data, experiments are usually designed to prevent predictors from being collinear.

GUIDED PRACTICE 9.13

The estimated value of the intercept is 1.925, and one might be tempted to make some interpretation of this coefficient, such as, it is the model's predicted price when each of the variables take value zero: income source is not verified, the borrower has no debt (debt-to-income and credit utilization are zero), and so on. Is this reasonable? Is there any value gained by making this interpretation?[5]

---

[5]Many of the variables do take a value 0 for at least one data point, and for those variables, it is reasonable. However, one variable never takes a value of zero: $\texttt{term}$, which describes the length of the loan, in months. If $\texttt{term}$ is set to zero, then the loan must be paid back immediately; the borrower must give the money back as soon as she receives it, which means it is not a real loan. Ultimately, the interpretation of the intercept in this setting is not insightful.

9.1.3 Adjusted R^2 as a better tool for multiple regression

We first used R^2 in Section 8.2 to determine the amount of variability in the response that was explained by the model:

$$R^2 = 1 - \frac{\text{variability in residuals}}{\text{variability in the outcome}} = 1 - \frac{Var(e_i)}{Var(y_i)}$$

where e_i represents the residuals of the model and y_i the outcomes. This equation remains valid in the multiple regression framework, but a small enhancement can make it even more informative when comparing models.

GUIDED PRACTICE 9.14

The variance of the residuals for the model given in Guided Practice 9.9 is 18.53, and the variance of the total price in all the auctions is 25.01. Calculate R^2 for this model.[6]

This strategy for estimating R^2 is acceptable when there is just a single variable. However, it becomes less helpful when there are many variables. The regular R^2 is a biased estimate of the amount of variability explained by the model when applied to a new sample of data. To get a better estimate, we use the adjusted R^2.

ADJUSTED R^2 AS A TOOL FOR MODEL ASSESSMENT

The **adjusted R^2** is computed as

$$R^2_{adj} = 1 - \frac{s^2_{\text{residuals}}/(n-k-1)}{s^2_{\text{outcome}}/(n-1)} = 1 - \frac{s^2_{\text{residuals}}}{s^2_{\text{outcome}}} \times \frac{n-1}{n-k-1}$$

where n is the number of cases used to fit the model and k is the number of predictor variables in the model. Remember that a categorical predictor with p levels will contribute $p - 1$ to the number of variables in the model.

Because k is never negative, the adjusted R^2 will be smaller – often times just a little smaller – than the unadjusted R^2. The reasoning behind the adjusted R^2 lies in the **degrees of freedom** associated with each variance, which is equal to $n - k - 1$ for the multiple regression context. If we were to make predictions for *new data* using our current model, we would find that the unadjusted R^2 would tend to be slightly overly optimistic, while the adjusted R^2 formula helps correct this bias.

GUIDED PRACTICE 9.15

There were $n = 10000$ auctions in the `loans` data set and $k = 9$ predictor variables in the model. Use n, k, and the variances from Guided Practice 9.14 to calculate R^2_{adj} for the interest rate model.[7]

GUIDED PRACTICE 9.16

Suppose you added another predictor to the model, but the variance of the errors $Var(e_i)$ didn't go down. What would happen to the R^2? What would happen to the adjusted R^2?[8]

Adjusted R^2 could have been used in Chapter 8. However, when there is only $k = 1$ predictors, adjusted R^2 is very close to regular R^2, so this nuance isn't typically important when the model has only one predictor.

---

[6] $R^2 = 1 - \frac{18.53}{25.01} = 0.2591$.

[7] $R^2_{adj} = 1 - \frac{18.53}{25.01} \times \frac{10000-1}{1000-9-1} = 0.2584$. While the difference is very small, it will be important when we fine tune the model in the next section.

[8] The unadjusted R^2 would stay the same and the adjusted R^2 would go down.

Exercises

9.1 Baby weights, Part I. The Child Health and Development Studies investigate a range of topics. One study considered all pregnancies between 1960 and 1967 among women in the Kaiser Foundation Health Plan in the San Francisco East Bay area. Here, we study the relationship between smoking and weight of the baby. The variable `smoke` is coded 1 if the mother is a smoker, and 0 if not. The summary table below shows the results of a linear regression model for predicting the average birth weight of babies, measured in ounces, based on the smoking status of the mother.[9]

	Estimate	Std. Error	t value	Pr($>$|t|)
(Intercept)	123.05	0.65	189.60	0.0000
smoke	-8.94	1.03	-8.65	0.0000

The variability within the smokers and non-smokers are about equal and the distributions are symmetric. With these conditions satisfied, it is reasonable to apply the model. (Note that we don't need to check linearity since the predictor has only two levels.)

(a) Write the equation of the regression model.

(b) Interpret the slope in this context, and calculate the predicted birth weight of babies born to smoker and non-smoker mothers.

(c) Is there a statistically significant relationship between the average birth weight and smoking?

9.2 Baby weights, Part II. Exercise 9.1 introduces a data set on birth weight of babies. Another variable we consider is `parity`, which is 1 if the child is the first born, and 0 otherwise. The summary table below shows the results of a linear regression model for predicting the average birth weight of babies, measured in ounces, from `parity`.

	Estimate	Std. Error	t value	Pr($>$|t|)
(Intercept)	120.07	0.60	199.94	0.0000
parity	-1.93	1.19	-1.62	0.1052

(a) Write the equation of the regression model.

(b) Interpret the slope in this context, and calculate the predicted birth weight of first borns and others.

(c) Is there a statistically significant relationship between the average birth weight and parity?

---

[9]Child Health and Development Studies, Baby weights data set.

9.3 Baby weights, Part III. We considered the variables `smoke` and `parity`, one at a time, in modeling birth weights of babies in Exercises 9.1 and 9.2. A more realistic approach to modeling infant weights is to consider all possibly related variables at once. Other variables of interest include length of pregnancy in days (`gestation`), mother's age in years (`age`), mother's height in inches (`height`), and mother's pregnancy weight in pounds (`weight`). Below are three observations from this data set.

	bwt	gestation	parity	age	height	weight	smoke
1	120	284	0	27	62	100	0
2	113	282	0	33	64	135	0
⋮	⋮	⋮	⋮	⋮	⋮	⋮	⋮
1236	117	297	0	38	65	129	0

The summary table below shows the results of a regression model for predicting the average birth weight of babies based on all of the variables included in the data set.

	Estimate	Std. Error	t value	$Pr(>|t|)$
(Intercept)	-80.41	14.35	-5.60	0.0000
gestation	0.44	0.03	15.26	0.0000
parity	-3.33	1.13	-2.95	0.0033
age	-0.01	0.09	-0.10	0.9170
height	1.15	0.21	5.63	0.0000
weight	0.05	0.03	1.99	0.0471
smoke	-8.40	0.95	-8.81	0.0000

(a) Write the equation of the regression model that includes all of the variables.

(b) Interpret the slopes of `gestation` and `age` in this context.

(c) The coefficient for `parity` is different than in the linear model shown in Exercise 9.2. Why might there be a difference?

(d) Calculate the residual for the first observation in the data set.

(e) The variance of the residuals is 249.28, and the variance of the birth weights of all babies in the data set is 332.57. Calculate the R^2 and the adjusted R^2. Note that there are 1,236 observations in the data set.

9.4 Absenteeism, Part I. Researchers interested in the relationship between absenteeism from school and certain demographic characteristics of children collected data from 146 randomly sampled students in rural New South Wales, Australia, in a particular school year. Below are three observations from this data set.

	eth	sex	lrn	days
1	0	1	1	2
2	0	1	1	11
⋮	⋮	⋮	⋮	⋮
146	1	0	0	37

The summary table below shows the results of a linear regression model for predicting the average number of days absent based on ethnic background (`eth`: 0 - aboriginal, 1 - not aboriginal), sex (`sex`: 0 - female, 1 - male), and learner status (`lrn`: 0 - average learner, 1 - slow learner).[10]

	Estimate	Std. Error	t value	$Pr(>|t|)$
(Intercept)	18.93	2.57	7.37	0.0000
eth	-9.11	2.60	-3.51	0.0000
sex	3.10	2.64	1.18	0.2411
lrn	2.15	2.65	0.81	0.4177

(a) Write the equation of the regression model.

(b) Interpret each one of the slopes in this context.

(c) Calculate the residual for the first observation in the data set: a student who is aboriginal, male, a slow learner, and missed 2 days of school.

(d) The variance of the residuals is 240.57, and the variance of the number of absent days for all students in the data set is 264.17. Calculate the R^2 and the adjusted R^2. Note that there are 146 observations in the data set.

9.5 GPA. A survey of 55 Duke University students asked about their GPA, number of hours they study at night, number of nights they go out, and their gender. Summary output of the regression model is shown below. Note that male is coded as 1.

	Estimate	Std. Error	t value	$Pr(>|t|)$
(Intercept)	3.45	0.35	9.85	0.00
studyweek	0.00	0.00	0.27	0.79
sleepnight	0.01	0.05	0.11	0.91
outnight	0.05	0.05	1.01	0.32
gender	-0.08	0.12	-0.68	0.50

(a) Calculate a 95% confidence interval for the coefficient of gender in the model, and interpret it in the context of the data.

(b) Would you expect a 95% confidence interval for the slope of the remaining variables to include 0? Explain

9.6 Cherry trees. Timber yield is approximately equal to the volume of a tree, however, this value is difficult to measure without first cutting the tree down. Instead, other variables, such as height and diameter, may be used to predict a tree's volume and yield. Researchers wanting to understand the relationship between these variables for black cherry trees collected data from 31 such trees in the Allegheny National Forest, Pennsylvania. Height is measured in feet, diameter in inches (at 54 inches above ground), and volume in cubic feet.[11]

	Estimate	Std. Error	t value	$Pr(>|t|)$
(Intercept)	-57.99	8.64	-6.71	0.00
height	0.34	0.13	2.61	0.01
diameter	4.71	0.26	17.82	0.00

(a) Calculate a 95% confidence interval for the coefficient of height, and interpret it in the context of the data.

(b) One tree in this sample is 79 feet tall, has a diameter of 11.3 inches, and is 24.2 cubic feet in volume. Determine if the model overestimates or underestimates the volume of this tree, and by how much.

---

[10]W. N. Venables and B. D. Ripley. *Modern Applied Statistics with S*. Fourth Edition. Data can also be found in the R MASS package. New York: Springer, 2002.

[11]D.J. Hand. *A handbook of small data sets*. Chapman & Hall/CRC, 1994.

9.2 Model selection

The best model is not always the most complicated. Sometimes including variables that are not evidently important can actually reduce the accuracy of predictions. In this section, we discuss model selection strategies, which will help us eliminate variables from the model that are found to be less important. It's common (and hip, at least in the statistical world) to refer to models that have undergone such variable pruning as **parsimonious**.

In practice, the model that includes all available explanatory variables is often referred to as the **full model**. The full model may not be the best model, and if it isn't, we want to identify a smaller model that is preferable.

9.2.1 Identifying variables in the model that may not be helpful

Adjusted R^2 describes the strength of a model fit, and it is a useful tool for evaluating which predictors are adding value to the model, where *adding value* means they are (likely) improving the accuracy in predicting future outcomes.

Let's consider two models, which are shown in Tables 9.6 and 9.7. The first table summarizes the full model since it includes all predictors, while the second does not include the `issued` variable.

	Estimate	Std. Error	t value	Pr(>|t|)
(Intercept)	1.9251	0.2102	9.16	<0.0001
income_ver: *source_only*	0.9750	0.0991	9.83	<0.0001
income_ver: *verified*	2.5374	0.1172	21.65	<0.0001
debt_to_income	0.0211	0.0029	7.18	<0.0001
credit_util	4.8959	0.1619	30.24	<0.0001
bankruptcy	0.3864	0.1324	2.92	0.0035
term	0.1537	0.0039	38.96	<0.0001
issued: *Jan2018*	0.0276	0.1081	0.26	0.7981
issued: *Mar2018*	-0.0397	0.1065	-0.37	0.7093
credit_checks	0.2282	0.0182	12.51	<0.0001
$R_{adj}^2 = 0.25843$				df = 9990

Figure 9.6: The fit for the full regression model, including the adjusted R^2.

	Estimate	Std. Error	t value	Pr(>|t|)
(Intercept)	1.9213	0.1982	9.69	<0.0001
income_ver: *source_only*	0.9740	0.0991	9.83	<0.0001
income_ver: *verified*	2.5355	0.1172	21.64	<0.0001
debt_to_income	0.0211	0.0029	7.19	<0.0001
credit_util	4.8958	0.1619	30.25	<0.0001
bankruptcy	0.3869	0.1324	2.92	0.0035
term	0.1537	0.0039	38.97	<0.0001
credit_checks	0.2283	0.0182	12.51	<0.0001
$R_{adj}^2 = 0.25854$				df = 9992

Figure 9.7: The fit for the regression model after dropping the `issued` variable.

EXAMPLE 9.17

Which of the two models is better?

We compare the adjusted R^2 of each model to determine which to choose. Since the first model has an R_{adj}^2 smaller than the R_{adj}^2 of the second model, we prefer the second model to the first.

Will the model without `issued` be better than the model with `issued`? We cannot know for sure, but based on the adjusted R^2, this is our best assessment.

9.2.2   Two model selection strategies

Two common strategies for adding or removing variables in a multiple regression model are called *backward elimination* and *forward selection*. These techniques are often referred to as **stepwise** model selection strategies, because they add or delete one variable at a time as they "step" through the candidate predictors.

Backward elimination starts with the model that includes all potential predictor variables. Variables are eliminated one-at-a-time from the model until we cannot improve the adjusted R^2. The strategy within each elimination step is to eliminate the variable that leads to the largest improvement in adjusted R^2.

EXAMPLE 9.18

Results corresponding to the *full model* for the `loans` data are shown in Figure 9.6. How should we proceed under the backward elimination strategy?

Our baseline adjusted R^2 from the full model is $R^2_{adj} = 0.25843$, and we need to determine whether dropping a predictor will improve the adjusted R^2. To check, we fit models that each drop a different predictor, and we record the adjusted R^2:

Exclude ...	income_ver	debt_to_income	credit_util	bankruptcy
	$R^2_{adj} = 0.22380$	$R^2_{adj} = 0.25468$	$R^2_{adj} = 0.19063$	$R^2_{adj} = 0.25787$

	term	issued	credit_checks
	$R^2_{adj} = 0.14581$	$R^2_{adj} = 0.25854$	$R^2_{adj} = 0.24689$

The model without `issued` has the highest adjusted R^2 of 0.25854, higher than the adjusted R^2 for the full model. Because eliminating `issued` leads to a model with a higher adjusted R^2, we drop `issued` from the model.

Since we eliminated a predictor from the model in the first step, we see whether we should eliminate any additional predictors. Our baseline adjusted R^2 is now $R^2_{adj} = 0.25854$. We now fit new models, which consider eliminating each of the remaining predictors in addition to `issued`:

Exclude issued and ...	income_ver	debt_to_income	credit_util
	$R^2_{adj} = 0.22395$	$R^2_{adj} = 0.25479$	$R^2_{adj} = 0.19074$

	bankruptcy	term	credit_checks
	$R^2_{adj} = 0.25798$	$R^2_{adj} = 0.14592$	$R^2_{adj} = 0.24701$

None of these models lead to an improvement in adjusted R^2, so we do not eliminate any of the remaining predictors. That is, after backward elimination, we are left with the model that keeps all predictors except `issued`, which we can summarize using the coefficients from Figure 9.7:

$$\widehat{rate} = 1.921 + 0.974 \times \texttt{income\_ver}_{\text{source\_only}} + 2.535 \times \texttt{income\_ver}_{\text{verified}}$$
$$+ 0.021 \times \texttt{debt\_to\_income} + 4.896 \times \texttt{credit\_util} + 0.387 \times \texttt{bankruptcy}$$
$$+ 0.154 \times \texttt{term} + 0.228 \times \texttt{credit\_check}$$

The **forward selection** strategy is the reverse of the backward elimination technique. Instead of eliminating variables one-at-a-time, we add variables one-at-a-time until we cannot find any variables that improve the model (as measured by adjusted R^2).

EXAMPLE 9.19

Construct a model for the `loans` data set using the forward selection strategy.

We start with the model that includes no variables. Then we fit each of the possible models with just one variable. That is, we fit the model including just `income_ver`, then the model including just `debt_to_income`, then a model with just `credit_util`, and so on. Then we examine the adjusted R^2 for each of these models:

Add ...	income_ver	debt_to_income	credit_util	bankruptcy
	$R^2_{adj} = 0.05926$	$R^2_{adj} = 0.01946$	$R^2_{adj} = 0.06452$	$R^2_{adj} = 0.00222$

	term	issued	credit_checks
	$R^2_{adj} = 0.12855$	$R^2_{adj} = -¡0.00018$	$R^2_{adj} = 0.01711$

In this first step, we compare the adjusted R^2 against a baseline model that has no predictors. The no-predictors model always has $R^2_{adj} = 0$. The model with one predictor that has the largest adjusted R^2 is the model with the `term` predictor, and because this adjusted R^2 is larger than the adjusted R^2 from the model with no predictors ($R^2_{adj} = 0$), we will add this variable to our model.

We repeat the process again, this time considering 2-predictor models where one of the predictors is `term` and with a new baseline of $R^2_{adj} = 0.12855$:

Add term and ...	income_ver	debt_to_income	credit_util
	$R^2_{adj} = 0.16851$	$R^2_{adj} = 0.14368$	$R^2_{adj} = 0.20046$

	bankruptcy	issued	credit_checks
	$R^2_{adj} = 0.13070$	$R^2_{adj} = 0.12840$	$R^2_{adj} = 0.14294$

The best second predictor, `credit_util`, has a higher adjusted R^2 (0.20046) than the baseline (0.12855), so we also add `credit_util` to the model.

Since we have again added a variable to the model, we continue and see whether it would be beneficial to add a third variable:

Add term, credit_util, and ...	income_ver	debt_to_income
	$R^2_{adj} = 0.24183$	$R^2_{adj} = 0.20810$

	bankruptcy	issued	credit_checks
	$R^2_{adj} = 0.20169$	$R^2_{adj} = 0.20031$	$R^2_{adj} = 0.21629$

The model adding `income_ver` improved adjusted R^2 (0.24183 to 0.20046), so we add `income_ver` to the model.

We continue on in this way, next adding `debt_to_income`, then `credit_checks`, and `bankruptcy`. At this point, we come again to the `issued` variable: adding this variable leads to $R^2_{adj} = 0.25843$, while keeping all the other variables but excluding `issued` leads to a higher $R^2_{adj} = 0.25854$. This means we do not add `issued`. In this example, we have arrived at the same model that we identified from backward elimination.

MODEL SELECTION STRATEGIES

Backward elimination begins with the model having the largest number of predictors and eliminates variables one-by-one until we are satisfied that all remaining variables are important to the model. Forward selection starts with no variables included in the model, then it adds in variables according to their importance until no other important variables are found.

Backward elimination and forward selection sometimes arrive at different final models. If trying both techniques and this happens, it's common to choose the model with the larger R^2_{adj}.

9.2.3   The p-value approach, an alternative to adjusted R^2

The p-value may be used as an alternative to R^2_{adj} for model selection:

Backward elimination with the p-value approach. In backward elimination, we would identify the predictor corresponding to the largest p-value. If the p-value is above the significance level, usually $\alpha = 0.05$, then we would drop that variable, refit the model, and repeat the process. If the largest p-value is less than $\alpha = 0.05$, then we would not eliminate any predictors and the current model would be our best-fitting model.

Forward selection with the p-value approach. In forward selection with p-values, we reverse the process. We begin with a model that has no predictors, then we fit a model for each possible predictor, identifying the model where the corresponding predictor's p-value is smallest. If that p-value is smaller than $\alpha = 0.05$, we add it to the model and repeat the process, considering whether to add more variables one-at-a-time. When none of the remaining predictors can be added to the model and have a p-value less than 0.05, then we stop adding variables and the current model would be our best-fitting model.

GUIDED PRACTICE 9.20

Examine Figure 9.7 on page 353, which considers the model including all variables except the variable for the month the loan was issued. If we were using the p-value approach with backward elimination and we were considering this model, which of these variables would be up for elimination? Would we drop that variable, or would we keep it in the model?[12]

While the adjusted R^2 and p-value approaches are similar, they sometimes lead to different models, with the R^2_{adj} approach tending to include more predictors in the final model.

ADJUSTED R^2 VS P-VALUE APPROACH

When the sole goal is to improve prediction accuracy, use R^2_{adj}. This is commonly the case in machine learning applications.

When we care about understanding which variables are statistically significant predictors of the response, or if there is interest in producing a simpler model at the potential cost of a little prediction accuracy, then the p-value approach is preferred.

Regardless of whether you use R^2_{adj} or the p-value approach, or if you use the backward elimination of forward selection strategy, our job is not done after variable selection. We must still verify the model conditions are reasonable.

---

[12]The `bankruptcy` predictor is up for elimination since it has the largest p-value. However, since that p-value is smaller than 0.05, we would still keep it in the model.

Exercises

9.7 Baby weights, Part IV. Exercise 9.3 considers a model that predicts a newborn's weight using several predictors (gestation length, parity, age of mother, height of mother, weight of mother, smoking status of mother). The table below shows the adjusted R-squared for the full model as well as adjusted R-squared values for all models we evaluate in the first step of the backwards elimination process.

	Model	Adjusted R^2
1	Full model	0.2541
2	No gestation	0.1031
3	No parity	0.2492
4	No age	0.2547
5	No height	0.2311
6	No weight	0.2536
7	No smoking status	0.2072

Which, if any, variable should be removed from the model first?

9.8 Absenteeism, Part II. Exercise 9.4 considers a model that predicts the number of days absent using three predictors: ethnic background (`eth`), gender (`sex`), and learner status (`lrn`). The table below shows the adjusted R-squared for the model as well as adjusted R-squared values for all models we evaluate in the first step of the backwards elimination process.

	Model	Adjusted R^2
1	Full model	0.0701
2	No ethnicity	-0.0033
3	No sex	0.0676
4	No learner status	0.0723

Which, if any, variable should be removed from the model first?

9.9 Baby weights, Part V. Exercise 9.3 provides regression output for the full model (including all explanatory variables available in the data set) for predicting birth weight of babies. In this exercise we consider a forward-selection algorithm and add variables to the model one-at-a-time. The table below shows the p-value and adjusted R^2 of each model where we include only the corresponding predictor. Based on this table, which variable should be added to the model first?

variable	gestation	parity	age	height	weight	smoke
p-value	2.2×10^{-16}	0.1052	0.2375	2.97×10^{-12}	8.2×10^{-8}	2.2×10^{-16}
R^2_{adj}	0.1657	0.0013	0.0003	0.0386	0.0229	0.0569

9.10 Absenteeism, Part III. Exercise 9.4 provides regression output for the full model, including all explanatory variables available in the data set, for predicting the number of days absent from school. In this exercise we consider a forward-selection algorithm and add variables to the model one-at-a-time. The table below shows the p-value and adjusted R^2 of each model where we include only the corresponding predictor. Based on this table, which variable should be added to the model first?

variable	ethnicity	sex	learner status
p-value	0.0007	0.3142	0.5870
R^2_{adj}	0.0714	0.0001	0

9.11 Movie lovers, Part I. Suppose a social scientist is interested in studying what makes audiences love or hate a movie. She collects a random sample of movies (genre, length, cast, director, budget, etc.) as well as a measure of the success of the movie (score on a film review aggregator website). If as part of her research she is interested in finding out which variables are significant predictors of movie success, what type of model selection method should she use?

9.12 Movie lovers, Part II. Suppose an online media streaming company is interested in building a movie recommendation system. The website maintains data on the movies in their database (genre, length, cast, director, budget, etc.) and additionally collects data from their subscribers (demographic information, previously watched movies, how they rated previously watched movies, etc.). The recommendation system will be deemed successful if subscribers actually watch, and rate highly, the movies recommended to them. Should the company use the adjusted R^2 or the p-value approach in selecting variables for their recommendation system?

9.3   Checking model conditions using graphs

Multiple regression methods using the model

$$\hat{y} = \beta_0 + \beta_1 x_1 + \beta_2 x_2 + \cdots + \beta_k x_k$$

generally depend on the following four conditions:

1. the residuals of the model are nearly normal (less important for larger data sets),
2. the variability of the residuals is nearly constant,
3. the residuals are independent, and
4. each variable is linearly related to the outcome.

9.3.1   Diagnostic plots

Diagnostic plots can be used to check each of these conditions. We will consider the model from the Lending Club loans data, and check whether there are any notable concerns:

$$\widehat{rate} = 1.921 + 0.974 \times \texttt{income\_ver}_{\texttt{source\_only}} + 2.535 \times \texttt{income\_ver}_{\texttt{verified}}$$
$$+ 0.021 \times \texttt{debt\_to\_income} + 4.896 \times \texttt{credit\_util} + 0.387 \times \texttt{bankruptcy}$$
$$+ 0.154 \times \texttt{term} + 0.228 \times \texttt{credit\_check}$$

Check for outliers. In theory, the distribution of the residuals should be nearly normal; in practice, normality can be relaxed for most applications. Instead, we examine a histogram of the residuals to check if there are any outliers: Figure 9.8 is a histogram of these outliers. Since this is a very large data set, only particularly extreme observations would be a concern in this particular case. There are no extreme observations that might cause a concern.

If we intended to construct what are called **prediction intervals** for future observations, we would be more strict and require the residuals to be nearly normal. Prediction intervals are further discussed in an online extra on the OpenIntro website:

www.openintro.org/d?id=stat_extra_linear_regression_supp

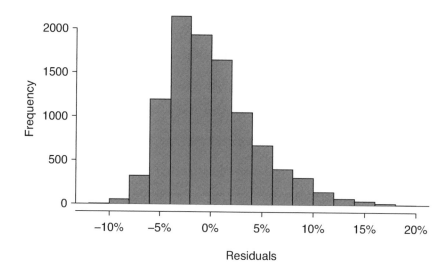

Figure 9.8: A histogram of the residuals.

Absolute values of residuals against fitted values. A plot of the absolute value of the residuals against their corresponding fitted values $(\hat{y}_i)$ is shown in Figure 9.9. This plot is helpful to check the condition that the variance of the residuals is approximately constant, and a smoothed line has been added to represent the approximate trend in this plot. There is more evident variability for fitted values that are larger, which we'll discuss further.

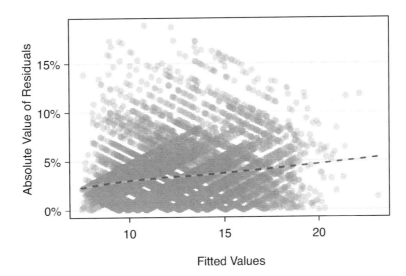

Figure 9.9: Comparing the absolute value of the residuals against the fitted values $(\hat{y}_i)$ is helpful in identifying deviations from the constant variance assumption.

Residuals in order of their data collection. This type of plot can be helpful when observations were collected in a sequence. Such a plot is helpful in identifying any connection between cases that are close to one another. The loans in this data set were issued over a 3 month period, and the month the loan was issued was not found to be important, suggesting this is not a concern for this data set. In cases where a data set does show some pattern for this check, **time series** methods may be useful.

Residuals against each predictor variable. We consider a plot of the residuals against each of the predictors in Figure 9.10. For those instances where there are only 2-3 groups, box plots are shown. For the numerical outcomes, a smoothed line has been fit to the data to make it easier to review. Ultimately, we are looking for any notable change in variability between groups or pattern in the data.

Here are the things of importance from these plots:

- There is some minor differences in variability between the verified income groups.

- There is a very clear pattern for the debt-to-income variable. What also stands out is that this variable is very strongly right skewed: there are few observations with very high debt-to-income ratios.

- The downward curve on the right side of the credit utilization and credit check plots suggests some minor misfitting for those larger values.

Having reviewed the diagnostic plots, there are two options. The first option is to, if we're not concerned about the issues observed, use this as the final model; if going this route, it is important to still note any abnormalities observed in the diagnostics. The second option is to try to improve the model, which is what we'll try to do with this particular model fit.

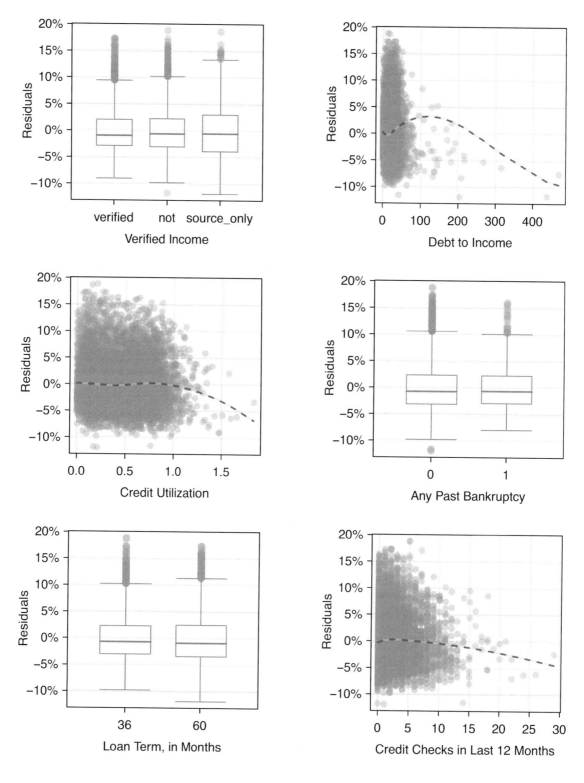

Figure 9.10: Diagnostic plots for residuals against each of the predictors. For the box plots, we're looking for notable differences in variability. For numerical predictors, we also check for trends or other structure in the data.

9.3.2   Options for improving the model fit

There are several options for improvement of a model, including transforming variables, seeking out additional variables to fill model gaps, or using more advanced methods that would account for challenges around inconsistent variability or nonlinear relationships between predictors and the outcome.

The main concern for the initial model is that there is a notable nonlinear relationship between the debt-to-income variable observed in Figure 9.10. To resolve this issue, we're going to consider a couple strategies for adjusting the relationship between the predictor variable and the outcome.

Let's start by taking a look at a histogram of debt_to_income in Figure 9.11. The variable is extremely skewed, and upper values will have a lot of leverage on the fit. Below are several options:

- log transformation $(\log x)$,

- square root transformation $(\sqrt{x})$,

- inverse transformation $(1/x)$,

- truncation (cap the max value possible)

If we inspected the data more closely, we'd observe some instances where the variable takes a value of 0, and since $\log(0)$ and $1/x$ are undefined when $x = 0$, we'll exclude these transformations from further consideration.[13] A square root transformation is valid for all values the variable takes, and truncating some of the larger observations is also a valid approach. We'll consider both of these approaches.

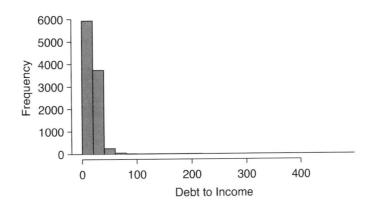

Figure 9.11: Histogram of debt_to_income, where extreme skew is evident.

To try transforming the variable, we make two new variables representing the transformed versions:

Square root. We create a new variable, sqrt_debt_to_income, where all the values are simply the square roots of the values in debt_to_income, and then refit the model as before. The result is shown in the left panel of Figure 9.12. The square root pulled in the higher values a bit, but the fit still doesn't look great since the smoothed line is still wavy.

Truncate at 50. We create a new variable, debt_to_income_50, where any values in debt_to_income that are greater than 50 are shrunk to exactly 50. Refitting the model once more, the diagnostic plot for this new variable is shown in the right panel of Figure 9.12. Here the fit looks much more reasonable, so this appears to be a reasonable approach.

The downside of using transformations is that it reduces the ease of interpreting the results. Fortunately, since the truncation transformation only affects a relatively small number of cases, the interpretation isn't dramatically impacted.

---

[13]There are ways to make them work, but we'll leave those options to a later course.

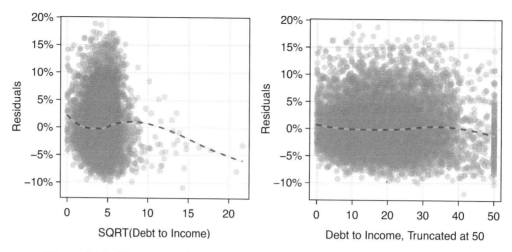

Figure 9.12: Histogram of `debt_to_income`, where extreme skew is evident.

As a next step, we'd evaluate the new model using the truncated version of `debt_to_income`, we would complete all the same procedures as before. The other two issues noted while inspecting diagnostics in Section 9.3.1 are still present in the updated model. If we choose to report this model, we would want to also discuss these shortcomings to be transparent in our work. Depending on what the model will be used, we could either try to bring those under control, or we could stop since those issues aren't severe. Had the non-constant variance been a little more dramatic, it would be a higher priority. Ultimately we decided that the model was reasonable, and we report its final form here:

$$\widehat{rate} = 1.562 + 1.002 \times \texttt{income\_ver}_{\texttt{source\_only}} + 2.436 \times \texttt{income\_ver}_{\texttt{verified}}$$
$$+ 0.048 \times \texttt{debt\_to\_income\_50} + 4.694 \times \texttt{credit\_util} + 0.394 \times \texttt{bankruptcy}$$
$$+ 0.153 \times \texttt{term} + 0.223 \times \texttt{credit\_check}$$

A sharp eye would notice that the coefficient for `debt_to_income_50` is more than twice as large as what the coefficient had been for the `debt_to_income` variable in the earlier model. This suggests those larger values not only were points with high leverage, but they were influential points that were dramatically impacting the coefficient.

"ALL MODELS ARE WRONG, BUT SOME ARE USEFUL" -GEORGE E.P. BOX

The truth is that no model is perfect. However, even imperfect models can be useful. Reporting a flawed model can be reasonable so long as we are clear and report the model's shortcomings.

Don't report results when conditions are grossly violated. While there is a little leeway in model conditions, don't go too far. If model conditions are very clearly violated, consider a new model, even if it means learning more statistical methods or hiring someone who can help. To help you get started, we've developed a couple additional sections that you may find on OpenIntro's website. These sections provide a light introduction to what are called **interaction terms** and to fitting **nonlinear curves** to data, respectively:

www.openintro.org/d?file=stat_extra_interaction_effects

www.openintro.org/d?file=stat_extra_nonlinear_relationships

Exercises

9.13 Baby weights, Part VI. Exercise 9.3 presents a regression model for predicting the average birth weight of babies based on length of gestation, parity, height, weight, and smoking status of the mother. Determine if the model assumptions are met using the plots below. If not, describe how to proceed with the analysis.

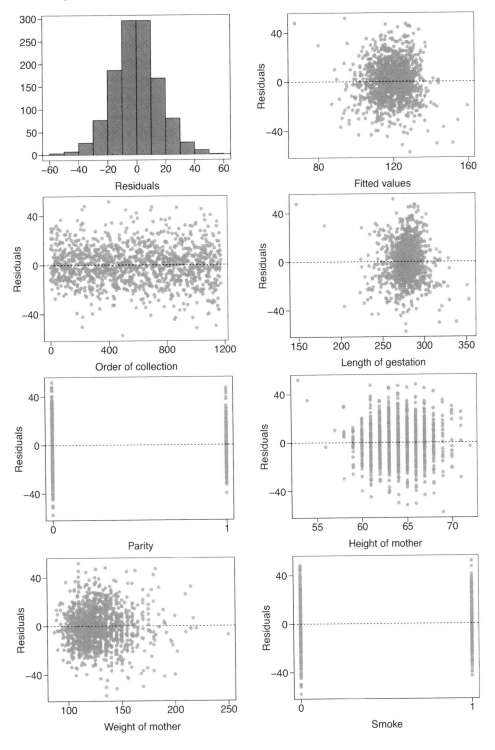

9.14   Movie returns, Part I. A FiveThirtyEight.com article reports that "Horror movies get nowhere near as much draw at the box office as the big-time summer blockbusters or action/adventure movies ... but there's a huge incentive for studios to continue pushing them out. The return-on-investment potential for horror movies is absurd." To investigate how the return-on-investment compares between genres and how this relationship has changed over time, an introductory statistics student fit a model predicting the ratio of gross revenue of movies from genre and release year for 1,070 movies released between 2000 and 2018. Using the plots given below, determine if this regression model is appropriate for these data.[14]

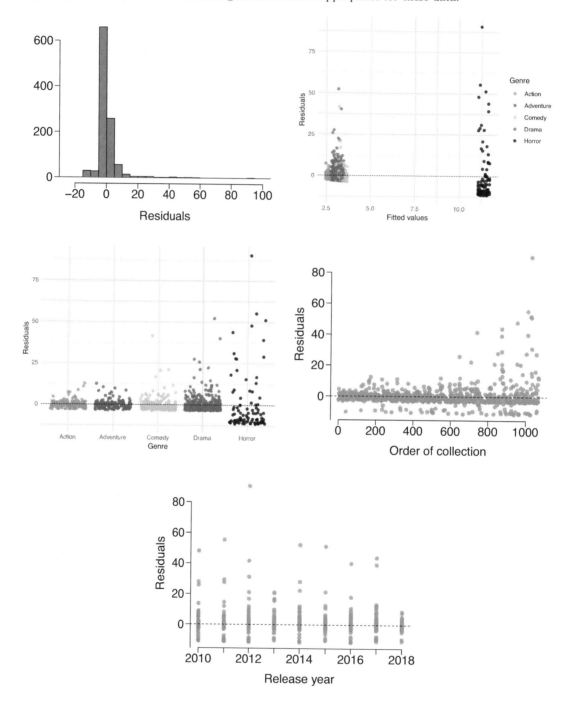

[14]FiveThirtyEight, Scary Movies Are The Best Investment In Hollywood.

9.4 Multiple regression case study: Mario Kart

We'll consider Ebay auctions of a video game called *Mario Kart* for the Nintendo Wii. The outcome variable of interest is the total price of an auction, which is the highest bid plus the shipping cost. We will try to determine how total price is related to each characteristic in an auction while simultaneously controlling for other variables. For instance, all other characteristics held constant, are longer auctions associated with higher or lower prices? And, on average, how much more do buyers tend to pay for additional Wii wheels (plastic steering wheels that attach to the Wii controller) in auctions? Multiple regression will help us answer these and other questions.

9.4.1 Data set and the full model

The `mariokart` data set includes results from 141 auctions. Four observations from this data set are shown in Figure 9.13, and descriptions for each variable are shown in Figure 9.14. Notice that the condition and stock photo variables are indicator variables, similar to `bankruptcy` in the `loan` data set.

	price	cond_new	stock_photo	duration	wheels
1	51.55	1	1	3	1
2	37.04	0	1	7	1
⋮	⋮	⋮	⋮	⋮	⋮
140	38.76	0	0	7	0
141	54.51	1	1	1	2

Figure 9.13: Four observations from the `mariokart` data set.

variable	description
price	Final auction price plus shipping costs, in US dollars.
cond_new	Indicator variable for if the game is new (1) or used (0).
stock_photo	Indicator variable for if the auction's main photo is a stock photo.
duration	The length of the auction, in days, taking values from 1 to 10.
wheels	The number of Wii wheels included with the auction. A *Wii wheel* is an optional steering wheel accessory that holds the Wii controller.

Figure 9.14: Variables and their descriptions for the `mariokart` data set.

GUIDED PRACTICE 9.21

We fit a linear regression model with the game's condition as a predictor of auction price. Results of this model are summarized below:

	Estimate	Std. Error	t value	Pr(>|t|)
(Intercept)	42.8711	0.8140	52.67	<0.0001
cond_new	10.8996	1.2583	8.66	<0.0001
				df = 139

Write down the equation for the model, note whether the slope is statistically different from zero, and interpret the coefficient.[15]

Sometimes there are underlying structures or relationships between predictor variables. For instance, new games sold on Ebay tend to come with more Wii wheels, which may have led to higher prices for those auctions. We would like to fit a model that includes all potentially important variables simultaneously. This would help us evaluate the relationship between a predictor variable and the outcome while controlling for the potential influence of other variables.

We want to construct a model that accounts for not only the game condition, as in Guided Practice 9.21, but simultaneously accounts for three other variables:

$$\widehat{price} = \beta_0 + \beta_1 \times \texttt{cond\_new} + \beta_2 \times \texttt{stock\_photo}$$
$$+ \beta_3 \times \texttt{duration} + \beta_4 \times \texttt{wheels}$$

Figure 9.15 summarizes the full model. Using this output, we identify the point estimates of each coefficient.

	Estimate	Std. Error	t value	Pr(>|t|)
(Intercept)	36.2110	1.5140	23.92	<0.0001
cond_new	5.1306	1.0511	4.88	<0.0001
stock_photo	1.0803	1.0568	1.02	0.3085
duration	-0.0268	0.1904	-0.14	0.8882
wheels	7.2852	0.5547	13.13	<0.0001
				df = 136

Figure 9.15: Output for the regression model where `price` is the outcome and `cond_new`, `stock_photo`, `duration`, and `wheels` are the predictors.

GUIDED PRACTICE 9.22

Write out the model's equation using the point estimates from Figure 9.15. How many predictors are there in this model?[16]

GUIDED PRACTICE 9.23

What does β_4, the coefficient of variable x_4 (Wii wheels), represent? What is the point estimate of β_4?[17]

---

[15]The equation for the line may be written as

$$\widehat{price} = 42.87 + 10.90 \times cond\_new$$

Examining the regression output in Guided Practice 9.21, we can see that the p-value for `cond_new` is very close to zero, indicating there is strong evidence that the coefficient is different from zero when using this simple one-variable model.

The `cond_new` is a two-level categorical variable that takes value 1 when the game is new and value 0 when the game is used. This means the 10.90 model coefficient predicts an extra $10.90 for those games that are new versus those that are used.

[16]$\widehat{price} = 36.21 + 5.13 \times \texttt{cond\_new} + 1.08 \times \texttt{stock\_photo} - 0.03 \times \texttt{duration} + 7.29 \times \texttt{wheels}$, with the $k = 4$ predictors.

[17]It is the average difference in auction price for each additional Wii wheel included when holding the other variables constant. The point estimate is $b_4 = 7.29$.

GUIDED PRACTICE 9.24

Compute the residual of the first observation in Figure 9.13 using the equation identified in Guided Practice 9.22.[18]

EXAMPLE 9.25

We estimated a coefficient for `cond_new` in Section 9.21 of $b_1 = 10.90$ with a standard error of $SE_{b_1} = 1.26$ when using simple linear regression. Why might there be a difference between that estimate and the one in the multiple regression setting?

If we examined the data carefully, we would see that there is collinearity among some predictors. For instance, when we estimated the connection of the outcome `price` and predictor `cond_new` using simple linear regression, we were unable to control for other variables like the number of Wii wheels included in the auction. That model was biased by the confounding variable `wheels`. When we use both variables, this particular underlying and unintentional bias is reduced or eliminated (though bias from other confounding variables may still remain).

9.4.2 Model selection

Let's revisit the model for the Mario Kart auction and complete model selection using backwards selection. Recall that the full model took the following form:

$$\widehat{price} = 36.21 + 5.13 \times \texttt{cond\_new} + 1.08 \times \texttt{stock\_photo} - 0.03 \times \texttt{duration} + 7.29 \times \texttt{wheels}$$

EXAMPLE 9.26

Results corresponding to the full model for the `mariokart` data were shown in Figure 9.15 on the facing page. For this model, we consider what would happen if dropping each of the variables in the model:

Exclude ...	cond_new	stock_photo	duration	wheels
	$R^2_{adj} = 0.6626$	$R^2_{adj} = 0.7107$	$R^2_{adj} = 0.7128$	$R^2_{adj} = 0.3487$

For the full model, $R^2_{adj} = 0.7108$. How should we proceed under the backward elimination strategy?

The third model without `duration` has the highest R^2_{adj} of 0.7128, so we compare it to R^2_{adj} for the full model. Because eliminating `duration` leads to a model with a higher R^2_{adj}, we drop `duration` from the model.

GUIDED PRACTICE 9.27

In Example 9.26, we eliminated the `duration` variable, which resulted in a model with $R^2_{adj} = 0.7128$. Let's look at if we would eliminate another variable from the model using backwards elimination:

Exclude duration and ...	cond_new	stock_photo	wheels
	$R^2_{adj} = 0.6587$	$R^2_{adj} = 0.7124$	$R^2_{adj} = 0.3414$

Should we eliminate any additional variable, and if so, which variable should we eliminate?[19]

---

[18]$e_i = y_i - \hat{y}_i = 51.55 - 49.62 = 1.93$, where 49.62 was computed using the variables values from the observation and the equation identified in Guided Practice 9.22.

[19]Removing any of the three remaining variables would lead to a decrease in R^2_{adj}, so we should not remove any additional variables from the model after we removed `duration`.

368   CHAPTER 9. MULTIPLE AND LOGISTIC REGRESSION

GUIDED PRACTICE 9.28

After eliminating the auction's duration from the model, we are left with the following reduced model:

$$\widehat{price} = 36.05 + 5.18 \times \texttt{cond\_new} + 1.12 \times \texttt{stock\_photo} + 7.30 \times \texttt{wheels}$$

How much would you predict for the total price for the Mario Kart game if it was used, used a stock photo, and included two wheels and put up for auction during the time period that the Mario Kart data were collected?[20]

GUIDED PRACTICE 9.29

Would you be surprised if the seller from Guided Practice 9.28 didn't get the exact price predicted?[21]

9.4.3 Checking model conditions using graphs

Let's take a closer look at the diagnostics for the Mario Kart model to check if the model we have identified is reasonable.

Check for outliers. A histogram of the residuals is shown in Figure 9.16. With a data set well over a hundred, we're primarily looking for major outliers. While one minor outlier appears on the upper end, it is not a concern for this large of a data set.

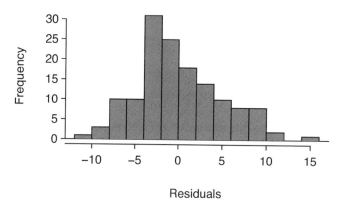

Figure 9.16: Histogram of the residuals. No clear outliers are evident.

Absolute values of residuals against fitted values. A plot of the absolute value of the residuals against their corresponding fitted values ($\hat{y}_i$) is shown in Figure 9.17. We don't see any obvious deviations from constant variance in this example.

Residuals in order of their data collection. A plot of the residuals in the order their corresponding auctions were observed is shown in Figure 9.18. Here we see no structure that indicates a problem.

Residuals against each predictor variable. We consider a plot of the residuals against the `cond_new` variable, the residuals against the `stock_photo` variable, and the residuals against the `wheels` variable. These plots are shown in Figure 9.19. For the two-level condition variable, we are guaranteed not to see any remaining trend, and instead we are checking that the variability doesn't fluctuate across groups, which it does not. However, looking at the stock

---

[20]We would plug in 0 for `cond_new` 1 for `stock_photo`, and 2 for `wheels` into the equation, which would return $51.77, which is the total price we would expect for the auction.
[21]No. The model provides the *average* auction price we would expect, and the price for one auction to the next will continue to vary a bit (but less than what our prediction would be without the model).

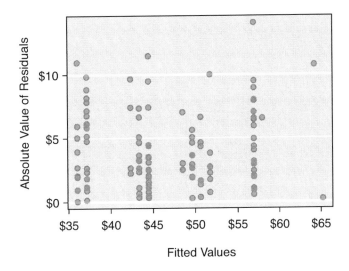

Figure 9.17: Absolute value of the residuals against the fitted values. No patterns are evident.

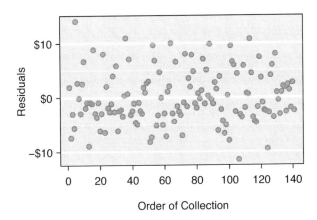

Figure 9.18: Residuals in the order that their corresponding observations were collected. There are no evident patterns.

photo variable, we find that there is some difference in the variability of the residuals in the two groups. Additionally, when we consider the residuals against the wheels variable, we see some possible structure. There appears to be curvature in the residuals, indicating the relationship is probably not linear.

As with the loans analysis, we would summarize diagnostics when reporting the model results. In the case of this auction data, we would report that there appears to be non-constant variance in the stock photo variable and that there may be a nonlinear relationship between the total price and the number of wheels included for an auction. This information would be important to buyers and sellers who may review the analysis, and omitting this information could be a setback to the very people who the model might assist.

Note: there are no exercises for this section.

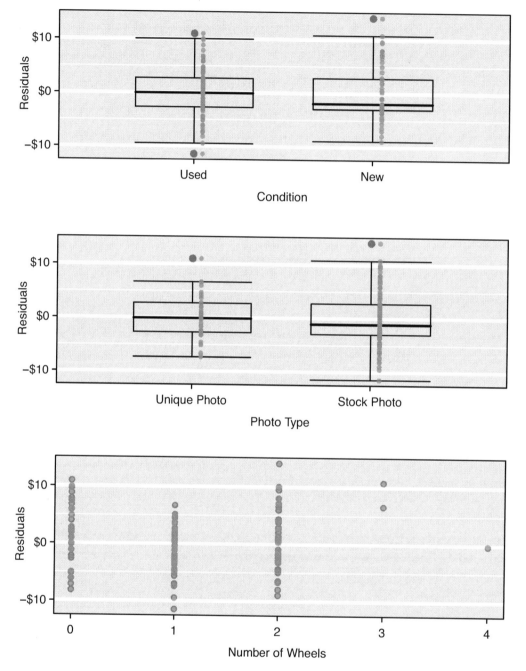

Figure 9.19: For the condition and stock photo variables, we check for differences in the distribution shape or variability of the residuals. In the case of the stock photos variable, we see a little less variability in the unique photo group than the stock photo group. For numerical predictors, we also check for trends or other structure. We see some slight bowing in the residuals against the `wheels` variable in the bottom plot.

9.5   Introduction to logistic regression

In this section we introduce **logistic regression** as a tool for building models when there is a categorical response variable with two levels, e.g. yes and no. Logistic regression is a type of **generalized linear model (GLM)** for response variables where regular multiple regression does not work very well. In particular, the response variable in these settings often takes a form where residuals look completely different from the normal distribution.

GLMs can be thought of as a two-stage modeling approach. We first model the response variable using a probability distribution, such as the binomial or Poisson distribution. Second, we model the parameter of the distribution using a collection of predictors and a special form of multiple regression. Ultimately, the application of a GLM will feel very similar to multiple regression, even if some of the details are different.

9.5.1   Resume data

We will consider experiment data from a study that sought to understand the effect of race and sex on job application callback rates; details of the study and a link to the data set may be found in Appendix B.9. To evaluate which factors were important, job postings were identified in Boston and Chicago for the study, and researchers created many fake resumes to send off to these jobs to see which would elicit a callback. The researchers enumerated important characteristics, such as years of experience and education details, and they used these characteristics to randomly generate the resumes. Finally, they randomly assigned a name to each resume, where the name would imply the applicant's sex and race.

The first names that were used and randomly assigned in this experiment were selected so that they would predominantly be recognized as belonging to Black or White individuals; other races were not considered in this study. While no name would definitively be inferred as pertaining to a Black individual or to a White individual, the researchers conducted a survey to check for racial association of the names; names that did not pass this survey check were excluded from usage in the experiment. You can find the full set of names that did pass the survey test and were ultimately used in the study in Figure 9.20. For example, Lakisha was a name that their survey indicated would be interpreted as a Black woman, while Greg was a name that would generally be interpreted to be associated with a White male.

first_name	race	sex	first_name	race	sex	first_name	race	sex
Aisha	black	female	Hakim	black	male	Laurie	white	female
Allison	white	female	Jamal	black	male	Leroy	black	male
Anne	white	female	Jay	white	male	Matthew	white	male
Brad	white	male	Jermaine	black	male	Meredith	white	female
Brendan	white	male	Jill	white	female	Neil	white	male
Brett	white	male	Kareem	black	male	Rasheed	black	male
Carrie	white	female	Keisha	black	female	Sarah	white	female
Darnell	black	male	Kenya	black	female	Tamika	black	female
Ebony	black	female	Kristen	white	female	Tanisha	black	female
Emily	white	female	Lakisha	black	female	Todd	white	male
Geoffrey	white	male	Latonya	black	female	Tremayne	black	male
Greg	white	male	Latoya	black	female	Tyrone	black	male

Figure 9.20: List of all 36 unique names along with the commonly inferred race and sex associated with these names.

The response variable of interest is whether or not there was a callback from the employer for the applicant, and there were 8 attributes that were randomly assigned that we'll consider, with special interest in the race and sex variables. Race and sex are **protected classes** in the United States, meaning they are not legally permitted factors for hiring or employment decisions. The full set of attributes considered is provided in Figure 9.21.

variable	description
callback	Specifies whether the employer called the applicant following submission of the application for the job.
job_city	City where the job was located: Boston or Chicago.
college_degree	An indicator for whether the resume listed a college degree.
years_experience	Number of years of experience listed on the resume.
honors	Indicator for the resume listing some sort of honors, e.g. employee of the month.
military	Indicator for if the resume listed any military experience.
email_address	Indicator for if the resume listed an email address for the applicant.
race	Race of the applicant, implied by their first name listed on the resume.
sex	Sex of the applicant (limited to only male and female in this study), implied by the first name listed on the resume.

Figure 9.21: Descriptions for the callback variable along with 8 other variables in the resume data set. Many of the variables are indicator variables, meaning they take the value 1 if the specified characteristic is present and 0 otherwise.

All of the attributes listed on each resume were randomly assigned. This means that no attributes that might be favorable or detrimental to employment would favor one demographic over another on these resumes. Importantly, due to the experimental nature of this study, we can infer causation between these variables and the callback rate, if the variable is statistically significant. Our analysis will allow us to compare the practical importance of each of the variables relative to each other.

9.5.2 Modeling the probability of an event

Logistic regression is a generalized linear model where the outcome is a two-level categorical variable. The outcome, Y_i, takes the value 1 (in our application, this represents a callback for the resume) with probability p_i and the value 0 with probability $1 - p_i$. Because each observation has a slightly different context, e.g. different education level or a different number of years of experience, the probability p_i will differ for each observation. Ultimately, it is this probability that we model in relation to the predictor variables: we will examine which resume characteristics correspond to higher or lower callback rates.

> **NOTATION FOR A LOGISTIC REGRESSION MODEL**
>
> The outcome variable for a GLM is denoted by Y_i, where the index i is used to represent observation i. In the resume application, Y_i will be used to represent whether resume i received a callback ($Y_i = 1$) or not ($Y_i = 0$).
>
> The predictor variables are represented as follows: $x_{1,i}$ is the value of variable 1 for observation i, $x_{2,i}$ is the value of variable 2 for observation i, and so on.

The logistic regression model relates the probability a resume would receive a callback (p_i) to the predictors $x_{1,i}, x_{2,i}, ..., x_{k,i}$ through a framework much like that of multiple regression:

$$transformation(p_i) = \beta_0 + \beta_1 x_{1,i} + \beta_2 x_{2,i} + \cdots + \beta_k x_{k,i} \tag{9.30}$$

We want to choose a transformation in the equation that makes practical and mathematical sense. For example, we want a transformation that makes the range of possibilities on the left hand side of the equation equal to the range of possibilities for the right hand side; if there was no transformation for this equation, the left hand side could only take values between 0 and 1, but the right hand side could take values outside of this range. A common transformation for p_i is the **logit transformation**, which may be written as

$$logit(p_i) = \log_e \left(\frac{p_i}{1 - p_i} \right)$$

The logit transformation is shown in Figure 9.22. Below, we rewrite the equation relating Y_i to its

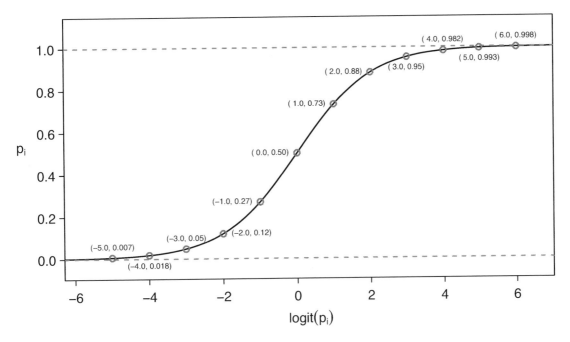

Figure 9.22: Values of p_i against values of $logit(p_i)$.

predictors using the logit transformation of p_i:

$$\log_e\left(\frac{p_i}{1-p_i}\right) = \beta_0 + \beta_1 x_{1,i} + \beta_2 x_{2,i} + \cdots + \beta_k x_{k,i}$$

In our resume example, there are 8 predictor variables, so $k = 8$. While the precise choice of a logit function isn't intuitive, it is based on theory that underpins generalized linear models, which is beyond the scope of this book. Fortunately, once we fit a model using software, it will start to feel like we're back in the multiple regression context, even if the interpretation of the coefficients is more complex.

EXAMPLE 9.31

We start by fitting a model with a single predictor: `honors`. This variable indicates whether the applicant had any type of honors listed on their resume, such as employee of the month. The following logistic regression model was fit using statistical software:

$$\log_e\left(\frac{p_i}{1-p_i}\right) = -2.4998 + 0.8668 \times \texttt{honors}$$

(a) If a resume is randomly selected from the study and it does not have any honors listed, what is the probability resulted in a callback?

(b) What would the probability be if the resume did list some honors?

———

(a) If a randomly chosen resume from those sent out is considered, and it does not list honors, then `honors` takes value 0 and the right side of the model equation equals -2.4998. Solving for p_i: $\frac{e^{-2.4998}}{1+e^{-2.4998}} = 0.076$. Just as we labeled a fitted value of y_i with a "hat" in single-variable and multiple regression, we do the same for this probability: $\hat{p}_i = 0.076$.

(b) If the resume had listed some honors, then the right side of the model equation is $-2.4998 + 0.8668 \times 1 = -1.6330$, which corresponds to a probability $\hat{p}_i = 0.163$.

Notice that we could examine -2.4998 and -1.6330 in Figure 9.22 to estimate the probability before formally calculating the value.

To convert from values on the logistic regression scale (e.g. -2.4998 and -1.6330 in Example 9.31), use the following formula, which is the result of solving for p_i in the regression model:

$$p_i = \frac{e^{\beta_0 + \beta_1 x_{1,i} + \cdots + \beta_k x_{k,i}}}{1 + e^{\beta_0 + \beta_1 x_{1,i} + \cdots + \beta_k x_{k,i}}}$$

As with most applied data problems, we substitute the point estimates for the parameters (the β_i) so that we can make use of this formula. In Example 9.31, the probabilities were calculated as

$$\frac{e^{-2.4998}}{1 + e^{-2.4998}} = 0.076 \qquad \frac{e^{-2.4998+0.8668}}{1 + e^{-2.4998+0.8668}} = 0.163$$

While knowing whether a resume listed honors provides some signal when predicting whether or not the employer would call, we would like to account for many different variables at once to understand how each of the different resume characteristics affected the chance of a callback.

9.5.3 Building the logistic model with many variables

We used statistical software to fit the logistic regression model with all 8 predictors described in Figure 9.21. Like multiple regression, the result may be presented in a summary table, which is shown in Figure 9.23. The structure of this table is almost identical to that of multiple regression; the only notable difference is that the p-values are calculated using the normal distribution rather than the t-distribution.

	Estimate	Std. Error	z value	$\Pr(>|z|)$
(Intercept)	-2.6632	0.1820	-14.64	<0.0001
job_city: *Chicago*	-0.4403	0.1142	-3.85	0.0001
college_degree	-0.0666	0.1211	-0.55	0.5821
years_experience	0.0200	0.0102	1.96	0.0503
honors	0.7694	0.1858	4.14	<0.0001
military	-0.3422	0.2157	-1.59	0.1127
email_address	0.2183	0.1133	1.93	0.0541
race: *white*	0.4424	0.1080	4.10	<0.0001
sex: *male*	-0.1818	0.1376	-1.32	0.1863

Figure 9.23: Summary table for the full logistic regression model for the resume callback example.

Just like multiple regression, we could trim some variables from the model. Here we'll use a statistic called **Akaike information criterion (AIC)**, which is an analog to how we used adjusted R-squared in multiple regression, and we look for models with a lower AIC through a backward elimination strategy. After using this criteria, the `college_degree` variable is eliminated, giving the smaller model summarized in Figure 9.24, which is what we'll rely on for the remainder of this section.

	Estimate	Std. Error	z value	$\Pr(>|z|)$
(Intercept)	-2.7162	0.1551	-17.51	<0.0001
job_city: *Chicago*	-0.4364	0.1141	-3.83	0.0001
years_experience	0.0206	0.0102	2.02	0.0430
honors	0.7634	0.1852	4.12	<0.0001
military	-0.3443	0.2157	-1.60	0.1105
email_address	0.2221	0.1130	1.97	0.0494
race: *white*	0.4429	0.1080	4.10	<0.0001
sex: *male*	-0.1959	0.1352	-1.45	0.1473

Figure 9.24: Summary table for the logistic regression model for the resume callback example, where variable selection has been performed using AIC.

EXAMPLE 9.32

The `race` variable had taken only two levels: `black` and `white`. Based on the model results, was race a meaningful factor for if a prospective employer would call back?

We see that the p-value for this coefficient is very small (very nearly zero), which implies that race played a statistically significant role in whether a candidate received a callback. Additionally, we see that the coefficient shown corresponds to the level of `white`, and it is positive. This positive coefficient reflects a positive gain in callback rate for resumes where the candidate's first name implied they were White. The data provide very strong evidence of racism by prospective employers that favors resumes where the first name is typically interpreted to be White.

The coefficient of `race`$_\texttt{white}$ in the full model in Figure 9.23, is nearly identical to the model shown in Figure 9.24. The predictors in this experiment were thoughtfully laid out so that the coefficient estimates would typically not be much influenced by which other predictors were in the model, which aligned with the motivation of the study to tease out which effects were important to getting a callback. In most observational data, it's common for point estimates to change a little, and sometimes a lot, depending on which other variables are included in the model.

EXAMPLE 9.33

Use the model summarized in Figure 9.24 to estimate the probability of receiving a callback for a job in Chicago where the candidate lists 14 years experience, no honors, no military experience, includes an email address, and has a first name that implies they are a White male.

We can start by writing out the equation using the coefficients from the model, then we can add in the corresponding values of each variable for this individual:

$$
\begin{aligned}
log_e&\left(\frac{p}{1-p}\right)\\
&= -2.7162 - 0.4364 \times \texttt{job\_city}_\texttt{Chicago} + 0.0206 \times \texttt{years\_experience} + 0.7634 \times \texttt{honors}\\
&\quad - 0.3443 \times \texttt{military} + 0.2221 \times \texttt{email} + 0.4429 \times \texttt{race}_\texttt{white} - 0.1959 \times \texttt{sex}_\texttt{male}\\
&= -2.7162 - 0.4364 \times 1 + 0.0206 \times 14 + 0.7634 \times 0\\
&\quad - 0.3443 \times 0 + 0.2221 \times 1 + 0.4429 \times 1 - 0.1959 \times 1\\
&= -2.3955
\end{aligned}
$$

We can now back-solve for p: the chance such an individual will receive a callback is about 8.35%.

EXAMPLE 9.34

Compute the probability of a callback for an individual with a name commonly inferred to be from a Black male but who otherwise has the same characteristics as the one described in Example 9.33.

We can complete the same steps for an individual with the same characteristics who is Black, where the only difference in the calculation is that the indicator variable `race`$_\texttt{white}$ will take a value of 0. Doing so yields a probability of 0.0553. Let's compare the results with those of Example 9.33.

In practical terms, an individual perceived as White based on their first name would need to apply to $\frac{1}{0.0835} \approx 12$ jobs on average to receive a callback, while an individual perceived as Black based on their first name would need to apply to $\frac{1}{0.0553} \approx 18$ jobs on average to receive a callback. That is, applicants who are perceived as Black need to apply to 50% more employers to receive a callback than someone who is perceived as White based on their first name for jobs like those in the study.

What we've quantified in this section is alarming and disturbing. However, one aspect that makes this racism so difficult to address is that the experiment, as well-designed as it is, cannot send us much signal about which employers are discriminating. It is only possible to say that discrimination is happening, even if we cannot say which particular callbacks – or non-callbacks – represent discrimination. Finding strong evidence of racism for individual cases is a persistent challenge in enforcing anti-discrimination laws.

9.5.4 Diagnostics for the callback rate model

> **LOGISTIC REGRESSION CONDITIONS**
>
> There are two key conditions for fitting a logistic regression model:
>
> 1. Each outcome Y_i is independent of the other outcomes.
> 2. Each predictor x_i is linearly related to $\text{logit}(p_i)$ if all other predictors are held constant.

The first logistic regression model condition – independence of the outcomes – is reasonable for the experiment since characteristics of resumes were randomly assigned to the resumes that were sent out.

The second condition of the logistic regression model is not easily checked without a fairly sizable amount of data. Luckily, we have 4870 resume submissions in the data set! Let's first visualize these data by plotting the true classification of the resumes against the model's fitted probabilities, as shown in Figure 9.25.

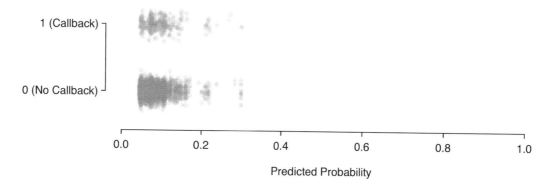

Figure 9.25: The predicted probability that each of the 4870 resumes results in a callback. Noise (small, random vertical shifts) have been added to each point so points with nearly identical values aren't plotted exactly on top of one another.

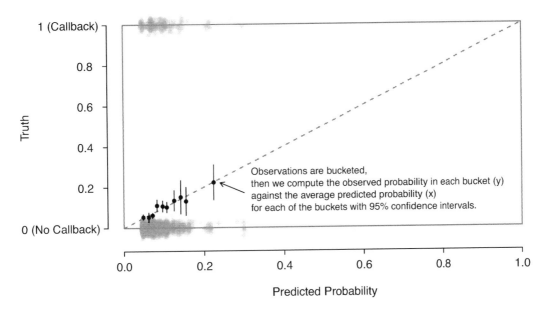

Figure 9.26: The dashed line is within the confidence bound of the 95% confidence intervals of each of the buckets, suggesting the logistic fit is reasonable.

We'd like to assess the quality of the model. For example, we might ask: if we look at resumes that we modeled as having a 10% chance of getting a callback, do we find about 10% of them actually receive a callback? We can check this for groups of the data by constructing a plot as follows:

1. Bucket the data into groups based on their predicted probabilities.

2. Compute the average predicted probability for each group.

3. Compute the observed probability for each group, along with a 95% confidence interval.

4. Plot the observed probabilities (with 95% confidence intervals) against the average predicted probabilities for each group.

The points plotted should fall close to the line $y = x$, since the predicted probabilities should be similar to the observed probabilities. We can use the confidence intervals to roughly gauge whether anything might be amiss. Such a plot is shown in Figure 9.26.

Additional diagnostics may be created that are similar to those featured in Section 9.3. For instance, we could compute residuals as the observed outcome minus the expected outcome ($e_i = Y_i - \hat{p}_i$), and then we could create plots of these residuals against each predictor. We might also create a plot like that in Figure 9.26 to better understand the deviations.

9.5.5   Exploring discrimination between groups of different sizes

Any form of discrimination is concerning, and this is why we decided it was so important to discuss this topic using data. The resume study also only examined discrimination in a single aspect: whether a prospective employer would call a candidate who submitted their resume. There was a 50% higher barrier for resumes simply when the candidate had a first name that was perceived to be from a Black individual. It's unlikely that discrimination would stop there.

EXAMPLE 9.35

Let's consider a sex-imbalanced company that consists of 20% women and 80% men,[22] and we'll suppose that the company is very large, consisting of perhaps 20,000 employees. Suppose when someone goes up for promotion at this company, 5 of their colleagues are randomly chosen to provide feedback on their work.

Now let's imagine that 10% of the people in the company are prejudiced against the other sex. That is, 10% of men are prejudiced against women, and similarly, 10% of women are prejudiced against men.

Who is discriminated against more at the company, men or women?

Let's suppose we took 100 men who have gone up for promotion in the past few years. For these men, $5 \times 100 = 500$ random colleagues will be tapped for their feedback, of which about 20% will be women (100 women). Of these 100 women, 10 are expected to be biased against the man they are reviewing. Then, of the 500 colleagues reviewing them, men will experience discrimination by about 2% of their colleagues when they go up for promotion.

Let's do a similar calculation for 100 women who have gone up for promotion in the last few years. They will also have 500 random colleagues providing feedback, of which about 400 (80%) will be men. Of these 400 men, about 40 (10%) hold a bias against women. Of the 500 colleagues providing feedback on the promotion packet for these women, 8% of the colleagues hold a bias against the women.

Example 9.35 highlights something profound: even in a hypothetical setting where each demographic has the same degree of prejudice against the other demographic, the smaller group experiences the negative effects more frequently. Additionally, if we would complete a handful of examples like the one above with different numbers, we'd learn that the greater the imbalance in the population groups, the more the smaller group is disproportionately impacted.[23]

Of course, there are other considerable real-world omissions from the hypothetical example. For example, studies have found instances where people from an oppressed group also discriminate against others within their own oppressed group. As another example, there are also instances where a majority group can be oppressed, with apartheid in South Africa being one such historic example. Ultimately, discrimination is complex, and there are many factors at play beyond the mathematics property we observed in Example 9.35.

We close this book on this serious topic, and we hope it inspires you to think about the power of reasoning with data. Whether it is with a formal statistical model or by using critical thinking skills to structure a problem, we hope the ideas you have learned will help you do more and do better in life.

---

[22]A more thoughtful example would include non-binary individuals.

[23]If a proportion p of a company are women and the rest of the company consists of men, then under the hypothetical situation the ratio of rates of discrimination against women vs men would be given by $\frac{1-p}{p}$; this ratio is always greater than 1 when $p < 0.5$.

Exercises

9.15 Possum classification, Part I. The common brushtail possum of the Australia region is a bit cuter than its distant cousin, the American opossum (see Figure 8.4 on page 307). We consider 104 brushtail possums from two regions in Australia, where the possums may be considered a random sample from the population. The first region is Victoria, which is in the eastern half of Australia and traverses the southern coast. The second region consists of New South Wales and Queensland, which make up eastern and northeastern Australia. We use logistic regression to differentiate between possums in these two regions. The outcome variable, called `population`, takes value 1 when a possum is from Victoria and 0 when it is from New South Wales or Queensland. We consider five predictors: `sex_male` (an indicator for a possum being male), `head_length`, `skull_ width`, `total_length`, and `tail_length`. Each variable is summarized in a histogram. The full logistic regression model and a reduced model after variable selection are summarized in the table.

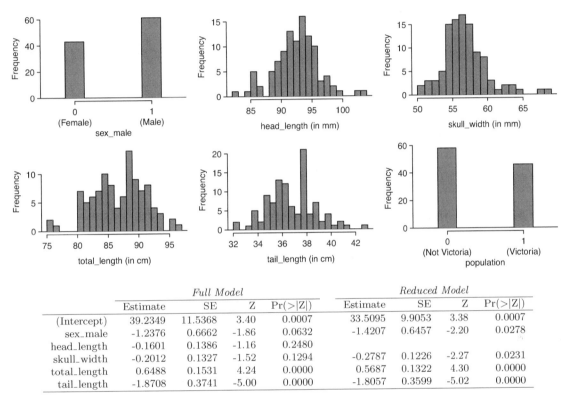

	Full Model				Reduced Model			
	Estimate	SE	Z	Pr(>|Z|)	Estimate	SE	Z	Pr(>|Z|)
(Intercept)	39.2349	11.5368	3.40	0.0007	33.5095	9.9053	3.38	0.0007
sex_male	-1.2376	0.6662	-1.86	0.0632	-1.4207	0.6457	-2.20	0.0278
head_length	-0.1601	0.1386	-1.16	0.2480				
skull_width	-0.2012	0.1327	-1.52	0.1294	-0.2787	0.1226	-2.27	0.0231
total_length	0.6488	0.1531	4.24	0.0000	0.5687	0.1322	4.30	0.0000
tail_length	-1.8708	0.3741	-5.00	0.0000	-1.8057	0.3599	-5.02	0.0000

(a) Examine each of the predictors. Are there any outliers that are likely to have a very large influence on the logistic regression model?

(b) The summary table for the full model indicates that at least one variable should be eliminated when using the p-value approach for variable selection: `head_length`. The second component of the table summarizes the reduced model following variable selection. Explain why the remaining estimates change between the two models.

9.16 Challenger disaster, Part I. On January 28, 1986, a routine launch was anticipated for the Challenger space shuttle. Seventy-three seconds into the flight, disaster happened: the shuttle broke apart, killing all seven crew members on board. An investigation into the cause of the disaster focused on a critical seal called an O-ring, and it is believed that damage to these O-rings during a shuttle launch may be related to the ambient temperature during the launch. The table below summarizes observational data on O-rings for 23 shuttle missions, where the mission order is based on the temperature at the time of the launch. *Temp* gives the temperature in Fahrenheit, *Damaged* represents the number of damaged O- rings, and *Undamaged* represents the number of O-rings that were not damaged.

Shuttle Mission	1	2	3	4	5	6	7	8	9	10	11	12
Temperature	53	57	58	63	66	67	67	67	68	69	70	70
Damaged	5	1	1	1	0	0	0	0	0	0	1	0
Undamaged	1	5	5	5	6	6	6	6	6	6	5	6

Shuttle Mission	13	14	15	16	17	18	19	20	21	22	23
Temperature	70	70	72	73	75	75	76	76	78	79	81
Damaged	1	0	0	0	0	1	0	0	0	0	0
Undamaged	5	6	6	6	6	5	6	6	6	6	6

(a) Each column of the table above represents a different shuttle mission. Examine these data and describe what you observe with respect to the relationship between temperatures and damaged O-rings.

(b) Failures have been coded as 1 for a damaged O-ring and 0 for an undamaged O-ring, and a logistic regression model was fit to these data. A summary of this model is given below. Describe the key components of this summary table in words.

	Estimate	Std. Error	z value	$\Pr(>|z|)$
(Intercept)	11.6630	3.2963	3.54	0.0004
Temperature	-0.2162	0.0532	-4.07	0.0000

(c) Write out the logistic model using the point estimates of the model parameters.

(d) Based on the model, do you think concerns regarding O-rings are justified? Explain.

9.17 Possum classification, Part II. A logistic regression model was proposed for classifying common brushtail possums into their two regions in Exercise 9.15. The outcome variable took value 1 if the possum was from Victoria and 0 otherwise.

	Estimate	SE	Z	$\Pr(>|Z|)$
(Intercept)	33.5095	9.9053	3.38	0.0007
sex_male	-1.4207	0.6457	-2.20	0.0278
skull_width	-0.2787	0.1226	-2.27	0.0231
total_length	0.5687	0.1322	4.30	0.0000
tail_length	-1.8057	0.3599	-5.02	0.0000

(a) Write out the form of the model. Also identify which of the variables are positively associated when controlling for other variables.

(b) Suppose we see a brushtail possum at a zoo in the US, and a sign says the possum had been captured in the wild in Australia, but it doesn't say which part of Australia. However, the sign does indicate that the possum is male, its skull is about 63 mm wide, its tail is 37 cm long, and its total length is 83 cm. What is the reduced model's computed probability that this possum is from Victoria? How confident are you in the model's accuracy of this probability calculation?

9.18 Challenger disaster, Part II. Exercise 9.16 introduced us to O-rings that were identified as a plausible explanation for the breakup of the Challenger space shuttle 73 seconds into takeoff in 1986. The investigation found that the ambient temperature at the time of the shuttle launch was closely related to the damage of O-rings, which are a critical component of the shuttle. See this earlier exercise if you would like to browse the original data.

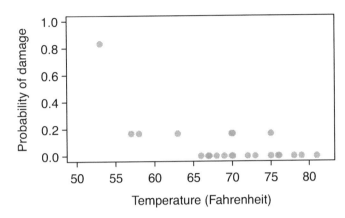

(a) The data provided in the previous exercise are shown in the plot. The logistic model fit to these data may be written as

$$\log\left(\frac{\hat{p}}{1-\hat{p}}\right) = 11.6630 - 0.2162 \times Temperature$$

where $\hat{p}$ is the model-estimated probability that an O-ring will become damaged. Use the model to calculate the probability that an O-ring will become damaged at each of the following ambient temperatures: 51, 53, and 55 degrees Fahrenheit. The model-estimated probabilities for several additional ambient temperatures are provided below, where subscripts indicate the temperature:

$\hat{p}_{57} = 0.341$	$\hat{p}_{59} = 0.251$	$\hat{p}_{61} = 0.179$	$\hat{p}_{63} = 0.124$
$\hat{p}_{65} = 0.084$	$\hat{p}_{67} = 0.056$	$\hat{p}_{69} = 0.037$	$\hat{p}_{71} = 0.024$

(b) Add the model-estimated probabilities from part (a) on the plot, then connect these dots using a smooth curve to represent the model-estimated probabilities.

(c) Describe any concerns you may have regarding applying logistic regression in this application, and note any assumptions that are required to accept the model's validity.

Chapter exercises

9.19 Multiple regression fact checking. Determine which of the following statements are true and false. For each statement that is false, explain why it is false.

(a) If predictors are collinear, then removing one variable will have no influence on the point estimate of another variable's coefficient.

(b) Suppose a numerical variable x has a coefficient of $b_1 = 2.5$ in the multiple regression model. Suppose also that the first observation has $x_1 = 7.2$, the second observation has a value of $x_1 = 8.2$, and these two observations have the same values for all other predictors. Then the predicted value of the second observation will be 2.5 higher than the prediction of the first observation based on the multiple regression model.

(c) If a regression model's first variable has a coefficient of $b_1 = 5.7$, then if we are able to influence the data so that an observation will have its x_1 be 1 larger than it would otherwise, the value y_1 for this observation would increase by 5.7.

(d) Suppose we fit a multiple regression model based on a data set of 472 observations. We also notice that the distribution of the residuals includes some skew but does not include any particularly extreme outliers. Because the residuals are not nearly normal, we should not use this model and require more advanced methods to model these data.

9.20 Logistic regression fact checking. Determine which of the following statements are true and false. For each statement that is false, explain why it is false.

(a) Suppose we consider the first two observations based on a logistic regression model, where the first variable in observation 1 takes a value of $x_1 = 6$ and observation 2 has $x_1 = 4$. Suppose we realized we made an error for these two observations, and the first observation was actually $x_1 = 7$ (instead of 6) and the second observation actually had $x_1 = 5$ (instead of 4). Then the predicted probability from the logistic regression model would increase the same amount for each observation after we correct these variables.

(b) When using a logistic regression model, it is impossible for the model to predict a probability that is negative or a probability that is greater than 1.

(c) Because logistic regression predicts probabilities of outcomes, observations used to build a logistic regression model need not be independent.

(d) When fitting logistic regression, we typically complete model selection using adjusted R^2.

9.21 Spam filtering, Part I. Spam filters are built on principles similar to those used in logistic regression. We fit a probability that each message is spam or not spam. We have several email variables for this problem: `to_multiple`, `cc`, `attach`, `dollar`, `winner`, `inherit`, `password`, `format`, `re_subj`, `exclaim_subj`, and `sent_email`. We won't describe what each variable means here for the sake of brevity, but each is either a numerical or indicator variable.

(a) For variable selection, we fit the full model, which includes all variables, and then we also fit each model where we've dropped exactly one of the variables. In each of these reduced models, the AIC value for the model is reported below. Based on these results, which variable, if any, should we drop as part of model selection? Explain.

Variable Dropped	AIC
None Dropped	1863.50
`to_multiple`	2023.50
`cc`	1863.18
`attach`	1871.89
`dollar`	1879.70
`winner`	1885.03
`inherit`	1865.55
`password`	1879.31
`format`	2008.85
`re_subj`	1904.60
`exclaim_subj`	1862.76
`sent_email`	1958.18

See the next page for part (b).

(b) Consider the following model selection stage. Here again we've computed the AIC for each leave-one-variable-out model. Based on the results, which variable, if any, should we drop as part of model selection? Explain.

Variable Dropped	AIC
None Dropped	1862.41
to_multiple	2019.55
attach	1871.17
dollar	1877.73
winner	1884.95
inherit	1864.52
password	1878.19
format	2007.45
re_subj	1902.94
sent_email	1957.56

9.22 Movie returns, Part II. The student from Exercise 9.14 analyzed return-on-investment (ROI) for movies based on release year and genre of movies. The plots below show the predicted ROI vs. actual ROI for each of the genres separately. Do these figures support the comment in the FiveThirtyEight.com article that states, "The return-on-investment potential for horror movies is absurd." Note that the x-axis range varies for each plot.

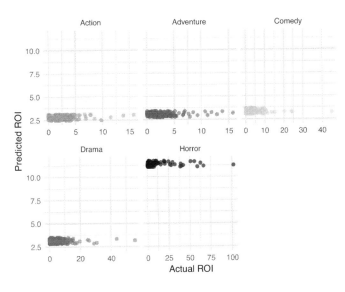

9.23 Spam filtering, Part II. In Exercise 9.21, we encountered a data set where we applied logistic regression to aid in spam classification for individual emails. In this exercise, we've taken a small set of these variables and fit a formal model with the following output:

	Estimate	Std. Error	z value	Pr(>|z|)
(Intercept)	-0.8124	0.0870	-9.34	0.0000
to_multiple	-2.6351	0.3036	-8.68	0.0000
winner	1.6272	0.3185	5.11	0.0000
format	-1.5881	0.1196	-13.28	0.0000
re_subj	-3.0467	0.3625	-8.40	0.0000

(a) Write down the model using the coefficients from the model fit.

(b) Suppose we have an observation where to_multiple = 0, winner = 1, format = 0, and re_subj = 0. What is the predicted probability that this message is spam?

(c) Put yourself in the shoes of a data scientist working on a spam filter. For a given message, how high must the probability a message is spam be before you think it would be reasonable to put it in a *spambox* (which the user is unlikely to check)? What tradeoffs might you consider? Any ideas about how you might make your spam-filtering system even better from the perspective of someone using your email service?

Appendix A

Exercise solutions

1 Introduction to data

1.1 (a) Treatment: $10/43 = 0.23 \rightarrow 23\%$.
(b) Control: $2/46 = 0.04 \rightarrow 4\%$. (c) A higher percentage of patients in the treatment group were pain free 24 hours after receiving acupuncture. (d) It is possible that the observed difference between the two group percentages is due to chance.

1.3 (a) "Is there an association between air pollution exposure and preterm births?" (b) 143,196 births in Southern California between 1989 and 1993. (c) Measurements of carbon monoxide, nitrogen dioxide, ozone, and particulate matter less than $10\mu g/m^3$ (PM$_{10}$) collected at air-quality-monitoring stations as well as length of gestation. Continuous numerical variables.

1.5 (a) "Does explicitly telling children not to cheat affect their likelihood to cheat?". (b) 160 children between the ages of 5 and 15. (c) Four variables: (1) age (numerical, continuous), (2) sex (categorical), (3) whether they were an only child or not (categorical), (4) whether they cheated or not (categorical).

1.7 Explanatory: acupuncture or not. Response: if the patient was pain free or not.

1.9 (a) $50 \times 3 = 150$. (b) Four continuous numerical variables: sepal length, sepal width, petal length, and petal width. (c) One categorical variable, species, with three levels: *setosa*, *versicolor*, and *virginica*.

1.11 (a) Airport ownership status (public/private), airport usage status (public/private), latitude, and longitude. (b) Airport ownership status: categorical, not ordinal. Airport usage status: categorical, not ordinal. Latitude: numerical, continuous. Longitude: numerical, continuous.

1.13 (a) Population: all births, sample: 143,196 births between 1989 and 1993 in Southern California. (b) If births in this time span at the geography can be considered to be representative of all births, then the results are generalizable to the population of Southern California. However, since the study is

observational the findings cannot be used to establish causal relationships.

1.15 (a) Population: all asthma patients aged 18-69 who rely on medication for asthma treatment. Sample: 600 such patients. (b) If the patients in this sample, who are likely not randomly sampled, can be considered to be representative of all asthma patients aged 18-69 who rely on medication for asthma treatment, then the results are generalizable to the population defined above. Additionally, since the study is experimental, the findings can be used to establish causal relationships.

1.17 (a) Observation. (b) Variable. (c) Sample statistic (mean). (d) Population parameter (mean).

1.19 (a) Observational. (b) Use stratified sampling to randomly sample a fixed number of students, say 10, from each section for a total sample size of 40 students.

1.21 (a) Positive, non-linear, somewhat strong. Countries in which a higher percentage of the population have access to the internet also tend to have higher average life expectancies, however rise in life expectancy trails off before around 80 years old. (b) Observational. (c) Wealth: countries with individuals who can widely afford the internet can probably also afford basic medical care. (Note: Answers may vary.)

1.23 (a) Simple random sampling is okay. In fact, it's rare for simple random sampling to not be a reasonable sampling method! (b) The student opinions may vary by field of study, so the stratifying by this variable makes sense and would be reasonable. (c) Students of similar ages are probably going to have more similar opinions, and we want clusters to be diverse with respect to the outcome of interest, so this would **not** be a good approach. (Additional thought: the clusters in this case may also have very different numbers of people, which can also create unexpected sample sizes.)

1.25 (a) The cases are 200 randomly sampled men and women. (b) The response variable is attitude towards a fictional microwave oven. (c) The explanatory variable is dispositional attitude. (d) Yes, the cases are sampled randomly. (e) This is an observational study since there is no random assignment to treatments. (f) No, we cannot establish a causal link between the explanatory and response variables since the study is observational. (g) Yes, the results of the study can be generalized to the population at large since the sample is random.

1.27 (a) Simple random sample. Non-response bias, if only those people who have strong opinions about the survey responds his sample may not be representative of the population. (b) Convenience sample. Under coverage bias, his sample may not be representative of the population since it consists only of his friends. It is also possible that the study will have non-response bias if some choose to not bring back the survey. (c) Convenience sample. This will have a similar issues to handing out surveys to friends. (d) Multi-stage sampling. If the classes are similar to each other with respect to student composition this approach should not introduce bias, other than potential non-response bias.

1.29 (a) Exam performance. (b) Light level: fluorescent overhead lighting, yellow overhead lighting, no overhead lighting (only desk lamps). (c) Sex: man, woman.

1.31 (a) Exam performance. (b) Light level (overhead lighting, yellow overhead lighting, no overhead lighting) and noise level (no noise, construction noise, and human chatter noise). (c) Since the researchers want to ensure equal gender representation, sex will be a blocking variable.

1.33 Need randomization and blinding. One possible outline: (1) Prepare two cups for each participant, one containing regular Coke and the other containing Diet Coke. Make sure the cups are identical and contain equal amounts of soda. Label the cups A (regular) and B (diet). (Be sure to randomize A and B for each trial!) (2) Give each participant the two cups, one cup at a time, in random order, and ask the participant to record a value that indicates how much she liked the beverage. Be sure that neither the participant nor the person handing out the cups knows the identity of the beverage to make this a double-blind experiment. (Answers may vary.)

1.35 (a) Observational study. (b) Dog: Lucy. Cat: Luna. (c) Oliver and Lily. (d) Positive, as the popularity of a name for dogs increases, so does the popularity of that name for cats.

1.37 (a) Experiment. (b) Treatment: 25 grams of chia seeds twice a day, control: placebo. (c) Yes, gender. (d) Yes, single blind since the patients were blinded to the treatment they received. (e) Since this is an experiment, we can make a causal statement. However, since the sample is not random, the causal statement cannot be generalized to the population at large.

1.39 (a) Non-responders may have a different response to this question, e.g. parents who returned the surveys likely don't have difficulty spending time with their children. (b) It is unlikely that the women who were reached at the same address 3 years later are a random sample. These missing responders are probably renters (as opposed to homeowners) which means that they might be in a lower socio-economic status than the respondents. (c) There is no control group in this study, this is an observational study, and there may be confounding variables, e.g. these people may go running because they are generally healthier and/or do other exercises.

1.41 (a) Randomized controlled experiment. (b) Explanatory: treatment group (categorical, with 3 levels). Response variable: Psychological well-being. (c) No, because the participants were volunteers. (d) Yes, because it was an experiment. (e) The statement should say "evidence" instead of "proof".

1.43 (a) County, state, driver's race, whether the car was searched or not, and whether the driver was arrested or not. (b) All categorical, non-ordinal. (c) Response: whether the car was searched or not. Explanatory: race of the driver.

2 Summarizing data

2.1 (a) Positive association: mammals with longer gestation periods tend to live longer as well. (b) Association would still be positive. (c) No, they are not independent. See part (a).

2.3 The graph below shows a ramp up period. There may also be a period of exponential growth at the start before the size of the petri dish becomes a factor in slowing growth.

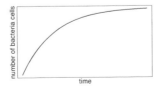

2.5 (a) Population mean, $\mu_{2007} = 52$; sample mean, $\bar{x}_{2008} = 58$. (b) Population mean, $\mu_{2001} = 3.37$; sample mean, $\bar{x}_{2012} = 3.59$.

2.7 Any 10 employees whose average number of days off is between the minimum and the mean number of days off for the entire workforce at this plant.

2.9 (a) Dist 2 has a higher mean since $20 > 13$, and a higher standard deviation since 20 is further from the rest of the data than 13. (b) Dist 1 has a higher mean since $-20 > -40$, and Dist 2 has a higher standard deviation since -40 is farther away from the rest of the data than -20. (c) Dist 2 has a higher mean since all values in this distribution are higher than those in Dist 1, but both distribution have the same standard deviation since they are equally variable around their respective means. (d) Both distributions have the same mean since they're both centered at 300, but Dist 2 has a higher standard deviation since the observations are farther from the mean than in Dist 1.

2.11 (a) About 30. (b) Since the distribution is right skewed the mean is higher than the median. (c) Q1: between 15 and 20, Q3: between 35 and 40, IQR: about 20. (d) Values that are considered to be unusually low or high lie more than $1.5 \times$IQR away from the quartiles. Upper fence: Q3 $+ 1.5 \times$ IQR $= 37.5 + 1.5 \times 20 = 67.5$; Lower fence: Q1 - $1.5 \times$ IQR $= 17.5 - 1.5 \times 20 = -12.5$; The lowest AQI recorded is not lower than 5 and the highest AQI recorded is not higher than 65, which are both within the fences. Therefore none of the days in this sample would be considered to have an unusually low or high AQI.

2.13 The histogram shows that the distribution is bimodal, which is not apparent in the box plot. The box plot makes it easy to identify more precise values of observations outside of the whiskers.

2.15 (a) The distribution of number of pets per household is likely right skewed as there is a natural boundary at 0 and only a few people have many pets. Therefore the center would be best described by the median, and variability would be best described by the IQR. (b) The distribution of number of distance to work is likely right skewed as there is a natural boundary at 0 and only a few people live a very long distance from work. Therefore the center would be best described by the median, and variability would be best described by the IQR. (c) The distribution of heights of males is likely symmetric. Therefore the center would be best described by the mean, and variability would be best described by the standard deviation.

2.17 (a) The median is a much better measure of the typical amount earned by these 42 people. The mean is much higher than the income of 40 of the 42 people. This is because the mean is an arithmetic average and gets affected by the two extreme observations. The median does not get effected as much since it is robust to outliers. (b) The IQR is a much better measure of variability in the amounts earned by nearly all of the 42 people. The standard deviation gets affected greatly by the two high salaries, but the IQR is robust to these extreme observations.

2.19 (a) The distribution is unimodal and symmetric with a mean of about 25 minutes and a standard deviation of about 5 minutes. There does not appear to be any counties with unusually high or low mean travel times. Since the distribution is already unimodal and symmetric, a log transformation is not necessary. (b) Answers will vary. There are pockets of longer travel time around DC, Southeastern NY, Chicago, Minneapolis, Los Angeles, and many other big cities. There is also a large section of shorter average commute times that overlap with farmland in the Midwest. Many farmers' homes are adjacent to their farmland, so their commute would be brief, which may explain why the average commute time for these counties is relatively low.

2.21 (a) We see the order of the categories and the relative frequencies in the bar plot. (b) There are no features that are apparent in the pie chart but not in the bar plot. (c) We usually prefer to use a bar plot as we can also see the relative frequencies of the categories in this graph.

2.23 The vertical locations at which the ideological groups break into the Yes, No, and Not Sure categories differ, which indicates that likelihood of supporting the DREAM act varies by political ideology. This suggests that the two variables may be dependent.

387

2.25 (a) (i) False. Instead of comparing counts, we should compare percentages of people in each group who suffered cardiovascular problems. (ii) True. (iii) False. Association does not imply causation. We cannot infer a causal relationship based on an observational study. The difference from part (ii) is subtle. (iv) True.

(b) Proportion of all patients who had cardiovascular problems: $\frac{7,979}{227,571} \approx 0.035$

(c) The expected number of heart attacks in the rosiglitazone group, if having cardiovascular problems and treatment were independent, can be calculated as the number of patients in that group multiplied by the overall cardiovascular problem rate in the study: $67,593 * \frac{7,979}{227,571} \approx 2370$.

(d) (i) H_0: The treatment and cardiovascular problems are independent. They have no relationship, and the difference in incidence rates between the rosiglitazone and pioglitazone groups is due to chance. H_A: The treatment and cardiovascular problems are not independent. The difference in the incidence rates between the rosiglitazone and pioglitazone groups is not due to chance and rosiglitazone is associated with an increased risk of serious cardiovascular problems. (ii) A higher number of patients with cardiovascular problems than expected under the assumption of independence would provide support for the alternative hypothesis as this would suggest that rosiglitazone increases the risk of such problems. (iii) In the actual study, we observed 2,593 cardiovascular events in the rosiglitazone group. In the 1,000 simulations under the independence model, we observed somewhat less than 2,593 in every single simulation, which suggests that the actual results did not come from the independence model. That is, the variables do not appear to be independent, and we reject the independence model in favor of the alternative. The study's results provide convincing evidence that rosiglitazone is associated with an increased risk of cardiovascular problems.

2.27 (a) Decrease: the new score is smaller than the mean of the 24 previous scores. (b) Calculate a weighted mean. Use a weight of 24 for the old mean and 1 for the new mean: $(24 \times 74 + 1 \times 64)/(24+1) = 73.6$. (c) The new score is more than 1 standard deviation away from the previous mean, so increase.

2.29 No, we would expect this distribution to be right skewed. There are two reasons for this: (1) there is a natural boundary at 0 (it is not possible to watch less than 0 hours of TV), (2) the standard deviation of the distribution is very large compared to the mean.

2.31 The distribution of ages of best actress winners are right skewed with a median around 30 years. The distribution of ages of best actor winners is also right skewed, though less so, with a median around 40 years. The difference between the peaks of these distributions suggest that best actress winners are typically younger than best actor winners. The ages of best actress winners are more variable than the ages of best actor winners. There are potential outliers on the higher end of both of the distributions.

2.33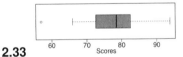

3 Probability

3.1 (a) False. These are independent trials. (b) False. There are red face cards. (c) True. A card cannot be both a face card and an ace.

3.3 (a) 10 tosses. Fewer tosses mean more variability in the sample fraction of heads, meaning there's a better chance of getting at least 60% heads. (b) 100 tosses. More flips means the observed proportion of heads would often be closer to the average, 0.50, and therefore also above 0.40. (c) 100 tosses. With more flips, the observed proportion of heads would often be closer to the average, 0.50. (d) 10 tosses. Fewer flips would increase variability in the fraction of tosses that are heads.

3.5 (a) $0.5^{10} = 0.00098$. (b) $0.5^{10} = 0.00098$. (c) $P(\text{at least one tails}) = 1 - P(\text{no tails}) = 1 - (0.5^{10}) \approx 1 - 0.001 = 0.999$.

3.7 (a) No, there are voters who are both independent and swing voters.

(b)

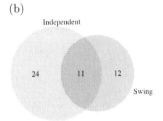

(c) Each Independent voter is either a swing voter or not. Since 35% of voters are Independents and 11% are both Independent and swing voters, the other 24% must not be swing voters. (d) 0.47. (e) 0.53. (f) P(Independent) × P(swing) = 0.35 × 0.23 = 0.08, which does not equal P(Independent and swing) = 0.11, so the events are dependent.

3.9 (a) If the class is not graded on a curve, they are independent. If graded on a curve, then neither independent nor disjoint – unless the instructor will only give one A, which is a situation we will ignore in parts (b) and (c). (b) They are probably not independent: if you study together, your study habits would be related, which suggests your course performances are also related. (c) No. See the answer to part (a) when the course is not graded on a curve. More generally: if two things are unrelated (independent), then one occurring does not preclude the other from occurring.

3.11 (a) $0.16 + 0.09 = 0.25$. (b) $0.17 + 0.09 = 0.26$. (c) Assuming that the education level of the husband and wife are independent: $0.25 \times 0.26 = 0.065$. You might also notice we actually made a second assumption: that the decision to get married is unrelated to education level. (d) The husband/wife independence assumption is probably not reasonable, because people often marry another person with a comparable level of education. We will leave it to you to think about whether the second assumption noted in part (c) is reasonable.

3.13 (a) No, but we could if A and B are independent. (b-i) 0.21. (b-ii) 0.79. (b-iii) 0.3. (c) No, because $0.1 \neq 0.21$, where 0.21 was the value computed under independence from part (a). (d) 0.143.

3.15 (a) No, 0.18 of respondents fall into this combination. (b) $0.60 + 0.20 - 0.18 = 0.62$. (c) $0.18/0.20 = 0.9$. (d) $0.11/0.33 \approx 0.33$. (e) No, otherwise the answers to (c) and (d) would be the same. (f) $0.06/0.34 \approx 0.18$.

3.17 (a) No. There are 6 females who like Five Guys Burgers. (b) $162/248 = 0.65$. (c) $181/252 = 0.72$. (d) Under the assumption of a dating choices being independent of hamburger preference, which on the surface seems reasonable: $0.65 \times 0.72 = 0.468$. (e) $(252 + 6 - 1)/500 = 0.514$.

3.19 (a)

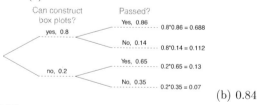

(b) 0.84

3.21 0.0714. Even when a patient tests positive for lupus, there is only a 7.14% chance that he actually has lupus. House may be right.

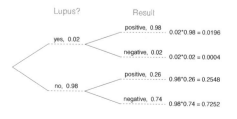

3.23 (a) 0.3. (b) 0.3. (c) 0.3. (d) $0.3 \times 0.3 = 0.09$. (e) Yes, the population that is being sampled from is identical in each draw.

3.25 (a) $2/9 \approx 0.22$. (b) $3/9 \approx 0.33$. (c) $\frac{3}{10} \times \frac{2}{9} \approx 0.067$. (d) No, e.g. in this exercise, removing one chip meaningfully changes the probability of what might be drawn next.

3.27 $P(^1\text{leggings}, {}^2\text{jeans}, {}^3\text{jeans}) = \frac{5}{24} \times \frac{7}{23} \times \frac{6}{22} = 0.0173$. However, the person with leggings could have come 2nd or 3rd, and these each have this same probability, so $3 \times 0.0173 = 0.0519$.

3.29 (a) 13. (b) No, these 27 students are not a random sample from the university's student population. For example, it might be argued that the proportion of smokers among students who go to the gym at 9 am on a Saturday morning would be lower than the proportion of smokers in the university as a whole.

3.31 (a) E(X) = 3.59. SD(X) = 9.64. (b) E(X) = -1.41. SD(X) = 9.64. (c) No, the expected net profit is negative, so on average you expect to lose money.

3.33 5% increase in value.

3.35 E = -0.0526. SD = 0.9986.

3.37 Approximate answers are OK. (a) $(29 + 32)/144 = 0.42$. (b) $21/144 = 0.15$. (c) $(26 + 12 + 15)/144 = 0.37$.

3.39 (a) Invalid. Sum is greater than 1. (b) Valid. Probabilities are between 0 and 1, and they sum to 1. In this class, every student gets a C. (c) Invalid. Sum is less than 1. (d) Invalid. There is a negative probability. (e) Valid. Probabilities are between 0 and 1, and they sum to 1. (f) Invalid. There is a negative probability.

3.41 0.8247.

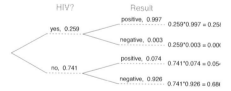

3.43 (a) E = \$3.90. SD = \$0.34. (b) E = \$27.30. SD = \$0.89.

3.45 $Var\left(\frac{X_1 + X_2}{2}\right)$
$= Var\left(\frac{X_1}{2} + \frac{X_2}{2}\right)$
$= \frac{Var(X_1)}{2^2} + \frac{Var(X_2)}{2^2}$
$= \frac{\sigma^2}{4} + \frac{\sigma^2}{4}$
$= \sigma^2/2$

3.47 $Var\left(\frac{X_1 + X_2 + \cdots + X_n}{n}\right)$
$= Var\left(\frac{X_1}{n} + \frac{X_2}{n} + \cdots + \frac{X_n}{n}\right)$
$= \frac{Var(X_1)}{n^2} + \frac{Var(X_2)}{n^2} + \cdots + \frac{Var(X_n)}{n^2}$
$= \frac{\sigma^2}{n^2} + \frac{\sigma^2}{n^2} + \cdots + \frac{\sigma^2}{n^2}$ (there are n of these terms)
$= n\frac{\sigma^2}{n^2}$
$= \sigma^2/n$

4 Distributions of random variables

4.1 (a) 8.85%. (b) 6.94%. (c) 58.86%. (d) 4.56%.

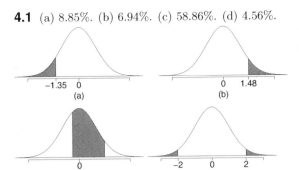

4.3 (a) Verbal: $N(\mu = 151, \sigma = 7)$, Quant: $N(\mu = 153, \sigma = 7.67)$. (b) $Z_{VR} = 1.29$, $Z_{QR} = 0.52$.

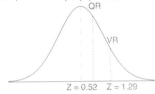

(c) She scored 1.29 standard deviations above the mean on the Verbal Reasoning section and 0.52 standard deviations above the mean on the Quantitative Reasoning section. (d) She did better on the Verbal Reasoning section since her Z-score on that section was higher. (e) $Perc_{VR} = 0.9007 \approx 90\%$, $Perc_{QR} = 0.6990 \approx 70\%$. (f) $100\% - 90\% = 10\%$ did better than her on VR, and $100\% - 70\% = 30\%$ did better than her on QR. (g) We cannot compare the raw scores since they are on different scales. Comparing her percentile scores is more appropriate when comparing her performance to others. (h) Answer to part (b) would not change as Z-scores can be calculated for distributions that are not normal. However, we could not answer parts (d)-(f) since we cannot use the normal probability table to calculate probabilities and percentiles without a normal model.

4.5 (a) $Z = 0.84$, which corresponds to approximately 159 on QR. (b) $Z = -0.52$, which corresponds to approximately 147 on VR.

4.7 (a) $Z = 1.2$, $P(Z > 1.2) = 0.1151$.
(b) $Z = -1.28 \rightarrow 70.6°$F or colder.

4.9 (a) $N(25, 2.78)$. (b) $Z = 1.08$, $P(Z > 1.08) = 0.1401$. (c) The answers are very close because only the units were changed. (The only reason why they differ at all because 28°C is 82.4°F, not precisely 83°F.) (d) Since $IQR = Q3 - Q1$, we first need to find $Q3$ and $Q1$ and take the difference between the two. Remember that $Q3$ is the 75th and $Q1$ is the 25th percentile of a distribution. Q1 = 23.13, Q3 = 26.86, IQR = 26. 86 - 23.13 = 3.73.

4.11 (a) No. The cards are not independent. For example, if the first card is an ace of clubs, that implies the second card cannot be an ace of clubs. Additionally, there are many possible categories, which would need to be simplified. (b) No. There are six events under consideration. The Bernoulli distribution allows for only two events or categories. Note that rolling a die could be a Bernoulli trial if we simply to two events, e.g. rolling a 6 and not rolling a 6, though specifying such details would be necessary.

4.13 (a) $0.875^2 \times 0.125 = 0.096$. (b) $\mu = 8$, $\sigma = 7.48$.

4.15 If p is the probability of a success, then the mean of a Bernoulli random variable X is given by
$\mu = E[X] = P(X = 0) \times 0 + P(X = 1) \times 1$
$= (1 - p) \times 0 + p \times 1 = 0 + p = p$

4.17 (a) Binomial conditions are met: (1) Independent trials: In a random sample, whether or not one 18-20 year old has consumed alcohol does not depend on whether or not another one has. (2) Fixed number of trials: $n = 10$. (3) Only two outcomes at each trial: Consumed or did not consume alcohol. (4) Probability of a success is the same for each trial: $p = 0.697$. (b) 0.203. (c) 0.203. (d) 0.167. (e) 0.997.

4.19 (a) $\mu 34.85$, $\sigma = 3.25$ (b) $Z = \frac{45-34.85}{3.25} = 3.12$. 45 is more than 3 standard deviations away from the mean, we can assume that it is an unusual observation. Therefore yes, we would be surprised. (c) Using the normal approximation, 0.0009. With 0.5 correction, 0.0015.

4.21 (a) $1 - 0.75^3 = 0.5781$. (b) 0.1406. (c) 0.4219. (d) $1 - 0.25^3 = 0.9844$.

4.23 (a) Geometric distribution: 0.109. (b) Binomial: 0.219. (c) Binomial: 0.137. (d) $1 - 0.875^6 = 0.551$. (e) Geometric: 0.084. (f) Using a binomial distribution with $n = 6$ and $p = 0.75$, we see that $\mu = 4.5$, $\sigma = 1.06$, and $Z = 2.36$. Since this is not within 2 SD, it may be considered unusual.

4.25 (a) $\overset{Anna}{1/5} \times \overset{Ben}{1/4} \times \overset{Carl}{1/3} \times \overset{Damian}{1/2} \times \overset{Eddy}{1/1} = 1/5! = 1/120$. (b) Since the probabilities must add to 1, there must be 5! = 120 possible orderings. (c) 8! = 40,320.

4.27 (a) 0.0804. (b) 0.0322. (c) 0.0193.

4.29 (a) Negative binomial with $n = 4$ and $p = 0.55$, where a success is defined here as a female student. The negative binomial setting is appropriate since the last trial is fixed but the order of the first 3 trials is unknown. (b) 0.1838. (c) $\binom{3}{1} = 3$. (d) In the binomial model there are no restrictions on the outcome of the last trial. In the negative binomial model the last trial is fixed. Therefore we are interested in the number of ways of orderings of the other $k - 1$ successes in the first $n - 1$ trials.

4.31 (a) Poisson with $\lambda = 75$. (b) $\mu = \lambda = 75$, $\sigma = \sqrt{\lambda} = 8.66$. (c) $Z = -1.73$. Since 60 is within 2 standard deviations of the mean, it would not generally be considered unusual. Note that we often use this rule of thumb even when the normal model does not apply. (d) Using Poisson with $\lambda = 75$: 0.0402.

4.33 (a) $\frac{\lambda^k \times e^{-\lambda}}{k!} = \frac{6.5^5 \times e^{-6.5}}{5!} = 0.1454$
(b) The probability will come to $0.0015 + 0.0098 + 0.0318 = 0.0431$ (0.0430 if no rounding error).
(c) The number of people per car is $11.7/6.5 = 1.8$, meaning people are coming in small clusters. That is, if one person arrives, there's a chance that they brought one or more other people in their vehicle. This means individuals (the people) are not independent, even if the car arrivals are independent, and this breaks a core assumption for the Poisson distribution. That is, the number of people visiting between 2pm and 3pm would not follow a Poisson distribution.

4.35 0 wins (-\$3): 0.1458. 1 win (-\$1): 0.3936. 2 wins (+\$1): 0.3543. 3 wins (+\$3): 0.1063.

4.37 Want to find the probability that there will be 1,786 or more enrollees. Using the normal approximation, with $\mu = np = 2{,}500 \times 0.7 = 1750$ and $\sigma = \sqrt{np(1-p)} = \sqrt{2{,}500 \times 0.7 \times 0.3} \approx 23$, $Z = 1.61$, and $P(Z > 1.61) = 0.0537$. With a 0.5 correction: 0.0559.

4.39 (a) $Z = 0.67$. (b) $\mu = \$1650$, $x = \$1800$. (c) $0.67 = \frac{1800 - 1650}{\sigma} \rightarrow \sigma = \$223.88$.

4.41 (a) $(1 - 0.471)^2 \times 0.471 = 0.1318$. (b) $0.471^3 = $ 0.1045. (c) $\mu = 1/0.471 = 2.12$, $\sigma = \sqrt{2.38} = 1.54$. (d) $\mu = 1/0.30 = 3.33$, $\sigma = 2.79$. (e) When p is smaller, the event is rarer, meaning the expected number of trials before a success and the standard deviation of the waiting time are higher.

4.43 $Z = 1.56$, $P(Z > 1.56) = 0.0594$, i.e. 6%.

4.45 (a) $Z = 0.73$, $P(Z > 0.73) = 0.2327$. (b) If you are bidding on only one auction and set a low maximum bid price, someone will probably outbid you. If you set a high maximum bid price, you may win the auction but pay more than is necessary. If bidding on more than one auction, and you set your maximum bid price very low, you probably won't win any of the auctions. However, if the maximum bid price is even modestly high, you are likely to win multiple auctions. (c) An answer roughly equal to the 10th percentile would be reasonable. Regrettably, no percentile cutoff point guarantees beyond any possible event that you win at least one auction. However, you may pick a higher percentile if you want to be more sure of winning an auction. (d) Answers will vary a little but should correspond to the answer in part (c). We use the 10th percentile: $Z = -1.28 \rightarrow \$69.80$.

4.47 (a) $Z = 3.5$, upper tail is 0.0002. (More precise value: 0.000233, but we'll use 0.0002 for the calculations here.)
(b) $0.0002 \times 2000 = 0.4$. We would expect about 0.4 10 year olds who are 76 inches or taller to show up.
(c) $\binom{2000}{0}(0.0002)^0(1 - 0.0002)^{2000} = 0.67029$.
(d) $\frac{0.4^0 \times e^{-0.4}}{0!} = \frac{1 \times e^{-0.4}}{1} = 0.67032$.

5 Foundations for inference

5.1 (a) Mean. Each student reports a numerical value: a number of hours. (b) Mean. Each student reports a number, which is a percentage, and we can average over these percentages. (c) Proportion. Each student reports Yes or No, so this is a categorical variable and we use a proportion. (d) Mean. Each student reports a number, which is a percentage like in part (b). (e) Proportion. Each student reports whether or not s/he expects to get a job, so this is a categorical variable and we use a proportion.

5.3 (a) The sample is from all computer chips manufactured at the factory during the week of production. We might be tempted to generalize the population to represent all weeks, but we should exercise caution here since the rate of defects may change over time. (b) The fraction of computer chips manufactured at the factory during the week of production that had defects. (c) Estimate the parameter using the data: $\hat{p} = \frac{27}{212} = 0.127$. (d) *Standard error* (or *SE*). (e) Compute the *SE* using $\hat{p} = 0.127$ in place of p: $SE \approx \sqrt{\frac{\hat{p}(1-\hat{p})}{n}} = \sqrt{\frac{0.127(1-0.127)}{212}} = 0.023$. (f) The standard error is the standard deviation of $\hat{p}$. A value of 0.10 would be about one standard error away from the observed value, which would not represent a very uncommon deviation. (Usually beyond about 2 standard errors is a good rule of thumb.) The engineer should not be surprised. (g) Recomputed standard error using $p = 0.1$: $SE = \sqrt{\frac{0.1(1-0.1)}{212}} = 0.021$. This value isn't very different, which is typical when the standard error is computed using relatively similar proportions (and even sometimes when those proportions are quite different!).

5.5 (a) Sampling distribution. (b) If the population proportion is in the 5-30% range, the success-failure condition would be satisfied and the sampling distribution would be symmetric. (c) We use the formula for the standard error: $SE = \sqrt{\frac{p(1-p)}{n}} = \sqrt{\frac{0.08(1-0.08)}{800}} = 0.0096$. (d) Standard error. (e) The distribution will tend to be more variable when we have fewer observations per sample.

5.7 Recall that the general formula is *point estimate* $\pm z^\star \times SE$. First, identify the three different values. The point estimate is 45%, $z^\star = 1.96$ for a 95% confidence level, and $SE = 1.2\%$. Then, plug the values into the formula: $45\% \pm 1.96 \times 1.2\%$ $\rightarrow$ $(42.6\%, 47.4\%)$ We are 95% confident that the proportion of US adults who live with one or more chronic conditions is between 42.6% and 47.4%.

5.9 (a) False. Confidence intervals provide a range of plausible values, and sometimes the truth is missed. A 95% confidence interval "misses" about 5% of the time. (b) True. Notice that the description focuses on the true population value. (c) True. If we examine the 95% confidence interval computed in Exercise 5.9, we can see that 50% is not included in this interval. This means that in a hypothesis test, we would reject the null hypothesis that the proportion is 0.5. (d) False. The standard error describes the uncertainty in the overall estimate from natural fluctuations due to randomness, not the uncertainty corresponding to individuals' responses.

5.11 (a) False. Inference is made on the population parameter, not the point estimate. The point estimate is always in the confidence interval. (b) True. (c) False. The confidence interval is not about a sample mean. (d) False. To be more confident that we capture the parameter, we need a wider interval. Think about needing a bigger net to be more sure of catching a fish in a murky lake. (e) True. Optional explanation: This is true since the normal model was used to model the sample mean. The margin of error is half the width of the interval, and the sample mean is the midpoint of the interval. (f) False. In the calculation of the standard error, we divide the standard deviation by the square root of the sample size. To cut the SE (or margin of error) in half, we would need to sample $2^2 = 4$ times the number of people in the initial sample.

5.13 (a) The visitors are from a simple random sample, so independence is satisfied. The success-failure condition is also satisfied, with both 64 and $752 - 64 = 688$ above 10. Therefore, we can use a normal distribution to model $\hat{p}$ and construct a confidence interval. (b) The sample proportion is $\hat{p} = \frac{64}{752} = 0.085$. The standard error is

$$SE = \sqrt{\frac{p(1-p)}{n}} \approx \sqrt{\frac{\hat{p}(1-\hat{p})}{n}}$$
$$= \sqrt{\frac{0.085(1 - 0.085)}{752}} = 0.010$$

(c) For a 90% confidence interval, use $z^\star = 1.65$. The confidence interval is $0.085 \pm 1.65 \times 0.010$ $\rightarrow$ $(0.0685, 0.1015)$. We are 90% confident that 6.85% to 10.15% of first-time site visitors will register using the new design.

5.15 (a) $H_0 : p = 0.5$ (Neither a majority nor minority of students' grades improved) $H_A : p \neq 0.5$ (Either a majority or a minority of students' grades improved)
(b) $H_0 : \mu = 15$ (The average amount of company time each employee spends not working is 15 minutes for March Madness.) $H_A : \mu \neq 15$ (The average amount of company time each employee spends not working is different than 15 minutes for March Madness.)

5.17 (1) The hypotheses should be about the population proportion (p), not the sample proportion. (2) The null hypothesis should have an equal sign. (3) The alternative hypothesis should have a not-equals sign, and (4) it should reference the null value, $p_0 = 0.6$, not the observed sample proportion. The correct way to set up these hypotheses is: $H_0 : p = 0.6$ and $H_A : p \neq 0.6$.

5.19 (a) This claim is reasonable, since the entire interval lies above 50%. (b) The value of 70% lies outside of the interval, so we have convincing evidence that the researcher's conjecture is wrong. (c) A 90% confidence interval will be narrower than a 95% confidence interval. Even without calculating the interval, we can tell that 70% would not fall in the interval, and we would reject the researcher's conjecture based on a 90% confidence level as well.

5.21 (i) Set up hypotheses. H_0: $p = 0.5$, H_A: $p \neq 0.5$. We will use a significance level of $\alpha = 0.05$. (ii) Check conditions: simple random sample gets us independence, and the success-failure conditions is satisfied since $0.5 \times 1000 = 500$ for each group is at least 10. (iii) Next, we calculate: $SE = \sqrt{0.5(1 - 0.5)/1000} = 0.016$. $Z = \frac{0.42 - 0.5}{0.016} = -5$, which has a one-tail area of about 0.0000003, so the p-value is twice this one-tail area at 0.0000006. (iv) Make a conclusion: Because the p-value is less than $\alpha = 0.05$, we reject the null hypothesis and conclude that the fraction of US adults who believe raising the minimum wage will help the economy is not 50%. Because the observed value is less than 50% and we have rejected the null hypothesis, we can conclude that this belief is held by fewer than 50% of US adults. (For reference, the survey also explores support for changing the minimum wage, which is a different question than if it will help the economy.)

5.23 If the p-value is 0.05, this means the test statistic would be either $Z = -1.96$ or $Z = 1.96$. We'll show the calculations for $Z = 1.96$. Standard error: $SE = \sqrt{0.3(1 - 0.3)/90} = 0.048$. Finally, set up the test statistic formula and solve for $\hat{p}$: $1.96 = \frac{\hat{p} - 0.3}{0.048}$ $\rightarrow$ $\hat{p} = 0.394$ Alternatively, if $Z = -1.96$ was used: $\hat{p} = 0.206$.

5.25 (a) H_0: Anti-depressants do not affect the symptoms of Fibromyalgia. H_A: Anti-depressants do affect the symptoms of Fibromyalgia (either helping or harming). (b) Concluding that anti-depressants either help or worsen Fibromyalgia symptoms when they actually do neither. (c) Concluding that anti-depressants do not affect Fibromyalgia symptoms when they actually do.

5.27 (a) We are 95% confident that Americans spend an average of 1.38 to 1.92 hours per day relaxing or pursuing activities they enjoy. (b) Their confidence level must be higher as the width of the confidence interval increases as the confidence level increases. (c) The new margin of error will be smaller, since as the sample size increases, the standard error decreases, which will decrease the margin of error.

5.29 (a) H_0: The restaurant meets food safety and sanitation regulations. H_A: The restaurant does not meet food safety and sanitation regulations. (b) The food safety inspector concludes that the restaurant does not meet food safety and sanitation regulations and shuts down the restaurant when the restaurant is actually safe. (c) The food safety inspector concludes that the restaurant meets food safety and sanitation regulations and the restaurant stays open when the restaurant is actually not safe. (d) A Type 1 Error may be more problematic for the restaurant owner since his restaurant gets shut down even though it meets the food safety and sanitation regulations. (e) A Type 2 Error may be more problematic for diners since the restaurant deemed safe by the inspector is actually not. (f) Strong evidence. Diners would rather a restaurant that meet the regulations get shut down than a restaurant that doesn't meet the regulations not get shut down.

5.31 (a) H_0 : $p_{unemp} = p_{underemp}$: The proportions of unemployed and underemployed people who are having relationship problems are equal. H_A : $p_{unemp} \neq p_{underemp}$: The proportions of unemployed and underemployed people who are having relationship problems are different. (b) If in fact the two population proportions are equal, the probability of observing at least a 2% difference between the sample proportions is approximately 0.35. Since this is a high probability we fail to reject the null hypothesis. The data do not provide convincing evidence that the proportion of of unemployed and underemployed people who are having relationship problems are different.

5.33 Because 130 is inside the confidence interval, we do not have convincing evidence that the true average is any different than what the nutrition label suggests.

5.35 True. If the sample size gets ever larger, then the standard error will become ever smaller. Eventually, when the sample size is large enough and the standard error is tiny, we can find statistically significant yet very small differences between the null value and point estimate (assuming they are not exactly equal).

5.37 (a) In effect, we're checking whether men are paid more than women (or vice-versa), and we'd expect these outcomes with either chance under the null hypothesis:

$$H_0 : p = 0.5 \qquad H_A : p \neq 0.5$$

We'll use p to represent the fraction of cases where men are paid more than women.
(b) Below is the completion of the hypothesis test.

- There isn't a good way to check independence here since the jobs are not a simple random sample. However, independence doesn't seem unreasonable, since the individuals in each job are different from each other. The success-failure condition is met since we check it using the null proportion: $p_0 n = (1 - p_0)n = 10.5$ is greater than 10.

- We can compute the sample proportion, SE, and test statistic:

$$\hat{p} = 19/21 = 0.905$$
$$SE = \sqrt{\frac{0.5 \times (1 - 0.5)}{21}} = 0.109$$
$$Z = \frac{0.905 - 0.5}{0.109} = 3.72$$

  The test statistic Z corresponds to an upper tail area of about 0.0001, so the p-value is 2 times this value: 0.0002.

- Because the p-value is smaller than 0.05, we reject the notion that all these gender pay disparities are due to chance. Because we observe that men are paid more in a higher proportion of cases and we have rejected H_0, we can conclude that men are being paid higher amounts in ways not explainable by chance alone.

If you're curious for more info around this topic, including a discussion about adjusting for additional factors that affect pay, please see the following video by Healthcare Triage: youtu.be/aVhgKSULNQA.

6 Inference for categorical data

6.1 (a) False. Doesn't satisfy success-failure condition. (b) True. The success-failure condition is not satisfied. In most samples we would expect $\hat{p}$ to be close to 0.08, the true population proportion. While $\hat{p}$ can be much above 0.08, it is bound below by 0, suggesting it would take on a right skewed shape. Plotting the sampling distribution would confirm this suspicion. (c) False. $SE_{\hat{p}} = 0.0243$, and $\hat{p} = 0.12$ is only $\frac{0.12-0.08}{0.0243} = 1.65$ SEs away from the mean, which would not be considered unusual. (d) True. $\hat{p} = 0.12$ is 2.32 standard errors away from the mean, which is often considered unusual. (e) False. Decreases the SE by a factor of $1/\sqrt{2}$.

6.3 (a) True. See the reasoning of 6.1(b). (b) True. We take the square root of the sample size in the SE formula. (c) True. The independence and success-failure conditions are satisfied. (d) True. The independence and success-failure conditions are satisfied.

6.5 (a) False. A confidence interval is constructed to estimate the population proportion, not the sample proportion. (b) True. 95% CI: 82% ± 2%. (c) True. By the definition of the confidence level. (d) True. Quadrupling the sample size decreases the SE and ME by a factor of $1/\sqrt{4}$. (e) True. The 95% CI is entirely above 50%.

6.7 With a random sample, independence is satisfied. The success-failure condition is also satisfied. $ME = z^{\star}\sqrt{\frac{\hat{p}(1-\hat{p})}{n}} = 1.96\sqrt{\frac{0.56 \times 0.44}{600}} = 0.0397 \approx 4\%$

6.9 (a) No. The sample only represents students who took the SAT, and this was also an online survey. (b) (0.5289, 0.5711). We are 90% confident that 53% to 57% of high school seniors who took the SAT are fairly certain that they will participate in a study abroad program in college. (c) 90% of such random samples would produce a 90% confidence interval that includes the true proportion. (d) Yes. The interval lies entirely above 50%.

6.11 (a) We want to check for a majority (or minority), so we use the following hypotheses:

$$H_0 : p = 0.5 \qquad H_A : p \neq 0.5$$

We have a sample proportion of $\hat{p} = 0.55$ and a sample size of $n = 617$ independents.
Since this is a random sample, independence is satisfied. The success-failure condition is also satisfied: 617×0.5 and $617 \times (1 - 0.5)$ are both at least 10 (we use the null proportion $p_0 = 0.5$ for this check in a one-proportion hypothesis test).
Therefore, we can model $\hat{p}$ using a normal distribution with a standard error of

$$SE = \sqrt{\frac{p(1-p)}{n}} = 0.02$$

(We use the null proportion $p_0 = 0.5$ to compute the standard error for a one-proportion hypothesis test.)

Next, we compute the test statistic:

$$Z = \frac{0.55 - 0.5}{0.02} = 2.5$$

This yields a one-tail area of 0.0062, and a p-value of $2 \times 0.0062 = 0.0124$.
Because the p-value is smaller than 0.05, we reject the null hypothesis. We have strong evidence that the support is different from 0.5, and since the data provide a point estimate above 0.5, we have strong evidence to support this claim by the TV pundit.
(b) No. Generally we expect a hypothesis test and a confidence interval to align, so we would expect the confidence interval to show a range of plausible values entirely above 0.5. However, if the confidence level is misaligned (e.g. a 99% confidence level and a $\alpha = 0.05$ significance level), then this is no longer generally true.

6.13 (a) $H_0 : p = 0.5$. $H_A : p \neq 0.5$. Independence (random sample) is satisfied, as is the success-failure conditions (using $p_0 = 0.5$, we expect 40 successes and 40 failures). $Z = 2.91 \rightarrow$ the one tail area is 0.0018, so the p-value is 0.0036. Since the p-value < 0.05, we reject the null hypothesis. Since we rejected H_0 and the point estimate suggests people are better than random guessing, we can conclude the rate of correctly identifying a soda for these people is significantly better than just by random guessing. (b) If in fact people cannot tell the difference between diet and regular soda and they were randomly guessing, the probability of getting a random sample of 80 people where 53 or more identify a soda correctly (or 53 or more identify a soda incorrectly) would be 0.0036.

6.15 Since a sample proportion ($\hat{p} = 0.55$) is available, we use this for the sample size calculations. The margin of error for a 90% confidence interval is $1.65 \times SE = 1.65 \times \sqrt{\frac{p(1-p)}{n}}$. We want this to be less than 0.01, where we use $\hat{p}$ in place of p:

$$1.65 \times \sqrt{\frac{0.55(1-0.55)}{n}} \leq 0.01$$
$$1.65^2 \frac{0.55(1-0.55)}{0.01^2} \leq n$$

From this, we get that n must be at least 6739.

6.17 This is not a randomized experiment, and it is unclear whether people would be affected by the behavior of their peers. That is, independence may not hold. Additionally, there are only 5 interventions under the provocative scenario, so the success-failure condition does not hold. Even if we consider a hypothesis test where we pool the proportions, the success-failure condition will not be satisfied. Since one condition is questionable and the other is not satisfied, the difference in sample proportions will not follow a nearly normal distribution.

6.19 (a) False. The entire confidence interval is above 0. (b) True. (c) True. (d) True. (e) False. It is simply the negated and reordered values: (-0.06,-0.02).

6.21 (a) Standard error:

$$SE = \sqrt{\frac{0.79(1-0.79)}{347} + \frac{0.55(1-0.55)}{617}} = 0.03$$

Using $z^\star = 1.96$, we get:

$$0.79 - 0.55 \pm 1.96 \times 0.03 \to (0.181, 0.299)$$

We are 95% confident that the proportion of Democrats who support the plan is 18.1% to 29.9% higher than the proportion of Independents who support the plan. (b) True.

6.23 (a) College grads: 23.7%. Non-college grads: 33.7%. (b) Let p_{CG} and p_{NCG} represent the proportion of college graduates and non-college graduates who responded "do not know". $H_0 : p_{CG} = p_{NCG}$. $H_A : p_{CG} \neq p_{NCG}$. Independence is satisfied (random sample), and the success-failure condition, which we would check using the pooled proportion ($\hat{p}_{pool} = 235/827 = 0.284$), is also satisfied. $Z = -3.18 \to$ p-value = 0.0014. Since the p-value is very small, we reject H_0. The data provide strong evidence that the proportion of college graduates who do not have an opinion on this issue is different than that of non-college graduates. The data also indicate that fewer college grads say they "do not know" than non-college grads (i.e. the data indicate the direction after we reject H_0).

6.25 (a) College grads: 35.2%. Non-college grads: 33.9%. (b) Let p_{CG} and p_{NCG} represent the proportion of college graduates and non-college grads who support offshore drilling. $H_0 : p_{CG} = p_{NCG}$. $H_A : p_{CG} \neq p_{NCG}$. Independence is satisfied (random sample), and the success-failure condition, which we would check using the pooled proportion ($\hat{p}_{pool} = 286/827 = 0.346$), is also satisfied. $Z = 0.39 \to$ p-value = 0.6966. Since the p-value > α (0.05), we fail to reject H_0. The data do not provide strong evidence of a difference between the proportions of college graduates and non-college graduates who support off-shore drilling in California.

6.27 Subscript C means control group. Subscript T means truck drivers. $H_0 : p_C = p_T$. $H_A : p_C \neq p_T$. Independence is satisfied (random samples), as is the success-failure condition, which we would check using the pooled proportion ($\hat{p}_{pool} = 70/495 = 0.141$). $Z = -1.65 \to$ p-value = 0.0989. Since the p-value is high (default to alpha = 0.05), we fail to reject H_0. The data do not provide strong evidence that the rates of sleep deprivation are different for non-transportation workers and truck drivers.

6.29 (a) Summary of the study:

		Virol. failure		
		Yes	No	Total
Treatment	Nevaripine	26	94	120
	Lopinavir	10	110	120
	Total	36	204	240

(b) $H_0 : p_N = p_L$. There is no difference in virologic failure rates between the Nevaripine and Lopinavir groups. $H_A : p_N \neq p_L$. There is some difference in virologic failure rates between the Nevaripine and Lopinavir groups. (c) Random assignment was used, so the observations in each group are independent. If the patients in the study are representative of those in the general population (something impossible to check with the given information), then we can also confidently generalize the findings to the population. The success-failure condition, which we would check using the pooled proportion ($\hat{p}_{pool} = 36/240 = 0.15$), is satisfied. $Z = 2.89 \to$ p-value = 0.0039. Since the p-value is low, we reject H_0. There is strong evidence of a difference in virologic failure rates between the Nevaripine and Lopinavir groups. Treatment and virologic failure do not appear to be independent.

6.31 (a) False. The chi-square distribution has one parameter called degrees of freedom. (b) True. (c) True. (d) False. As the degrees of freedom increases, the shape of the chi-square distribution becomes more symmetric.

6.33 (a) H_0: The distribution of the format of the book used by the students follows the professor's predictions. H_A: The distribution of the format of the book used by the students does not follow the professor's predictions. (b) $E_{hard\ copy} = 126 \times 0.60 = 75.6$. $E_{print} = 126 \times 0.25 = 31.5$. $E_{online} = 126 \times 0.15 = 18.9$. (c) Independence: The sample is not random. However, if the professor has reason to believe that the proportions are stable from one term to the next and students are not affecting each other's study habits, independence is probably reasonable. Sample size: All expected counts are at least 5. (d) $\chi^2 = 2.32$, $df = 2$, p-value = 0.313. (e) Since the p-value is large, we fail to reject H_0. The data do not provide strong evidence indicating the professor's predictions were statistically inaccurate.

6.35 (a) Two-way table:

	Quit		
Treatment	Yes	No	Total
Patch + support group	40	110	150
Only patch	30	120	150
Total	70	230	300

(b-i) $E_{row_1,col_1} = \frac{(row\ 1\ total) \times (col\ 1\ total)}{table\ total} = 35$. This is lower than the observed value.
(b-ii) $E_{row_2,col_2} = \frac{(row\ 2\ total) \times (col\ 2\ total)}{table\ total} = 115$. This is lower than the observed value.

6.37 H_0: The opinion of college grads and non-grads is not different on the topic of drilling for oil and natural gas off the coast of California. H_A: Opinions regarding the drilling for oil and natural gas off the coast of California has an association with earning a college degree.

$$E_{row\ 1,col\ 1} = 151.5 \qquad E_{row\ 1,col\ 2} = 134.5$$
$$E_{row\ 2,col\ 1} = 162.1 \qquad E_{row\ 2,col\ 2} = 143.9$$
$$E_{row\ 3,col\ 1} = 124.5 \qquad E_{row\ 3,col\ 2} = 110.5$$

Independence: The samples are both random, unrelated, and from less than 10% of the population, so independence between observations is reasonable. Sample size: All expected counts are at least 5. $\chi^2 = 11.47$, $df = 2 \rightarrow$ p-value $= 0.003$. Since the p-value $< \alpha$, we reject H_0. There is strong evidence that there is an association between support for offshore drilling and having a college degree.

6.39 No. The samples at the beginning and at the end of the semester are not independent since the survey is conducted on the same students.

6.41 (a) H_0: The age of Los Angeles residents is independent of shipping carrier preference variable. H_A: The age of Los Angeles residents is associated with the shipping carrier preference variable. (b) The conditions are not satisfied since some expected counts are below 5.

6.43 (a) Independence is satisfied (random sample), as is the success-failure condition (40 smokers, 160 non-smokers). The 95% CI: (0.145, 0.255). We are 95% confident that 14.5% to 25.5% of all students at this university smoke. (b) We want $z^\star SE$ to be no larger than 0.02 for a 95% confidence level. We use $z^\star = 1.96$ and plug in the point estimate $\hat{p} = 0.2$ within the SE formula: $1.96\sqrt{0.2(1-0.2)/n} \leq 0.02$. The sample size n should be at least 1,537.

6.45 (a) Proportion of graduates from this university who found a job within one year of graduating. $\hat{p} = 348/400 = 0.87$. (b) This is a random sample, so the observations are independent. Success-failure condition is satisfied: 348 successes, 52 failures, both well above 10. (c) (0.8371, 0.9029). We are 95% confident that approximately 84% to 90% of graduates from this university found a job within one year of completing their undergraduate degree. (d) 95% of such random samples would produce a 95% confidence interval that includes the true proportion of students at this university who found a job within one year of graduating from college. (e) (0.8267, 0.9133). Similar interpretation as before. (f) 99% CI is wider, as we are more confident that the true proportion is within the interval and so need to cover a wider range.

6.47 Use a chi-squared goodness of fit test. H_0: Each option is equally likely. H_A: Some options are preferred over others. Total sample size: 99. Expected counts: $(1/3) * 99 = 33$ for each option. These are all above 5, so conditions are satisfied. $df = 3 - 1 = 2$ and $\chi^2 = \frac{(43-33)^2}{33} + \frac{(21-33)^2}{33} + \frac{(35-33)^2}{33} = 7.52 \rightarrow$ p-value $= 0.023$. Since the p-value is less than 5%, we reject H_0. The data provide convincing evidence that some options are preferred over others.

6.49 (a) $H_0 : p = 0.38$. $H_A : p \neq 0.38$. Independence (random sample) and the success-failure condition are satisfied. $Z = -20.5 \rightarrow$ p-value ≈ 0. Since the p-value is very small, we reject H_0. The data provide strong evidence that the proportion of Americans who only use their cell phones to access the internet is different than the Chinese proportion of 38%, and the data indicate that the proportion is lower in the US. (b) If in fact 38% of Americans used their cell phones as a primary access point to the internet, the probability of obtaining a random sample of 2,254 Americans where 17% or less or 59% or more use their only their cell phones to access the internet would be approximately 0. (c) (0.1545, 0.1855). We are 95% confident that approximately 15.5% to 18.6% of all Americans primarily use their cell phones to browse the internet.

7 Inference for numerical data

7.1 (a) $df = 6 - 1 = 5$, $t_5^\star = 2.02$ (column with two tails of 0.10, row with $df = 5$). (b) $df = 21 - 1 = 20$, $t_{20}^\star = 2.53$ (column with two tails of 0.02, row with $df = 20$). (c) $df = 28$, $t_{28}^\star = 2.05$. (d) $df = 11$, $t_{11}^\star = 3.11$.

7.3 (a) 0.085, do not reject H_0. (b) 0.003, reject H_0. (c) 0.438, do not reject H_0. (d) 0.042, reject H_0.

7.5 The mean is the midpoint: $\bar{x} = 20$. Identify the margin of error: $ME = 1.015$, then use $t_{35}^\star = 2.03$ and $SE = s/\sqrt{n}$ in the formula for margin of error to identify $s = 3$.

7.7 (a) H_0: $\mu = 8$ (New Yorkers sleep 8 hrs per night on average.) H_A: $\mu \neq 8$ (New Yorkers sleep less or more than 8 hrs per night on average.) (b) Independence: The sample is random. The min/max suggest there are no concerning outliers. $T = -1.75$. $df = 25 - 1 = 24$. (c) p-value $= 0.093$. If in fact the true population mean of the amount New Yorkers sleep per night was 8 hours, the probability of getting a random sample of 25 New Yorkers where the average amount of sleep is 7.73 hours per night or less (or 8.27 hours or more) is 0.093. (d) Since p-value > 0.05, do not reject H_0. The data do not provide strong evidence that New Yorkers sleep more or less than 8 hours per night on average. (e) No, since the p-value is smaller than $1 - 0.90 = 0.10$.

7.9 T is either -2.09 or 2.09. Then $\bar{x}$ is one of the following:

$$-2.09 = \frac{\bar{x} - 60}{\frac{8}{\sqrt{20}}} \rightarrow \bar{x} = 56.26$$

$$2.09 = \frac{\bar{x} - 60}{\frac{8}{\sqrt{20}}} \rightarrow \bar{x} = 63.74$$

7.11 (a) We will conduct a 1-sample t-test. H_0: $\mu = 5$. H_A: $\mu \neq 5$. We'll use $\alpha = 0.05$. This is a random sample, so the observations are independent. To proceed, we assume the distribution of years of piano lessons is approximately normal. $SE = 2.2/\sqrt{20} = 0.4919$. The test statistic is $T = (4.6 - 5)/SE = -0.81$. $df = 20 - 1 = 19$. The one-tail area is about 0.21, so the p-value is about 0.42, which is bigger than $\alpha = 0.05$ and we do not reject H_0. That is, we do not have sufficiently strong evidence to reject the notion that the average is 5 years.
(b) Using $SE = 0.4919$ and $t^\star_{df=19} = 2.093$, the confidence interval is $(3.57, 5.63)$. We are 95% confident that the average number of years a child takes piano lessons in this city is 3.57 to 5.63 years.
(c) They agree, since we did not reject the null hypothesis and the null value of 5 was in the t-interval.

7.13 If the sample is large, then the margin of error will be about $1.96 \times 100/\sqrt{n}$. We want this value to be less than 10, which leads to $n \geq 384.16$, meaning we need a sample size of at least 385 (round up for sample size calculations!).

7.15 Paired, data are recorded in the same cities at two different time points. The temperature in a city at one point is not independent of the temperature in the same city at another time point.

7.17 (a) Since it's the same students at the beginning and the end of the semester, there is a pairing between the datasets, for a given student their beginning and end of semester grades are dependent.
(b) Since the subjects were sampled randomly, each observation in the men's group does not have a special correspondence with exactly one observation in the other (women's) group. (c) Since it's the same subjects at the beginning and the end of the study, there is a pairing between the datasets, for a subject student their beginning and end of semester artery thickness are dependent. (d) Since it's the same subjects at the beginning and the end of the study, there is a pairing between the datasets, for a subject student their beginning and end of semester weights are dependent.

7.19 (a) For each observation in one data set, there is exactly one specially corresponding observation in the other data set for the same geographic location. The data are paired. (b) $H_0 : \mu_{\text{diff}} = 0$ (There is no difference in average number of days exceeding 90°F

in 1948 and 2018 for NOAA stations.) $H_A : \mu_{\text{diff}} \neq 0$ (There is a difference.) (c) Locations were randomly sampled, so independence is reasonable. The sample size is at least 30, so we're just looking for particularly extreme outliers: none are present (the observation off left in the histogram would be considered a clear outlier, but not a particularly extreme one). Therefore, the conditions are satisfied. (d) $SE = 17.2/\sqrt{197} = 1.23$. $T = \frac{2.9-0}{1.23} = 2.36$ with degrees of freedom $df = 197 - 1 = 196$. This leads to a one-tail area of 0.0096 and a p-value of about 0.019. (e) Since the p-value is less than 0.05, we reject H_0. The data provide strong evidence that NOAA stations observed more 90°F days in 2018 than in 1948. (f) Type 1 Error, since we may have incorrectly rejected H_0. This error would mean that NOAA stations did not actually observe a decrease, but the sample we took just so happened to make it appear that this was the case. (g) No, since we rejected H_0, which had a null value of 0.

7.21 (a) $SE = 1.23$ and $t^\star = 1.65$. $2.9 \pm 1.65 \times 1.23 \rightarrow (0.87, 4.93)$.
(b) We are 90% confident that there was an increase of 0.87 to 4.93 in the average number of days that hit 90°F in 2018 relative to 1948 for NOAA stations.
(c) Yes, since the interval lies entirely above 0.

7.23 (a) These data are paired. For example, the Friday the 13th in say, September 1991, would probably be more similar to the Friday the 6th in September 1991 than to Friday the 6th in another month or year.
(b) Let $\mu_{diff} = \mu_{sixth} - \mu_{thirteenth}$. $H_0 : \mu_{diff} = 0$. $H_A : \mu_{diff} \neq 0$.
(c) Independence: The months selected are not random. However, if we think these dates are roughly equivalent to a simple random sample of all such Friday 6th/13th date pairs, then independence is reasonable. To proceed, we must make this strong assumption, though we should note this assumption in any reported results. Normality: With fewer than 10 observations, we would need to see clear outliers to be concerned. There is a borderline outlier on the right of the histogram of the differences, so we would want to report this in formal analysis results.
(d) $T = 4.93$ for $df = 10 - 1 = 9 \rightarrow$ p-value = 0.001.
(e) Since p-value < 0.05, reject H_0. The data provide strong evidence that the average number of cars at the intersection is higher on Friday the 6th than on Friday the 13th. (We should exercise caution about generalizing the interpretation to all intersections or roads.)
(f) If the average number of cars passing the intersection actually was the same on Friday the 6th and 13th, then the probability that we would observe a test statistic so far from zero is less than 0.01.
(g) We might have made a Type 1 Error, i.e. incorrectly rejected the null hypothesis.

7.25 (a) $H_0 : \mu_{diff} = 0$. $H_A : \mu_{diff} \neq 0$. $T = -2.71$. $df = 5$. p-value = 0.042. Since p-value < 0.05, reject H_0. The data provide strong evidence that the average number of traffic accident related emergency room admissions are different between Friday the 6th and Friday the 13th. Furthermore, the data indicate that the direction of that difference is that accidents are lower on Friday the 6th relative to Friday the 13th.
(b) (-6.49, -0.17).
(c) This is an observational study, not an experiment, so we cannot so easily infer a causal intervention implied by this statement. It is true that there is a difference. However, for example, this does not mean that a responsible adult going out on Friday the 13th has a higher chance of harm than on any other night.

7.27 (a) Chicken fed linseed weighed an average of 218.75 grams while those fed horsebean weighed an average of 160.20 grams. Both distributions are relatively symmetric with no apparent outliers. There is more variability in the weights of chicken fed linseed.
(b) $H_0 : \mu_{ls} = \mu_{hb}$. $H_A : \mu_{ls} \neq \mu_{hb}$.
We leave the conditions to you to consider.
$T = 3.02$, $df = min(11,9) = 9 \rightarrow$ p-value = 0.014. Since p-value < 0.05, reject H_0. The data provide strong evidence that there is a significant difference between the average weights of chickens that were fed linseed and horsebean.
(c) Type 1 Error, since we rejected H_0.
(d) Yes, since p-value > 0.01, we would not have rejected H_0.

7.29 $H_0 : \mu_C = \mu_S$. $H_A : \mu_C \neq \mu_S$. $T = 3.27$, $df = 11 \rightarrow$ p-value = 0.007. Since p-value < 0.05, reject H_0. The data provide strong evidence that the average weight of chickens that were fed casein is different than the average weight of chickens that were fed soybean (with weights from casein being higher). Since this is a randomized experiment, the observed difference can be attributed to the diet.

7.31 Let $\mu_{diff} = \mu_{pre} - \mu_{post}$. $H_0 : \mu_{diff} = 0$: Treatment has no effect. $H_A : \mu_{diff} \neq 0$: Treatment has an effect on P.D.T. scores, either positive or negative. Conditions: The subjects are randomly assigned to treatments, so independence within and between groups is satisfied. All three sample sizes are smaller than 30, so we look for clear outliers. There is a borderline outlier in the first treatment group. Since it is borderline, we will proceed, but we should report this caveat with any results. For all three groups: $df = 13$. $T_1 = 1.89 \rightarrow$ p-value = 0.081, $T_2 = 1.35 \rightarrow$ p-value = 0.200), $T_3 = -1.40 \rightarrow$ (p-value = 0.185). We do not reject the null hypothesis for any of these groups. As earlier noted, there is some uncertainty about if the method applied is reasonable for the first group.

7.33 Difference we care about: 40. Single tail of 90%: $1.28 \times SE$. Rejection region bounds: $\pm 1.96 \times SE$ (if 5% significance level). Setting $3.24 \times SE = 40$, subbing in $SE = \sqrt{\frac{94^2}{n} + \frac{94^2}{n}}$, and solving for the sample size n gives 116 plots of land for each fertilizer.

7.35 Alternative.

7.37 $H_0: \mu_1 = \mu_2 = \cdots = \mu_6$. H_A: The average weight varies across some (or all) groups. Independence: Chicks are randomly assigned to feed types (presumably kept separate from one another), therefore independence of observations is reasonable. Approx. normal: the distributions of weights within each feed type appear to be fairly symmetric. Constant variance: Based on the side-by-side box plots, the constant variance assumption appears to be reasonable. There are differences in the actual computed standard deviations, but these might be due to chance as these are quite small samples. $F_{5,65} = 15.36$ and the p-value is approximately 0. With such a small p-value, we reject H_0. The data provide convincing evidence that the average weight of chicks varies across some (or all) feed supplement groups.

7.39 (a) H_0: The population mean of MET for each group is equal to the others. H_A: At least one pair of means is different. (b) Independence: We don't have any information on how the data were collected, so we cannot assess independence. To proceed, we must assume the subjects in each group are independent. In practice, we would inquire for more details. Normality: The data are bound below by zero and the standard deviations are larger than the means, indicating very strong skew. However, since the sample sizes are extremely large, even extreme skew is acceptable. Constant variance: This condition is sufficiently met, as the standard deviations are reasonably consistent across groups. (c) See below, with the last column omitted:

	Df	Sum Sq	Mean Sq	F value
coffee	4	10508	2627	5.2
Residuals	50734	25564819	504	
Total	50738	25575327		

(d) Since p-value is very small, reject H_0. The data provide convincing evidence that the average MET differs between at least one pair of groups.

7.41 (a) H_0: Average GPA is the same for all majors. H_A: At least one pair of means are different. (b) Since p-value > 0.05, fail to reject H_0. The data do not provide convincing evidence of a difference between the average GPAs across three groups of majors. (c) The total degrees of freedom is 195+2 = 197, so the sample size is 197 + 1 = 198.

7.43 (a) False. As the number of groups increases, so does the number of comparisons and hence the modified significance level decreases. (b) True. (c) True. (d) False. We need observations to be independent regardless of sample size.

7.45 (a) H_0: Average score difference is the same for all treatments. H_A: At least one pair of means are different. (b) We should check conditions. If we look back to the earlier exercise, we will see that the patients were randomized, so independence is satisfied. There are some minor concerns about skew, especially with the third group, though this may be acceptable. The standard deviations across the groups are reasonably similar. Since the p-value is less than 0.05, reject H_0. The data provide convincing evidence of a difference between the average reduction in score among treatments. (c) We determined that at least two means are different in part (b), so we now conduct $K = 3 \times 2/2 = 3$ pairwise t-tests that each use $\alpha = 0.05/3 = 0.0167$ for a significance level. Use the following hypotheses for each pairwise test. H_0: The two means are equal. H_A: The two means are different. The sample sizes are equal and we use the pooled SD, so we can compute $SE = 3.7$ with the pooled $df = 39$. The p-value for Trmt 1 vs. Trmt 3 is the only one under 0.05: p-value = 0.035 (or 0.024 if using s_{pooled} in place of s_1 and s_3, though this won't affect the final conclusion). The p-value is larger than $0.05/3 = 1.67$, so we do not have strong evidence to conclude that it is this particular pair of groups that are different. That is, we cannot identify if which particular pair of groups are actually different, even though we've rejected the notion that they are all the same!

7.47 $H_0 : \mu_T = \mu_C$. $H_A : \mu_T \neq \mu_C$. $T = 2.24$, $df = 21 \rightarrow$ p-value = 0.036. Since p-value < 0.05, reject H_0. The data provide strong evidence that the average food consumption by the patients in the treatment and control groups are different. Furthermore, the data indicate patients in the distracted eating (treatment) group consume more food than patients in the control group.

7.49 False. While it is true that paired analysis requires equal sample sizes, only having the equal sample sizes isn't, on its own, sufficient for doing a paired test. Paired tests require that there be a special correspondence between each pair of observations in the two groups.

7.51 (a) We are building a distribution of sample statistics, in this case the sample mean. Such a distribution is called a sampling distribution. (b) Because we are dealing with the distribution of sample means, we need to check to see if the Central Limit Theorem applies. Our sample size is greater than 30, and we are told that random sampling is employed. With these conditions met, we expect that the distribution of the sample mean will be nearly normal and therefore symmetric. (c) Because we are dealing with a sampling distribution, we measure its variability with the standard error. $SE = 18.2/\sqrt{45} = 2.713$. (d) The sample means will be more variable with the smaller sample size.

7.53 (a) We should set 1.0% equal to 2.84 standard errors: $2.84 \times SE_{desired} = 1.0\%$ (see Example 7.37 on page 282 for details). This means the standard error should be about $SE = 0.35\%$ to achieve the desired statistical power.
(b) The margin of error was $0.5 \times (2.6\% - (-0.2\%)) = 1.4\%$, so the standard error in the experiment must have been $1.96 \times SE_{original} = 1.4\% \rightarrow SE_{original} = 0.71\%$.
(c) The standard error decreases with the square root of the sample size, so we should increase the sample size by a factor of $2.03^2 = 4.12$.
(d) The team should run an experiment 4.12 times larger, so they should have a random sample of 4.12% of their users in each of the experiment arms in the new experiment.

7.55 Independence: it is a random sample, so we can assume that the students in this sample are independent of each other with respect to number of exclusive relationships they have been in. Notice that there are no students who have had no exclusive relationships in the sample, which suggests some student responses are likely missing (perhaps only positive values were reported). The sample size is at least 30, and there are no particularly extreme outliers, so the normality condition is reasonable. 90% CI: (2.97, 3.43). We are 90% confident that undergraduate students have been in 2.97 to 3.43 exclusive relationships, on average.

7.57 The hypotheses should be about the population mean (μ), not the sample mean. The null hypothesis should have an equal sign and the alternative hypothesis should be about the null hypothesized value, not the observed sample mean. Correction:

$$H_0 : \mu = 10 \ hours$$
$$H_A : \mu \neq 10 \ hours$$

A two-sided test allows us to consider the possibility that the data show us something that we would find surprising.

8 Introduction to linear regression

8.1 (a) The residual plot will show randomly distributed residuals around 0. The variance is also approximately constant. (b) The residuals will show a fan shape, with higher variability for smaller x. There will also be many points on the right above the line. There is trouble with the model being fit here.

8.3 (a) Strong relationship, but a straight line would not fit the data. (b) Strong relationship, and a linear fit would be reasonable. (c) Weak relationship, and trying a linear fit would be reasonable. (d) Moderate relationship, but a straight line would not fit the data. (e) Strong relationship, and a linear fit would be reasonable. (f) Weak relationship, and trying a linear fit would be reasonable.

8.5 (a) Exam 2 since there is less of a scatter in the plot of final exam grade versus exam 2. Notice that the relationship between Exam 1 and the Final Exam appears to be slightly nonlinear. (b) Exam 2 and the final are relatively close to each other chronologically, or Exam 2 may be cumulative so has greater similarities in material to the final exam. Answers may vary.

8.7 (a) $r = -0.7 \rightarrow$ (4). (b) $r = 0.45 \rightarrow$ (3). (c) $r = 0.06 \rightarrow$ (1). (d) $r = 0.92 \rightarrow$ (2).

8.9 (a) The relationship is positive, weak, and possibly linear. However, there do appear to be some anomalous observations along the left where several students have the same height that is notably far from the cloud of the other points. Additionally, there are many students who appear not to have driven a car, and they are represented by a set of points along the bottom of the scatterplot. (b) There is no obvious explanation why simply being tall should lead a person to drive faster. However, one confounding factor is gender. Males tend to be taller than females on average, and personal experiences (anecdotal) may suggest they drive faster. If we were to follow-up on this suspicion, we would find that sociological studies confirm this suspicion. (c) Males are taller on average and they drive faster. The gender variable is indeed an important confounding variable.

8.11 (a) There is a somewhat weak, positive, possibly linear relationship between the distance traveled and travel time. There is clustering near the lower left corner that we should take special note of. (b) Changing the units will not change the form, direction or strength of the relationship between the two variables. If longer distances measured in miles are associated with longer travel time measured in minutes, longer distances measured in kilometers will be associated with longer travel time measured in hours. (c) Changing units doesn't affect correlation: $r = 0.636$.

8.13 (a) There is a moderate, positive, and linear relationship between shoulder girth and height. (b) Changing the units, even if just for one of the variables, will not change the form, direction or strength of the relationship between the two variables.

8.15 In each part, we can write the husband ages as a linear function of the wife ages. (a) $age_H = age_W + 3$. (b) $age_H = age_W - 2$. (c) $age_H = 2 \times age_W$. Since the slopes are positive and these are perfect linear relationships, the correlation will be exactly 1 in all three parts. An alternative way to gain insight into this solution is to create a mock data set, e.g. 5 women aged 26, 27, 28, 29, and 30, then find the husband ages for each wife in each part and create a scatterplot.

8.17 Correlation: no units. Intercept: kg. Slope: kg/cm.

8.19 Over-estimate. Since the residual is calculated as *observed − predicted*, a negative residual means that the predicted value is higher than the observed value.

8.21 (a) There is a positive, very strong, linear association between the number of tourists and spending. (b) Explanatory: number of tourists (in thousands). Response: spending (in millions of US dollars). (c) We can predict spending for a given number of tourists using a regression line. This may be useful information for determining how much the country may want to spend in advertising abroad, or to forecast expected revenues from tourism. (d) Even though the relationship appears linear in the scatterplot, the residual plot actually shows a nonlinear relationship. This is not a contradiction: residual plots can show divergences from linearity that can be difficult to see in a scatterplot. A simple linear model is inadequate for modeling these data. It is also important to consider that these data are observed sequentially, which means there may be a hidden structure not evident in the current plots but that is important to consider.

8.23 (a) First calculate the slope: $b_1 = R \times s_y/s_x = 0.636 \times 113/99 = 0.726$. Next, make use of the fact that the regression line passes through the point $(\bar{x}, \bar{y})$: $\bar{y} = b_0 + b_1 \times \bar{x}$. Plug in $\bar{x}$, $\bar{y}$, and b_1, and solve for b_0: 51. Solution: $\widehat{travel\ time} = 51 + 0.726 \times distance$. (b) b_1: For each additional mile in distance, the model predicts an additional 0.726 minutes in travel time. b_0: When the distance traveled is 0 miles, the travel time is expected to be 51 minutes. It does not make sense to have a travel distance of 0 miles in this context. Here, the y-intercept serves only to adjust the height of the line and is meaningless by itself. (c) $R^2 = 0.636^2 = 0.40$. About 40% of the variability in travel time is accounted for by the model, i.e. explained by the distance traveled. (d) $\widehat{travel\ time} = 51 + 0.726 \times distance = 51 + 0.726 \times 103 \approx 126$ minutes. (Note: we should be cautious in our predictions with this model since we have not yet evaluated whether it is a well-fit model.) (e) $e_i = y_i - \hat{y}_i = 168 - 126 = 42$ minutes. A positive residual means that the model underestimates the travel time. (f) No, this calculation would require extrapolation.

8.25 (a) $\widehat{murder} = -29.901 + 2.559 \times poverty\%$. (b) Expected murder rate in metropolitan areas with no poverty is -29. 901 per million. This is obviously not a meaningful value, it just serves to adjust the height of the regression line. (c) For each additional percentage increase in poverty, we expect murders per million to be higher on average by 2.559. (d) Poverty level explains 70.52% of the variability in murder rates in metropolitan areas. (e) $\sqrt{0.7052} = 0.8398$.

8.27 (a) There is an outlier in the bottom right. Since it is far from the center of the data, it is a point with high leverage. It is also an influential point since, without that observation, the regression line would have a very different slope.
(b) There is an outlier in the bottom right. Since it is far from the center of the data, it is a point with high leverage. However, it does not appear to be affecting the line much, so it is not an influential point.
(c) The observation is in the center of the data (in the x-axis direction), so this point does *not* have high leverage. This means the point won't have much effect on the slope of the line and so is not an influential point.

8.29 (a) There is a negative, moderate-to-strong, somewhat linear relationship between percent of families who own their home and the percent of the population living in urban areas in 2010. There is one outlier: a state where 100% of the population is urban. The variability in the percent of homeownership also increases as we move from left to right in the plot. (b) The outlier is located in the bottom right corner, horizontally far from the center of the other points, so it is a point with high leverage. It is an influen-

tial point since excluding this point from the analysis would greatly affect the slope of the regression line.

8.31 (a) The relationship is positive, moderate-to-strong, and linear. There are a few outliers but no points that appear to be influential.
(b) $\widehat{weight} = -105.0113 + 1.0176 \times height$.
Slope: For each additional centimeter in height, the model predicts the average weight to be 1.0176 additional kilograms (about 2.2 pounds).
Intercept: People who are 0 centimeters tall are expected to weigh - 105.0113 kilograms. This is obviously not possible. Here, the y- intercept serves only to adjust the height of the line and is meaningless by itself.
(c) H_0: The true slope coefficient of height is zero ($\beta_1 = 0$).
H_A: The true slope coefficient of height is different than zero ($\beta_1 \neq 0$).
The p-value for the two-sided alternative hypothesis ($\beta_1 \neq 0$) is incredibly small, so we reject H_0. The data provide convincing evidence that height and weight are positively correlated. The true slope parameter is indeed greater than 0.
(d) $R^2 = 0.72^2 = 0.52$. Approximately 52% of the variability in weight can be explained by the height of individuals.

8.33 (a) H_0: $\beta_1 = 0$. H_A: $\beta_1 \neq 0$. The p-value, as reported in the table, is incredibly small and is smaller than 0.05, so we reject H_0. The data provide convincing evidence that wives' and husbands' heights are positively correlated.
(b) $\widehat{height}_W = 43.5755 + 0.2863 \times height_H$.
(c) Slope: For each additional inch in husband's height, the average wife's height is expected to be an additional 0.2863 inches on average. Intercept: Men who are 0 inches tall are expected to have wives who are, on average, 43.5755 inches tall. The intercept here is meaningless, and it serves only to adjust the height of the line.
(d) The slope is positive, so r must also be positive. $r = \sqrt{0.09} = 0.30$.
(e) 63.33. Since R^2 is low, the prediction based on this regression model is not very reliable.
(f) No, we should avoid extrapolating.

8.35 (a) $H_0 : \beta_1 = 0; H_A : \beta_1 \neq 0$ (b) The p-value for this test is approximately 0, therefore we reject H_0. The data provide convincing evidence that poverty percentage is a significant predictor of murder rate. (c) $n = 20, df = 18, T_{18}^* = 2.10$; $2.559 \pm 2.10 \times 0.390 = (1.74, 3.378)$; For each percentage point poverty is higher, murder rate is expected to be higher on average by 1.74 to 3.378 per million. (d) Yes, we rejected H_0 and the confidence interval does not include 0.

8.37 (a) True. (b) False, correlation is a measure of the linear association between any two numerical variables.

8.39 (a) The point estimate and standard error are $b_1 = 0.9112$ and $SE = 0.0259$. We can compute a T-score: $T = (0.9112 - 1)/0.0259 = -3.43$. Using $df = 168$, the p-value is about 0.001, which is less than $\alpha = 0.05$. That is, the data provide strong evidence that the average difference between husbands' and wives' ages has actually changed over time. (b) $\widehat{age}_W = 1.5740 + 0.9112 \times age_H$. (c) Slope: For each additional year in husband's age, the model predicts an additional 0.9112 years in wife's age. This means that wives' ages tend to be lower for later ages, suggesting the average gap of husband and wife age is larger for older people. Intercept: Men who are 0 years old are expected to have wives who are on average 1.5740 years old. The intercept here is meaningless and serves only to adjust the height of the line. (d) $R = \sqrt{0.88} = 0.94$. The regression of wives' ages on husbands' ages has a positive slope, so the correlation coefficient will be positive. (e) $\widehat{age}_W = 1.5740 + 0.9112 \times 55 = 51.69$. Since R^2 is pretty high, the prediction based on this regression model is reliable. (f) No, we shouldn't use the same model to predict an 85 year old man's wife's age. This would require extrapolation. The scatterplot from an earlier exercise shows that husbands in this data set are approximately 20 to 65 years old. The regression model may not be reasonable outside of this range.

8.41 There is an upwards trend. However, the variability is higher for higher calorie counts, and it looks like there might be two clusters of observations above and below the line on the right, so we should be cautious about fitting a linear model to these data.

8.43 (a) $r = -0.72 \rightarrow (2)$ (b) $r = 0.07 \rightarrow (4)$ (c) $r = 0.86 \rightarrow (1)$ (d) $r = 0.99 \rightarrow (3)$

9 Multiple and logistic regression

9.1 (a) $\widehat{baby\_weight} = 123.05 - 8.94 \times smoke$ (b) The estimated body weight of babies born to smoking mothers is 8.94 ounces lower than babies born to non-smoking mothers. Smoker: $123.05 - 8.94 \times 1 = 114.11$ ounces. Non-smoker: $123.05 - 8.94 \times 0 = 123.05$ ounces. (c) $H_0: \beta_1 = 0$. $H_A: \beta_1 \neq 0$. $T = -8.65$, and the p-value is approximately 0. Since the p-value is very small, we reject H_0. The data provide strong evidence that the true slope parameter is different than 0 and that there is an association between birth weight and smoking. Furthermore, having rejected H_0, we can conclude that smoking is associated with lower birth weights.

9.3 (a) $\widehat{baby\_weight} = -80.41 + 0.44 \times gestation - 3.33 \times parity - 0.01 \times age + 1.15 \times height + 0.05 \times weight - 8.40 \times smoke$. (b) $\beta_{gestation}$: The model predicts a 0.44 ounce increase in the birth weight of the baby for each additional day of pregnancy, all else held constant. β_{age}: The model predicts a 0.01 ounce decrease in the birth weight of the baby for each additional year in mother's age, all else held constant. (c) Parity might be correlated with one of the other variables in the model, which complicates model estimation. (d) $\widehat{baby\_weight} = 120.58$. $e = 120 - 120.58 = -0.58$. The model over-predicts this baby's birth weight. (e) $R^2 = 0.2504$. $R_{adj}^2 = 0.2468$.

9.5 (a) (-0.32, 0.16). We are 95% confident that male students on average have GPAs 0.32 points lower to 0.16 points higher than females when controlling for the other variables in the model. (b) Yes, since the p-value is larger than 0.05 in all cases (not including the intercept).

9.7 Remove age.

9.9 Based on the p-value alone, either gestation or smoke should be added to the model first. However, since the adjusted R^2 for the model with gestation is higher, it would be preferable to add gestation in the first step of the forward- selection algorithm. (Other explanations are possible. For instance, it would be reasonable to only use the adjusted R^2.)

9.11 She should use p-value selection since she is interested in finding out about significant predictors, not just optimizing predictions.

9.13 Nearly normal residuals: With so many observations in the data set, we look for particularly extreme outliers in the histogram and do not see any. variability of residuals: The scatterplot of the residuals versus the fitted values does not show any overall structure. However, values that have very low or very high fitted values appear to also have somewhat larger outliers. In addition, the residuals do appear to have constant variability between the two parity and smoking status groups, though these items are relatively minor.
Independent residuals: The scatterplot of residuals versus the order of data collection shows a random scatter, suggesting that there is no apparent structures related to the order the data were collected.
Linear relationships between the response variable and numerical explanatory variables: The residuals vs. height and weight of mother are randomly distributed around 0. The residuals vs. length of gestation plot also does not show any clear or strong remaining structures, with the possible exception of very short or long gestations. The rest of the residuals do appear to be randomly distributed around 0. All concerns raised here are relatively mild. There are some outliers, but there is so much data that the influence of such observations will be minor.

9.15 (a) There are a few potential outliers, e.g. on the left in the `total_length` variable, but nothing that will be of serious concern in a data set this large. (b) When coefficient estimates are sensitive to which variables are included in the model, this typically indicates that some variables are collinear. For example, a possum's gender may be related to its head length, which would explain why the coefficient (and p-value) for `sex_male` changed when we removed the `head_length` variable. Likewise, a possum's skull width is likely to be related to its head length, probably even much more closely related than the head length was to gender.

9.17 (a) The logistic model relating $\hat{p}_i$ to the predictors may be written as $\log\left(\frac{\hat{p}_i}{1-\hat{p}_i}\right) = 33.5095 - 1.4207 \times sex\_male_i - 0.2787 \times skull\_width_i + 0.5687 \times total\_length_i - 1.8057 \times tail\_length_i$. Only `total_length` has a positive association with a possum being from Victoria. (b) $\hat{p} = 0.0062$. While the probability is very near zero, we have not run diagnostics on the model. We might also be a little skeptical that the model will remain accurate for a possum found in a US zoo. For example, perhaps the zoo selected a possum with specific characteristics but only looked in one region. On the other hand, it is encouraging that the possum was caught in the wild. (Answers regarding the reliability of the model probability will vary.)

9.19 (a) False. When predictors are collinear, it means they are correlated, and the inclusion of one variable can have a substantial influence on the point estimate (and standard error) of another. (b) True. (c) False. This would only be the case if the data was from an experiment and x_1 was one of the variables set by the researchers. (Multiple regression can be useful for forming hypotheses about causal relationships, but it offers zero guarantees.) (d) False. We should check normality like we would for inference for a single mean: we look for particularly extreme outliers if $n \geq 30$ or for clear outliers if $n < 30$.

9.21 (a) `exclaim_subj` should be removed, since it's removal reduces AIC the most (and the resulting model has lower AIC than the None Dropped model). (b) Removing any variable will increase AIC, so we should not remove any variables from this set.

9.23 (a) The equation is:

$$\log\left(\frac{p_i}{1-p_i}\right) = -0.8124$$
$$-2.6351 \times \texttt{to\_multiple}$$
$$+1.6272 \times \texttt{winner}$$
$$-1.5881 \times \texttt{format}$$
$$-3.0467 \times \texttt{re\_subj}$$

(b) First find $\log\left(\frac{p}{1-p}\right)$, then solve for p:

$$\log\left(\frac{p}{1-p}\right)$$
$$= -0.8124 - 2.6351 \times 0 + 1.6272 \times 1$$
$$-1.5881 \times 0 - 3.0467 \times 0$$
$$= 0.8148$$
$$\frac{p}{1-p} = e^{0.8148} \quad \rightarrow \quad p = 0.693$$

(c) It should probably be pretty high, since it could be very disruptive to the person using the email service if they are missing emails that aren't spam. Even only a 90% chance that a message is spam is probably enough to warrant keeping it in the inbox. Maybe a probability of 99% would be a reasonable cutoff. As for other ideas to make it even better, it may be worth building a second model that tries to classify the importance of an email message. If we have both the spam model and the importance model, we now have a better way to think about cost-benefit tradeoffs. For instance, perhaps we would be willing to have a lower probability-of-spam threshold for messages we were confident were not important, and perhaps we want an even higher probability threshold (e.g. 99.99%) for emails we are pretty sure are important.

Appendix B

Data sets within the text

Each data set within the text is described in this appendix, and there is a corresponding page for each of these data sets at **openintro.org/data**. This page also includes additional data sets that can be used for honing your skills. Each data set has its own page with the following information:

- List of the data set's variables.
- CSV download.
- R object file download.

B.1   Introduction to data

1.1 `stent30`, `stent365` → The stent data is split across two data sets, one for days 0-30 results and one for days 0-365 results.
Chimowitz MI, Lynn MJ, Derdeyn CP, et al. 2011. Stenting versus Aggressive Medical Therapy for Intracranial Arterial Stenosis. New England Journal of Medicine 365:993-1003. www.nejm.org/doi/full/10.1056/NEJMoa1105335.
NY Times article: www.nytimes.com/2011/09/08/health/research/08stent.html.

1.2 `loan50`, `loans_full_schema` → This data comes from Lending Club (lendingclub.com), which provides a large set of data on the people who received loans through their platform. The data used in the textbook comes from a sample of the loans made in Q1 (Jan, Feb, March) 2018.

1.2 `county`, `county_complete` → These data come from several government sources. For those variables included in the county data set, only the most recent data is reported, as of what was available in late 2018. Data prior to 2011 is all from census.gov, where the specific Quick Facts page providing the data is no longer available. The more recent data comes from USDA (ers.usda.gov), Bureau of Labor Statistics (bls.gov/lau), SAIPE (census.gov/did/www/saipe), and American Community Survey (census.gov/programs-surveys/acs).

1.3 Nurses' Health Study → For more information on this data set, see www.channing.harvard.edu/nhs

1.4 The study we had in mind when discussing the simple randomization (no blocking) study was Anturane Reinfarction Trial Research Group. 1980. *Sulfinpyrazone in the prevention of sudden death after myocardial infarction*. New England Journal of Medicine 302(5):250-256.

B.2   Summarizing data

2.1 `loan50`, `county` → These data sets are described in Data Appendix B.1.

2.2 `loan50`, `county` → These data sets are described in Data Appendix B.1.

2.3 `malaria` → Lyke et al. 2017. PfSPZ vaccine induces strain-transcending T cells and durable protection against heterologous controlled human malaria infection. PNAS 114(10):2711-2716. www.pnas.org/content/114/10/2711

B.3   Probability

3.1 `loan50`, `county` → These data sets are described in Data Appendix B.1.

3.1 `playing_cards` → Data set describing the 52 cards in a standard deck.

3.2 `family_college` → Simulated data based on real population summaries at nces.ed.gov/pubs2001/2001126.pdf.

3.2 `smallpox` → Fenner F. 1988. Smallpox and Its Eradication (History of International Public Health, No. 6). Geneva: World Health Organization. ISBN 92-4-156110-6.

3.2 Mammogram screening, probabilities → The probabilities reported were obtained using studies reported at www.breastcancer.org and www.ncbi.nlm.nih.gov/pmc/articles/PMC1173421.

3.2 Jose campus visits, probabilities → This example was made up.

3.3 No data sets were described in this section.

3.4 Course material purchases and probabilities → This example was made up.

3.4 Auctions for TV and toaster → This example was made up.

3.4 `stocks_18` → Monthly returns for Caterpillar, Exxon Mobil Corp, and Google for November 2015 to October 2018.

3.5 `fcid` → This sample can be considered a simple random sample from the US population. It relies on the USDA Food Commodity Intake Database.

B.4   Distributions of random variables

4.1 SAT and ACT score distributions → The SAT score data comes from the 2018 distribution, which is provided at
reports.collegeboard.org/pdf/2018-total-group-sat-suite-assessments-annual-report.pdf
The ACT score data is available at
act.org/content/dam/act/unsecured/documents/cccr2018/P_99_999999_N_S_N00_ACT-GCPR_National.pdf
We also acknowledge that the actual ACT score distribution is *not* nearly normal. However, since the topic is very accessible, we decided to keep the context and examples.

4.1 Male heights → The distribution is based on the USDA Food Commodity Intake Database.

4.1 `possum` → The distribution parameters are based on a sample of possums from Australia and New Guinea. The original source of this data is as follows. Lindenmayer DB, et al. 1995. *Morphological variation among columns of the mountain brushtail possum, Trichosurus caninus Ogilby (Phalangeridae: Marsupiala).* Australian Journal of Zoology 43: 449-458.

4.2 Exceeding insurance deductible → These statistics were made up but are possible values one might observe for low-deductible plans.

4.3 Exceeding insurance deductible → These statistics were made up but are possible values one might observe for low-deductible plans.

4.3 Smoking friends → Unfortunately, we don't currently have additional information on the source for the 30% statistic, so don't consider this one as fact since we cannot verify it was from a reputable source.

4.3 US smoking rate → The 15% smoking rate in the US figure is close to the value from the Centers for Disease Control and Prevention website, which reports a value of 14% as of the 2017 estimate:
cdc.gov/tobacco/data_statistics/fact_sheets/adult_data/cig_smoking/index.htm

4.4 Football kicker → This example was made up.

4.4 Heart attack admissions → This example was made up, though the heart attack admissions are realistic for some hospitals.

4.5 `ami_occurrences` → This is a simulated data set but resembles actual AMI data for New York City based on typical AMI incidence rates.

B.5 Foundations for inference

5.1 `pew_energy_2018` → The actual data has more observations than were referenced in this chapter. That is, we used a subsample since it helped smooth some of the examples to have a bit more variability. The `pew_energy_2018` data set represents the full data set for each of the different energy source questions, which covers solar, wind, offshore drilling, hydrolic fracturing, and nuclear energy. The statistics used to construct the data are from the following page:

www.pewinternet.org/2018/05/14/majorities-see-government-efforts-to-protect-the-environment-as-insufficient/

5.2 `pew_energy_2018` → See the details for this data set above in the Section 5.1 data section.

5.2 `ebola_survey` → In New York City on October 23rd, 2014, a doctor who had recently been treating Ebola patients in Guinea went to the hospital with a slight fever and was subsequently diagnosed with Ebola. Soon thereafter, an NBC 4 New York/The Wall Street Journal/Marist Poll found that 82% of New Yorkers favored a "mandatory 21-day quarantine for anyone who has come in contact with an Ebola patient". This poll included responses of 1,042 New York adults between Oct 26th and 28th, 2014. Poll ID NY141026 on maristpoll.marist.edu.

5.3 `pew_energy_2018` → See the details for this data set above in the Section 5.1 data section.

5.3 Rosling questions → We noted much smaller samples than the Roslings' describe in their book, Factfulness, The samples we describe are similar but not the same as the actual rates. The approximate rates for the correct answers for the two questions for (sometimes different) populations discussed in the book, as reported in Factfulness, are

- 80% of the world's 1 year olds have been vaccinated against some disease: 13% get this correct (17% in the US). gapm.io/q9
- Number of children in the world in 2100: 9% correct. gapm.io/q5

Here are a few more questions and a rough percent of people who get them correct:

- In all low-income countries across the world today, how many girls finish primary school: 20%, 40%, or 60%? Answer: 60%. About 7% of people get this question correct. gapm.io/q1
- What is the life expectancy of the world today: 50 years, 60 years, or 70 years? Answer: 70 years. In the US, about 43% of people get this question correct. gapm.io/q4
- In 1996, tigers, giant pandas, and black rhinos were all listed as endangered. How many of these three species are more critically endangered today: two of them, one of them, none of them? Answer: none of them. About 7% of people get this question correct. gapm.io/q11
- How many people in the world have some access to electricity? 20%, 50%, 80%. Answer: 80%. About 22% of people get this correct. gapm.io/q12

For more information, check out the book, Factfulness.

5.3 `pew_energy_2018` → See the details for this data set above in the Section 5.1 data section.

5.3 `nuclear_survey` → A simple random sample of 1,028 US adults in March 2013 found that 56% of US adults support nuclear arms reduction.
www.gallup.com/poll/161198/favor-russian-nuclear-arms-reductions.aspx

5.3 Car manufacturing → This example was made up.

5.3 `stent30`, `stent365` → These data sets are described in Data Appendix B.1.

B.6   Inference for categorical data

6.1 Payday loans → The statistics come from the following source:
pewtrusts.org/-/media/assets/2017/04/payday-loan-customers-want-more-protections-methodology.pdf

6.1 Tire factory → This example was made up.

6.2 cpr → Böttiger et al. *Efficacy and safety of thrombolytic therapy after initially unsuccessful cardiopulmonary resuscitation: a prospective clinical trial.* The Lancet, 2001.

6.2 fish_oil_18 → Manson JE, et al. 2018. *Marine n-3 Fatty Acids and Prevention of Cardiovascular Disease and Cancer.* NEJMoa1811403.

6.2 mammogram → Miller AB. 2014. *Twenty five year follow-up for breast cancer incidence and mortality of the Canadian National Breast Screening Study: randomised screening trial.* BMJ 2014;348:g366.

6.2 drone_blades → The quality control data set for quadcopter drone blades is a made-up data set for an example. We provide the simulated data in the drone_blades data set.

6.3 jury → The jury data set for examining discrimination is a made-up data set an example. We provide the simulated data in the jury data set.

6.3 sp500_1950_2018 → Data is sourced from finance.yahoo.com.

6.4 ask → Minson JA, Ruedy NE, Schweitzer ME. *There is such a thing as a stupid question: Question disclosure in strategic communication.*
opim.wharton.upenn.edu/DPlab/papers/workingPapers/
Minson_working_Ask%20(the%20Right%20Way)%20and%20You%20Shall%20Receive.pdf

6.4 diabetes2 → Zeitler P, et al. 2012. *A Clinical Trial to Maintain Glycemic Control in Youth with Type 2 Diabetes.* N Engl J Med.

B.7   Inference for numerical data

7.1 Risso's dolphins → Endo T and Haraguchi K. 2009. *High mercury levels in hair samples from residents of Taiji, a Japanese whaling town.* Marine Pollution Bulletin 60(5):743-747.

Taiji was featured in the movie *The Cove*, and it is a significant source of dolphin and whale meat in Japan. Thousands of dolphins pass through the Taiji area annually, and we assumes these 19 dolphins reasonably represent a simple random sample from those dolphins.

7.1 Croaker white fish → fda.gov/food/foodborneillnesscontaminants/metals/ucm115644.htm

7.1 run17 → www.cherryblossom.org

7.2 textbooks, ucla_textbooks_f18 → Data were collected by OpenIntro staff in 2010 and again in 2018. For the 2018 sample, we sampled 201 UCLA courses. Of those, 68 required books that could be found on Amazon. The websites where information was retrieved:
sa.ucla.edu/ro/public/soc, ucla.verbacompare.com, and amazon.com.

7.3 stem_cells → Menard C, et al. 2005. Transplantation of cardiac-committed mouse embryonic stem cells to infarcted sheep myocardium: a preclinical study. The Lancet: 366:9490, p1005-1012.

7.3 ncbirths → Birth records released by North Carolina in 2004. Unfortunately, we don't currently have additional information on the source for this data set.

7.3 Exam versions → This example was made up.

7.4 Blood pressure statistics → The blood pressure standard deviation for patients with blood pressure ranging from 140 to 180 mmHg is guessed and may be a little (but likely not dramatically) imprecise from what we'd observe in actual data.

7.5 toy_anova → Data used for Figure 7.19, where this data was made up.

7.5 mlb_players_18 → Data were retrieved from mlb.mlb.com/stats. Only players with at least 100 at bats were considered during the analysis.

7.5 classdata → This example was made up.

B.8 Introduction to linear regression

8.1 `simulated_scatter` → Fake data used for the first three plots. The perfect linear plot uses group 4 data, where `group` variable in the data set (Figure 8.1). The group of 3 imperfect linear plots use groups 1-3 (Figure 8.2). The sinusoidal curve uses group 5 data (Figure 8.3). The group of 3 scatterplots with residual plots use groups 6-8 (Figure 8.8). The correlation plots uses groups 9-19 data (Figures 8.9 and 8.10).

8.1 `possum` → This data set is described in Data Appendix B.4.

8.2 `elmhurst` → These data were sampled from a table of data for all freshman from the 2011 class at Elmhurst College that accompanied an article titled *What Students Really Pay to Go to College* published online by *The Chronicle of Higher Education*: chronicle.com/article/What-Students-Really-Pay-to-Go/131435.

8.2 `simulated_scatter` → The plots for things that can go wrong uses groups 20-23 (Figure 8.12).

8.2 `mariokart` → Auction data from Ebay (ebay.com) for the game Mario Kart for the Nintendo Wii. This data set was collected in early October, 2009.

8.3 `simulated_scatter` → The plots for types of outliers uses groups 24-29 (Figure 8.18).

8.4 `midterms_house` → Data was retrieved from Wikipedia.

B.9 Multiple and logistic regression

9.1 `loans_full_schema` → This data set is described in Data Appendix B.1.

9.2 `loans_full_schema` → This data set is described in Data Appendix B.1.

9.3 `loans_full_schema` → This data set is described in Data Appendix B.1.

9.4 `mariokart` → This data set is described in Data Appendix B.8.

9.5 `resume` → Bertrand M, Mullainathan S. 2004. *Are Emily and Greg More Employable than Lakisha and Jamal? A Field Experiment on Labor Market Discrimination.* The American Economic Review 94:4 (991-1013). www.nber.org/papers/w9873

We did omit discussion of some structure in the data for the analysis presented: the experiment design included blocking, where typically four resumes were sent to each job: one for each inferred race/sex combination (as inferred based on the first name). We did not worry about this blocking aspect, since accounting for the blocking would *reduce* the standard error without notably changing the point estimates for the `race` and `sex` variables versus the analysis performed in the section. That is, the most interesting conclusions in the study are unaffected even when completing a more sophisticated analysis.

Appendix C

Distribution tables

C.1 Normal Probability Table

A **normal probability table** may be used to find percentiles of a normal distribution using a Z-score, or vice-versa. Such a table lists Z-scores and the corresponding percentiles. An abbreviated probability table is provided in Figure C.1 that we'll use for the examples in this appendix. A full table may be found on page 410.

Z	Second decimal place of Z									
	0.00	0.01	0.02	0.03	**0.04**	0.05	0.06	0.07	0.08	0.09
0.0	0.5000	0.5040	0.5080	0.5120	0.5160	0.5199	0.5239	0.5279	0.5319	0.5359
0.1	0.5398	0.5438	0.5478	0.5517	0.5557	0.5596	0.5636	0.5675	0.5714	0.5753
0.2	0.5793	0.5832	0.5871	0.5910	0.5948	0.5987	0.6026	0.6064	0.6103	0.6141
0.3	0.6179	0.6217	0.6255	0.6293	0.6331	0.6368	0.6406	0.6443	0.6480	0.6517
0.4	0.6554	0.6591	0.6628	0.6664	0.6700	0.6736	0.6772	0.6808	0.6844	0.6879
0.5	0.6915	0.6950	0.6985	0.7019	0.7054	0.7088	0.7123	0.7157	0.7190	0.7224
0.6	0.7257	0.7291	0.7324	0.7357	0.7389	0.7422	0.7454	0.7486	0.7517	0.7549
0.7	0.7580	0.7611	0.7642	0.7673	0.7704	0.7734	0.7764	0.7794	0.7823	0.7852
0.8	0.7881	0.7910	0.7939	0.7967	**0.7995**	0.8023	0.8051	0.8078	0.8106	0.8133
0.9	0.8159	0.8186	0.8212	0.8238	0.8264	0.8289	0.8315	0.8340	0.8365	0.8389
1.0	*0.8413*	0.8438	0.8461	0.8485	0.8508	0.8531	0.8554	0.8577	0.8599	0.8621
1.1	0.8643	0.8665	0.8686	0.8708	0.8729	0.8749	0.8770	0.8790	0.8810	0.8830
⋮	⋮	⋮	⋮	⋮	⋮	⋮	⋮	⋮	⋮	⋮

Figure C.1: A section of the normal probability table. The percentile for a normal random variable with $Z = 1.00$ has been *highlighted*, and the percentile closest to 0.8000 has also been **highlighted**.

When using a normal probability table to find a percentile for Z (rounded to two decimals), identify the proper row in the normal probability table up through the first decimal, and then determine the column representing the second decimal value. The intersection of this row and column is the percentile of the observation. For instance, the percentile of $Z = 0.45$ is shown in row 0.4 and column 0.05 in Figure C.1: 0.6736, or the 67.36^{th} percentile.

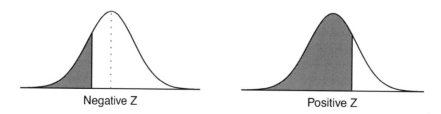

Negative Z   Positive Z

Figure C.2: The area to the left of Z represents the percentile of the observation.

EXAMPLE C.1

SAT scores follow a normal distribution, $N(1100, 200)$. Ann earned a score of 1300 on her SAT with a corresponding Z-score of $Z = 1$. She would like to know what percentile she falls in among all SAT test-takers.

Ann's **percentile** is the percentage of people who earned a lower SAT score than her. We shade the area representing those individuals in the following graph:

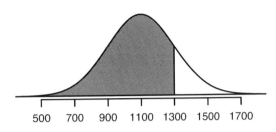

The total area under the normal curve is always equal to 1, and the proportion of people who scored below Ann on the SAT is equal to the *area* shaded in the graph. We find this area by looking in row 1.0 and column 0.00 in the normal probability table: 0.8413. In other words, Ann is in the 84th percentile of SAT takers.

EXAMPLE C.2

How do we find an upper tail area?

The normal probability table *always* gives the area to the left. This means that if we want the area to the right, we first find the lower tail and then subtract it from 1. For instance, 84.13% of SAT takers scored below Ann, which means 15.87% of test takers scored higher than Ann.

We can also find the Z-score associated with a percentile. For example, to identify Z for the 80th percentile, we look for the value closest to 0.8000 in the middle portion of the table: 0.7995. We determine the Z-score for the 80th percentile by combining the row and column Z values: 0.84.

EXAMPLE C.3

Find the SAT score for the 80th percentile.

We look for the are to the value in the table closest to 0.8000. The closest value is 0.7995, which corresponds to $Z = 0.84$, where 0.8 comes from the row value and 0.04 comes from the column value. Next, we set up the equation for the Z-score and the unknown value x as follows, and then we solve for x:

$$Z = 0.84 = \frac{x - 1100}{200} \quad \rightarrow \quad x = 1268$$

The College Board scales scores to increments of 10, so the 80th percentile is 1270. (Reporting 1268 would have been perfectly okay for our purposes.)

For additional details about working with the normal distribution and the normal probability table, see Section 4.1, which starts on page 133.

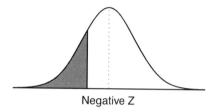

Negative Z

| Second decimal place of Z | | | | | | | | | | |
0.09	0.08	0.07	0.06	0.05	0.04	0.03	0.02	0.01	0.00	Z
0.0002	0.0003	0.0003	0.0003	0.0003	0.0003	0.0003	0.0003	0.0003	0.0003	−3.4
0.0003	0.0004	0.0004	0.0004	0.0004	0.0004	0.0004	0.0005	0.0005	0.0005	−3.3
0.0005	0.0005	0.0005	0.0006	0.0006	0.0006	0.0006	0.0006	0.0007	0.0007	−3.2
0.0007	0.0007	0.0008	0.0008	0.0008	0.0008	0.0009	0.0009	0.0009	0.0010	−3.1
0.0010	0.0010	0.0011	0.0011	0.0011	0.0012	0.0012	0.0013	0.0013	0.0013	−3.0
0.0014	0.0014	0.0015	0.0015	0.0016	0.0016	0.0017	0.0018	0.0018	0.0019	−2.9
0.0019	0.0020	0.0021	0.0021	0.0022	0.0023	0.0023	0.0024	0.0025	0.0026	−2.8
0.0026	0.0027	0.0028	0.0029	0.0030	0.0031	0.0032	0.0033	0.0034	0.0035	−2.7
0.0036	0.0037	0.0038	0.0039	0.0040	0.0041	0.0043	0.0044	0.0045	0.0047	−2.6
0.0048	0.0049	0.0051	0.0052	0.0054	0.0055	0.0057	0.0059	0.0060	0.0062	−2.5
0.0064	0.0066	0.0068	0.0069	0.0071	0.0073	0.0075	0.0078	0.0080	0.0082	−2.4
0.0084	0.0087	0.0089	0.0091	0.0094	0.0096	0.0099	0.0102	0.0104	0.0107	−2.3
0.0110	0.0113	0.0116	0.0119	0.0122	0.0125	0.0129	0.0132	0.0136	0.0139	−2.2
0.0143	0.0146	0.0150	0.0154	0.0158	0.0162	0.0166	0.0170	0.0174	0.0179	−2.1
0.0183	0.0188	0.0192	0.0197	0.0202	0.0207	0.0212	0.0217	0.0222	0.0228	−2.0
0.0233	0.0239	0.0244	0.0250	0.0256	0.0262	0.0268	0.0274	0.0281	0.0287	−1.9
0.0294	0.0301	0.0307	0.0314	0.0322	0.0329	0.0336	0.0344	0.0351	0.0359	−1.8
0.0367	0.0375	0.0384	0.0392	0.0401	0.0409	0.0418	0.0427	0.0436	0.0446	−1.7
0.0455	0.0465	0.0475	0.0485	0.0495	0.0505	0.0516	0.0526	0.0537	0.0548	−1.6
0.0559	0.0571	0.0582	0.0594	0.0606	0.0618	0.0630	0.0643	0.0655	0.0668	−1.5
0.0681	0.0694	0.0708	0.0721	0.0735	0.0749	0.0764	0.0778	0.0793	0.0808	−1.4
0.0823	0.0838	0.0853	0.0869	0.0885	0.0901	0.0918	0.0934	0.0951	0.0968	−1.3
0.0985	0.1003	0.1020	0.1038	0.1056	0.1075	0.1093	0.1112	0.1131	0.1151	−1.2
0.1170	0.1190	0.1210	0.1230	0.1251	0.1271	0.1292	0.1314	0.1335	0.1357	−1.1
0.1379	0.1401	0.1423	0.1446	0.1469	0.1492	0.1515	0.1539	0.1562	0.1587	−1.0
0.1611	0.1635	0.1660	0.1685	0.1711	0.1736	0.1762	0.1788	0.1814	0.1841	−0.9
0.1867	0.1894	0.1922	0.1949	0.1977	0.2005	0.2033	0.2061	0.2090	0.2119	−0.8
0.2148	0.2177	0.2206	0.2236	0.2266	0.2296	0.2327	0.2358	0.2389	0.2420	−0.7
0.2451	0.2483	0.2514	0.2546	0.2578	0.2611	0.2643	0.2676	0.2709	0.2743	−0.6
0.2776	0.2810	0.2843	0.2877	0.2912	0.2946	0.2981	0.3015	0.3050	0.3085	−0.5
0.3121	0.3156	0.3192	0.3228	0.3264	0.3300	0.3336	0.3372	0.3409	0.3446	−0.4
0.3483	0.3520	0.3557	0.3594	0.3632	0.3669	0.3707	0.3745	0.3783	0.3821	−0.3
0.3859	0.3897	0.3936	0.3974	0.4013	0.4052	0.4090	0.4129	0.4168	0.4207	−0.2
0.4247	0.4286	0.4325	0.4364	0.4404	0.4443	0.4483	0.4522	0.4562	0.4602	−0.1
0.4641	0.4681	0.4721	0.4761	0.4801	0.4840	0.4880	0.4920	0.4960	0.5000	−0.0

*For $Z \leq -3.50$, the probability is less than or equal to 0.0002.

411

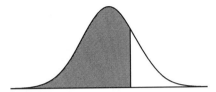

Positive Z

| | Second decimal place of Z | | | | | | | | | |
Z	0.00	0.01	0.02	0.03	0.04	0.05	0.06	0.07	0.08	0.09
0.0	0.5000	0.5040	0.5080	0.5120	0.5160	0.5199	0.5239	0.5279	0.5319	0.5359
0.1	0.5398	0.5438	0.5478	0.5517	0.5557	0.5596	0.5636	0.5675	0.5714	0.5753
0.2	0.5793	0.5832	0.5871	0.5910	0.5948	0.5987	0.6026	0.6064	0.6103	0.6141
0.3	0.6179	0.6217	0.6255	0.6293	0.6331	0.6368	0.6406	0.6443	0.6480	0.6517
0.4	0.6554	0.6591	0.6628	0.6664	0.6700	0.6736	0.6772	0.6808	0.6844	0.6879
0.5	0.6915	0.6950	0.6985	0.7019	0.7054	0.7088	0.7123	0.7157	0.7190	0.7224
0.6	0.7257	0.7291	0.7324	0.7357	0.7389	0.7422	0.7454	0.7486	0.7517	0.7549
0.7	0.7580	0.7611	0.7642	0.7673	0.7704	0.7734	0.7764	0.7794	0.7823	0.7852
0.8	0.7881	0.7910	0.7939	0.7967	0.7995	0.8023	0.8051	0.8078	0.8106	0.8133
0.9	0.8159	0.8186	0.8212	0.8238	0.8264	0.8289	0.8315	0.8340	0.8365	0.8389
1.0	0.8413	0.8438	0.8461	0.8485	0.8508	0.8531	0.8554	0.8577	0.8599	0.8621
1.1	0.8643	0.8665	0.8686	0.8708	0.8729	0.8749	0.8770	0.8790	0.8810	0.8830
1.2	0.8849	0.8869	0.8888	0.8907	0.8925	0.8944	0.8962	0.8980	0.8997	0.9015
1.3	0.9032	0.9049	0.9066	0.9082	0.9099	0.9115	0.9131	0.9147	0.9162	0.9177
1.4	0.9192	0.9207	0.9222	0.9236	0.9251	0.9265	0.9279	0.9292	0.9306	0.9319
1.5	0.9332	0.9345	0.9357	0.9370	0.9382	0.9394	0.9406	0.9418	0.9429	0.9441
1.6	0.9452	0.9463	0.9474	0.9484	0.9495	0.9505	0.9515	0.9525	0.9535	0.9545
1.7	0.9554	0.9564	0.9573	0.9582	0.9591	0.9599	0.9608	0.9616	0.9625	0.9633
1.8	0.9641	0.9649	0.9656	0.9664	0.9671	0.9678	0.9686	0.9693	0.9699	0.9706
1.9	0.9713	0.9719	0.9726	0.9732	0.9738	0.9744	0.9750	0.9756	0.9761	0.9767
2.0	0.9772	0.9778	0.9783	0.9788	0.9793	0.9798	0.9803	0.9808	0.9812	0.9817
2.1	0.9821	0.9826	0.9830	0.9834	0.9838	0.9842	0.9846	0.9850	0.9854	0.9857
2.2	0.9861	0.9864	0.9868	0.9871	0.9875	0.9878	0.9881	0.9884	0.9887	0.9890
2.3	0.9893	0.9896	0.9898	0.9901	0.9904	0.9906	0.9909	0.9911	0.9913	0.9916
2.4	0.9918	0.9920	0.9922	0.9925	0.9927	0.9929	0.9931	0.9932	0.9934	0.9936
2.5	0.9938	0.9940	0.9941	0.9943	0.9945	0.9946	0.9948	0.9949	0.9951	0.9952
2.6	0.9953	0.9955	0.9956	0.9957	0.9959	0.9960	0.9961	0.9962	0.9963	0.9964
2.7	0.9965	0.9966	0.9967	0.9968	0.9969	0.9970	0.9971	0.9972	0.9973	0.9974
2.8	0.9974	0.9975	0.9976	0.9977	0.9977	0.9978	0.9979	0.9979	0.9980	0.9981
2.9	0.9981	0.9982	0.9982	0.9983	0.9984	0.9984	0.9985	0.9985	0.9986	0.9986
3.0	0.9987	0.9987	0.9987	0.9988	0.9988	0.9989	0.9989	0.9989	0.9990	0.9990
3.1	0.9990	0.9991	0.9991	0.9991	0.9992	0.9992	0.9992	0.9992	0.9993	0.9993
3.2	0.9993	0.9993	0.9994	0.9994	0.9994	0.9994	0.9994	0.9995	0.9995	0.9995
3.3	0.9995	0.9995	0.9995	0.9996	0.9996	0.9996	0.9996	0.9996	0.9996	0.9997
3.4	0.9997	0.9997	0.9997	0.9997	0.9997	0.9997	0.9997	0.9997	0.9997	0.9998

*For $Z \geq 3.50$, the probability is greater than or equal to 0.9998.

C.2   *t*-Probability Table

A ***t*-probability table** may be used to find tail areas of a *t*-distribution using a T-score, or vice-versa. Such a table lists T-scores and the corresponding percentiles. A partial ***t*-table** is shown in Figure C.3, and the complete table starts on page 414. Each row in the *t*-table represents a *t*-distribution with different degrees of freedom. The columns correspond to tail probabilities. For instance, if we know we are working with the *t*-distribution with *df* = 18, we can examine row 18, which is highlighted in Figure C.3. If we want the value in this row that identifies the T-score (cutoff) for an upper tail of 10%, we can look in the column where *one tail* is 0.100. This cutoff is 1.33. If we had wanted the cutoff for the lower 10%, we would use -1.33. Just like the normal distribution, all *t*-distributions are symmetric.

one tail	0.100	0.050	0.025	0.010	0.005
two tails	0.200	0.100	0.050	0.020	0.010
df 1	3.08	6.31	12.71	31.82	63.66
2	1.89	2.92	4.30	6.96	9.92
3	1.64	2.35	3.18	4.54	5.84
⋮	⋮	⋮	⋮	⋮	⋮
17	1.33	1.74	2.11	2.57	2.90
18	**1.33**	**1.73**	**2.10**	**2.55**	**2.88**
19	1.33	1.73	2.09	2.54	2.86
20	1.33	1.72	2.09	2.53	2.85
⋮	⋮	⋮	⋮	⋮	
400	1.28	1.65	1.97	2.34	2.59
500	1.28	1.65	1.96	2.33	2.59
∞	1.28	1.64	1.96	2.33	2.58

Figure C.3: An abbreviated look at the *t*-table. Each row represents a different *t*-distribution. The columns describe the cutoffs for specific tail areas. The row with *df* = 18 has been **highlighted**.

EXAMPLE C.4

What proportion of the *t*-distribution with 18 degrees of freedom falls below -2.10?

Just like a normal probability problem, we first draw the picture and shade the area below -2.10:

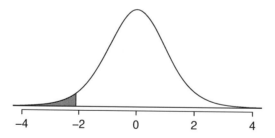

To find this area, we first identify the appropriate row: *df* = 18. Then we identify the column containing the absolute value of -2.10; it is the third column. Because we are looking for just one tail, we examine the top line of the table, which shows that a one tail area for a value in the third row corresponds to 0.025. That is, 2.5% of the distribution falls below -2.10.

In the next example we encounter a case where the exact T-score is not listed in the table.

EXAMPLE C.5

A t-distribution with 20 degrees of freedom is shown in the left panel of Figure C.4. Estimate the proportion of the distribution falling above 1.65.

We identify the row in the t-table using the degrees of freedom: $df = 20$. Then we look for 1.65; it is not listed. It falls between the first and second columns. Since these values bound 1.65, their tail areas will bound the tail area corresponding to 1.65. We identify the one tail area of the first and second columns, 0.050 and 0.10, and we conclude that between 5% and 10% of the distribution is more than 1.65 standard deviations above the mean. If we like, we can identify the precise area using statistical software: 0.0573.

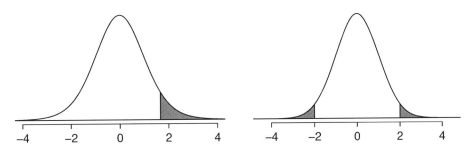

Figure C.4: Left: The t-distribution with 20 degrees of freedom, with the area above 1.65 shaded. Right: The t-distribution with 475 degrees of freedom, with the area further than 2 units from 0 shaded.

EXAMPLE C.6

A t-distribution with 475 degrees of freedom is shown in the right panel of Figure C.4. Estimate the proportion of the distribution falling more than 2 units from the mean (above or below).

As before, first identify the appropriate row: $df = 475$. This row does not exist! When this happens, we use the next smaller row, which in this case is $df = 400$. Next, find the columns that capture 2.00; because $1.97 < 3 < 2.34$, we use the third and fourth columns. Finally, we find bounds for the tail areas by looking at the two tail values: 0.02 and 0.05. We use the two tail values because we are looking for two symmetric tails in the t-distribution.

GUIDED PRACTICE C.7

What proportion of the t-distribution with 19 degrees of freedom falls above -1.79 units?[1]

EXAMPLE C.8

Find the value of $t^{\star}_{18}$ using the t-table, where $t^{\star}_{18}$ is the cutoff for the t-distribution with 18 degrees of freedom where 95% of the distribution lies between $-t^{\star}_{18}$ and $+t^{\star}_{18}$.

For a 95% confidence interval, we want to find the cutoff $t^{\star}_{18}$ such that 95% of the t-distribution is between $-t^{\star}_{18}$ and $t^{\star}_{18}$; this is the same as where the two tails have a total area of 0.05. We look in the t-table on page 412, find the column with area totaling 0.05 in the two tails (third column), and then the row with 18 degrees of freedom: $t^{\star}_{18} = 2.10$.

---

[1]We find the shaded area *above* -1.79 (we leave the picture to you). The small left tail is between 0.025 and 0.05, so the larger upper region must have an area between 0.95 and 0.975.

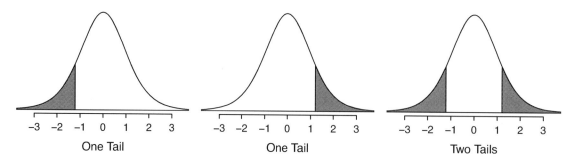

one tail	0.100	0.050	0.025	0.010	0.005
two tails	0.200	0.100	0.050	0.020	0.010
df 1	3.08	6.31	12.71	31.82	63.66
2	1.89	2.92	4.30	6.96	9.92
3	1.64	2.35	3.18	4.54	5.84
4	1.53	2.13	2.78	3.75	4.60
5	1.48	2.02	2.57	3.36	4.03
6	1.44	1.94	2.45	3.14	3.71
7	1.41	1.89	2.36	3.00	3.50
8	1.40	1.86	2.31	2.90	3.36
9	1.38	1.83	2.26	2.82	3.25
10	1.37	1.81	2.23	2.76	3.17
11	1.36	1.80	2.20	2.72	3.11
12	1.36	1.78	2.18	2.68	3.05
13	1.35	1.77	2.16	2.65	3.01
14	1.35	1.76	2.14	2.62	2.98
15	1.34	1.75	2.13	2.60	2.95
16	1.34	1.75	2.12	2.58	2.92
17	1.33	1.74	2.11	2.57	2.90
18	1.33	1.73	2.10	2.55	2.88
19	1.33	1.73	2.09	2.54	2.86
20	1.33	1.72	2.09	2.53	2.85
21	1.32	1.72	2.08	2.52	2.83
22	1.32	1.72	2.07	2.51	2.82
23	1.32	1.71	2.07	2.50	2.81
24	1.32	1.71	2.06	2.49	2.80
25	1.32	1.71	2.06	2.49	2.79
26	1.31	1.71	2.06	2.48	2.78
27	1.31	1.70	2.05	2.47	2.77
28	1.31	1.70	2.05	2.47	2.76
29	1.31	1.70	2.05	2.46	2.76
30	1.31	1.70	2.04	2.46	2.75

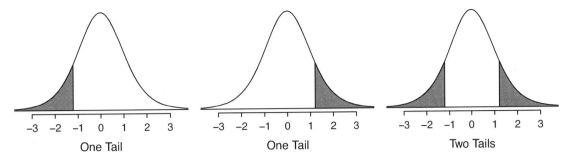

One Tail   One Tail   Two Tails

one tail	0.100	0.050	0.025	0.010	0.005
two tails	0.200	0.100	0.050	0.020	0.010
df 31	1.31	1.70	2.04	2.45	2.74
32	1.31	1.69	2.04	2.45	2.74
33	1.31	1.69	2.03	2.44	2.73
34	1.31	1.69	2.03	2.44	2.73
35	1.31	1.69	2.03	2.44	2.72
36	1.31	1.69	2.03	2.43	2.72
37	1.30	1.69	2.03	2.43	2.72
38	1.30	1.69	2.02	2.43	2.71
39	1.30	1.68	2.02	2.43	2.71
40	1.30	1.68	2.02	2.42	2.70
41	1.30	1.68	2.02	2.42	2.70
42	1.30	1.68	2.02	2.42	2.70
43	1.30	1.68	2.02	2.42	2.70
44	1.30	1.68	2.02	2.41	2.69
45	1.30	1.68	2.01	2.41	2.69
46	1.30	1.68	2.01	2.41	2.69
47	1.30	1.68	2.01	2.41	2.68
48	1.30	1.68	2.01	2.41	2.68
49	1.30	1.68	2.01	2.40	2.68
50	1.30	1.68	2.01	2.40	2.68
60	1.30	1.67	2.00	2.39	2.66
70	1.29	1.67	1.99	2.38	2.65
80	1.29	1.66	1.99	2.37	2.64
90	1.29	1.66	1.99	2.37	2.63
100	1.29	1.66	1.98	2.36	2.63
150	1.29	1.66	1.98	2.35	2.61
200	1.29	1.65	1.97	2.35	2.60
300	1.28	1.65	1.97	2.34	2.59
400	1.28	1.65	1.97	2.34	2.59
500	1.28	1.65	1.96	2.33	2.59
∞	1.28	1.65	1.96	2.33	2.58

C.3   Chi-Square Probability Table

A **chi-square probability table** may be used to find tail areas of a chi-square distribution. The **chi-square table** is partially shown in Figure C.5, and the complete table may be found on page 417. When using a chi-square table, we examine a particular row for distributions with different degrees of freedom, and we identify a range for the area (e.g. 0.025 to 0.05). Note that the chi-square table provides upper tail values, which is different than the normal and t-distribution tables.

Upper tail		0.3	0.2	0.1	0.05	0.02	0.01	0.005	0.001
df	2	2.41	**3.22**	**4.61**	5.99	7.82	9.21	10.60	13.82
	3	3.66	4.64	6.25	7.81	9.84	11.34	12.84	16.27
	4	4.88	5.99	7.78	9.49	11.67	13.28	14.86	18.47
	5	6.06	7.29	9.24	11.07	13.39	15.09	16.75	20.52
	6	7.23	8.56	10.64	12.59	15.03	16.81	18.55	22.46
	7	8.38	9.80	12.02	14.07	16.62	18.48	20.28	24.32

Figure C.5: A section of the chi-square table. A complete table is in Appendix C.3.

EXAMPLE C.9

Figure C.6(a) shows a chi-square distribution with 3 degrees of freedom and an upper shaded tail starting at 6.25. Use Figure C.5 to estimate the shaded area.

This distribution has three degrees of freedom, so only the row with 3 degrees of freedom (df) is relevant. This row has been italicized in the table. Next, we see that the value – 6.25 – falls in the column with upper tail area 0.1. That is, the shaded upper tail of Figure C.6(a) has area 0.1.

This example was unusual, in that we observed the *exact* value in the table. In the next examples, we encounter situations where we cannot precisely estimate the tail area and must instead provide a range of values.

EXAMPLE C.10

Figure C.6(b) shows the upper tail of a chi-square distribution with 2 degrees of freedom. The area above value 4.3 has been shaded; find this tail area.

The cutoff 4.3 falls between the second and third columns in the 2 degrees of freedom row. Because these columns correspond to tail areas of 0.2 and 0.1, we can be certain that the area shaded in Figure C.6(b) is between 0.1 and 0.2.

EXAMPLE C.11

Figure C.6(c) shows an upper tail for a chi-square distribution with 5 degrees of freedom and a cutoff of 5.1. Find the tail area.

Looking in the row with 5 df, 5.1 falls below the smallest cutoff for this row (6.06). That means we can only say that the area is *greater than* 0.3.

EXAMPLE C.12

Figure C.6(d) shows a cutoff of 11.7 on a chi-square distribution with 7 degrees of freedom. Find the area of the upper tail.

The value 11.7 falls between 9.80 and 12.02 in the 7 df row. Thus, the area is between 0.1 and 0.2.

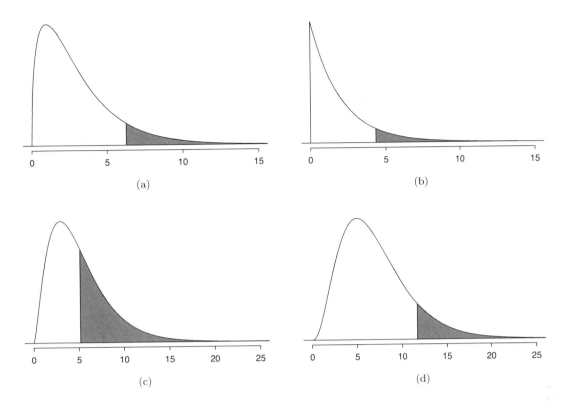

Figure C.6: **(a)** Chi-square distribution with 3 degrees of freedom, area above 6.25 shaded. **(b)** 2 degrees of freedom, area above 4.3 shaded. **(c)** 5 degrees of freedom, area above 5.1 shaded. **(d)** 7 degrees of freedom, area above 11.7 shaded.

Upper tail		0.3	0.2	0.1	0.05	0.02	0.01	0.005	0.001
df	1	1.07	1.64	2.71	3.84	5.41	6.63	7.88	10.83
	2	2.41	3.22	4.61	5.99	7.82	9.21	10.60	13.82
	3	3.66	4.64	6.25	7.81	9.84	11.34	12.84	16.27
	4	4.88	5.99	7.78	9.49	11.67	13.28	14.86	18.47
	5	6.06	7.29	9.24	11.07	13.39	15.09	16.75	20.52
	6	7.23	8.56	10.64	12.59	15.03	16.81	18.55	22.46
	7	8.38	9.80	12.02	14.07	16.62	18.48	20.28	24.32
	8	9.52	11.03	13.36	15.51	18.17	20.09	21.95	26.12
	9	10.66	12.24	14.68	16.92	19.68	21.67	23.59	27.88
	10	11.78	13.44	15.99	18.31	21.16	23.21	25.19	29.59
	11	12.90	14.63	17.28	19.68	22.62	24.72	26.76	31.26
	12	14.01	15.81	18.55	21.03	24.05	26.22	28.30	32.91
	13	15.12	16.98	19.81	22.36	25.47	27.69	29.82	34.53
	14	16.22	18.15	21.06	23.68	26.87	29.14	31.32	36.12
	15	17.32	19.31	22.31	25.00	28.26	30.58	32.80	37.70
	16	18.42	20.47	23.54	26.30	29.63	32.00	34.27	39.25
	17	19.51	21.61	24.77	27.59	31.00	33.41	35.72	40.79
	18	20.60	22.76	25.99	28.87	32.35	34.81	37.16	42.31
	19	21.69	23.90	27.20	30.14	33.69	36.19	38.58	43.82
	20	22.77	25.04	28.41	31.41	35.02	37.57	40.00	45.31
	25	28.17	30.68	34.38	37.65	41.57	44.31	46.93	52.62
	30	33.53	36.25	40.26	43.77	47.96	50.89	53.67	59.70
	40	44.16	47.27	51.81	55.76	60.44	63.69	66.77	73.40
	50	54.72	58.16	63.17	67.50	72.61	76.15	79.49	86.66

Index

Made in the USA
Middletown, DE
25 January 2022

59649767R00234